D1356775

Global Engineering and Construction

Global Engineering and Construction

Dr. J. K. Yates, PhD, BSCE

San Jose State University
San Jose, California

BICENTENNIAL
1807
WILEY
2007
BICENTENNIAL

JOHN WILEY & SONS, INC.

For general information about our other products and services, please contact our Customer Care Department within the United States at (800) 762-2974, outside the United States at (317) 572-3993 or fax (317) 572-4002.

Wiley also publishes its books in a variety of electronic formats. Some content that appears in print may not be available in electronic books. For more information about Wiley products, visit our web site at www.wiley.com.

Library of Congress Cataloging-in-Publication Data:

Yates, J. K., 1955–
    Global engineering and construction/J. K. Yates.
        p. cm.
    Includes index.
    ISBN-13: 978-0-471-74382-8 (cloth)
    ISBN-10: 0-471-74382-8 (cloth)
  1. Civil engineering.   2. Globalization.   I. Title.

    TA153.Y 38 2006
    624—dc22
2006010117

Printed in the United States of America

10 9 8 7 6 5 4 3 2 1

To my parents, Stanley and Amelia Yates,
who taught me the value of an education.
I wish they had lived long enough
to see the result of that lesson.

Contents

Chapter 3
Managing Global Engineering and Construction Projects | 36

Chapter 4
Global Competitiveness in the Engineering and Construction Industry | 77

Chapter 5
Global Engineering and Construction Alliances | 103

Chapter 6
Global Construction Financial Techniques | 121

Chapter 7
Global Legal Issues for Engineers and Constructors | **136**

Chapter 8
International Engineering and Construction Standards | **163**

Chapter 9
Global Environmental Issues of Concern to Engineers and Constructors | **181**

Chapter 10
Global Productivity Issues on Construction Projects | **197**

Chapter 11
Global Planning and Construction Delays | 214

Chapter 12
Global Terrorism: Kidnapping and Design Considerations | 236

Chapter 13
Preparing Engineers and Constructors to Work Globally | 263

Chapter 14
Country-Specific Information | 288

Preface

In the twenty-first century engineering and construction professionals are facing unique challenges due to the rapidly increasing globalization of the world that they did not have to address in the twentieth century. Telecommunications systems have irrevocably linked the world, and political, social, and economic events have changed how engineers and constructors now interface with the rest of the world. Modern educational systems have provided engineers and constructors with solid scientific and engineering backgrounds, but in the twenty-first century, engineering and construction (E&C) professionals need to be familiar with the eccentricities of other cultures and how to work effectively in the global arena. Organizations are facing many challenges that were not present before the twenty-first century because foreign investors are now buying domestic firms, and it is harder to distinguish between domestic and foreign firms.

Construction materials are not produced in every country in the world, and engineers and constructors have to rely on purchasing materials in the global marketplace, which requires an understanding of how to work effectively with members of foreign societies. Members of engineering and construction, firms have always competed against other domestic firms for construction materials but now they must compete against foreign firms for scarce materials. Substitute materials, such as fiber-reinforced composites, are only used on a small proportion of construction projects, and until they are universally adopted, and standards are developed and implemented for their use, firms will have to continue to compete for traditional construction materials in the global arena.

This book was written for both practicing engineers and constructors who are trying to adjust to working in a global environment and for college students who will soon be entering the global workforce. The majority of engineers and constructors who are released early from overseas work assignments are released because of cultural infractions rather than a lack of technical ability. If engineers and constructors are provided with books such as this one that provide information on how to work in global environments and what is and is not acceptable in other cultures, then hopefully the incidence of cultural infractions will be reduced, and global working environments will become more harmonious.

Engineering and construction professionals do not even have to leave their own country to experience cultural exchanges since many E&C firms are hiring foreign engineers to obtain the technical expertise required to complete contracts for clients. Some large industrial projects have personnel from 20 or 30 countries working together on projects either globally, using the Internet, or in the same office in order to complete projects on time. Cultural misunderstandings happen mainly because of ignorance rather than malice, and if E&C personnel are provided with information on what constitutes cultural infractions and how to work effectively with people from other cultures, they are able to adjust rapidly to working with foreign personnel.

The material in this book was written with operations-level engineers and constructors in mind rather than management personnel; therefore, it does not contain information on negotiating contracts. Operations-level E&C personnel are the professionals that design and construct projects after bid proposals have been accepted by clients. This book focuses on the daily issues that affect the lives of operations-level E&C professionals that result from cultural, social, or political differences between nations.

The topics included in this book were selected to help increase the awareness of engineers and constructors about the challenges they will face in the global E&C arena and to provide techniques and methods for addressing those challenges. The material in this book highlights which issues are the hardest to resolve when working on global E&C projects, and it provides suggestions on how to conquer the impediments created by working in multicultural environments.

When E&C personnel work on construction projects in foreign countries, they work not only with personnel from the host nation but also with people from other countries, because clients usually hire foreign nationals from several different nations to obtain the technical expertise required to design or build projects. Global project managers and project management team members may be managing multicultural personnel and laborers, and in some situations, they may be managing workers from every region of the world.

While E&C professionals are designing, constructing, or managing global projects, they do not have the time to conduct research investigations on how personnel from other countries operate or on cultural differences between their native country and host nations. Expatriates usually try adjusting to other cultures after they arrive in a foreign country rather than investigating cultural nuances before they accept an overseas assignment. This book is a compendium of both cultural and technical information that may be used by E&C personnel as a reference when they are assigned to work on overseas projects or with foreign nationals.

This book also was designed to be used in university programs as a textbook for courses on global engineering and construction to provide insight to students on what it will be like for them to design, construct, or manage foreign projects or to work with foreign nationals in their native country. The topics included in this book do not duplicate material that is taught in existing engineering and construction programs; they merely augment that material. Therefore, this book is appropriate for use at either the undergraduate or graduate level for a one-semester (or quarter) course.

This book is the result of twenty-three years of researching global engineering and construction topics with the assistance of global E&C personnel, living in an Islamic developing country, and traveling to over 25 countries throughout the world. In addition, engineers and constructors from every region of the world contributed material that is included in this book. Twenty-seven engineers and constructors conducted research into their native countries and provided information on what engineers and constructors need to know before working in their countries. A knowledge-acquisition system was used to collect data from E&C personnel in Southeast Asia, and hundreds of E&C professionals provided information on their experiences working with foreign nationals.

Extensive interviews were conducted at 58 global E&C companies with executives, including chief executive officers (CEOs), vice presidents, project managers, and construction managers, as well as with engineers, constructors, owners, and government officials. Government documents also were investigated along with manuscripts and refereed journal articles, PhD dissertations, and masters theses.

The chapters of this book are organized into specific topics based on the issue areas that affect E&C personnel when they work on global projects or with foreign nationals. The chapters are grouped into either topics that augment educational programs or global topics of interest to E&C personnel who work in multicultural environments. The subjects covered in the book are divided into the following two areas:

Global Topics and Chapters That Enhance the Education of Engineers and Constructors

- Project management (Chapter 3)
- Engineering design (Chapter 2)
- Legal issues (Chapter 7)
- Technical and quality standards (Chapter 8)
- Environmental issues (Chapter 9)
- Productivity improvement (Chapter 10)
- Planning and engineering delays and mitigation strategies (Chapter 11)

Global Topics of Interest to Engineering and Construction Professionals

- Concepts of culture and global issues (Chapter 2)
- Global competitiveness (Chapter 4)
- Global engineering and construction alliances (Chapter 5)
- Global financing techniques (Chapter 6)
- Preparing to work overseas (Chapter 13)
- Country-specific information (Chapter 14)

The main purpose of this book is to help foster multicultural working relationships and to provide E&C personnel with techniques and methods for designing, constructing, or managing global construction projects.

About the Author

Dr. J. K. Yates is the program coordinator for the Construction Engineering program in the Civil and Environmental Engineering Department at San Jose State University in San Jose, California. Dr. Yates was also a professor at Polytechnic University in Brooklyn, New York; Iowa State University, Ames, Iowa; and a visiting professor at the University of Colorado in Boulder, Colorado.

Dr. Yates has a Bachelor of Science degree in civil engineering from the University of Washington and Doctorate of Philosophy (Ph.D.) degree in civil engineering, with a concentration in construction engineering and management, from Texas A&M University, College Station, Texas with minors in international finance and management, international political science, business analysis, construction science, and archeology/anthropology.

Dr. Yates has worked for several domestic and global engineering and construction firms in the United States and globally. Dr. Yates is the author of forty-seven refereed articles and is a member of the American Society of Civil Engineers, the Project Management Institute, and the American Association of Cost Engineers International. Dr. Yates received the *Distinguished Professor* award for Polytechnic University in 1994, was the Associated General Contractor's *Outstanding Professor in America* in 1997, and was one of the Engineering News Record's *Those Who Made Marks on the Construction Industry* in 1991.

Acknowledgments

This book would not have been possible without the assistance of many people throughout the world. The idea for this book started germinating when a group of Indonesian engineers and constructors at Bontang Bay, in East Kalimantan, on the island of Borneo, in Indonesia, inquired as to whether I could do anything about the cultural clashes that were taking place between the Indonesian and expatriate engineers and constructors. My first response was that one person could not solve all the different cultural issues they were trying to address, but my second response was that maybe something could be done, and research for this book commenced. My research culminated two decades later in the writing of this book. It is to those 20 Indonesian engineers and constructors that I extend my appreciation, and I thank them for providing insights into why citizens from different nations have problems working together.

I also would like to acknowledge the assistance of Ir. Pandri Prabona (Indonesia), Toshio Kosugi (Japan), K. Y. Moon (South Korea), Yee-Wen Chen (Taiwan), Hsiung Yun-Lin (Taiwan), and Robert H. H. Shen (Taiwan) for distributing hundreds of questionnaires to their coworkers and for collecting and returning them to me, because their assistance provided invaluable data on the issues and challenges that face foreign engineers and constructors when they are working globally.

My former graduate students, from over 50 countries, contributed to this book by researching engineering, construction, and project management issues or by providing country-specific information to guide expatriate engineers and constructors when they are working on global engineering and construction projects. I want to specifically thank the following former students (next to each of their names is a citation of the chapter that contains their research):

- Alan Epstein and Joseph A. Smith (Chapter 7, global legal issues)
- Amin Ahmadi (Chapter 14, Iran)
- Swagata Guhthakurta, Basim Yacob Al Hamer and Amit Vakharia (Chapter 10, global productivity improvement)
- Angelo Perez (Chapter 14, the Philippines)
- Atta Elatta (Chapter 14, Yemen)
- Bogdan Clainau (Chapter 12 and Chapter 14, terrorism and Romania)
- Carlos Sotelo (Chapter 14, Peru)
- Jeffrey Herbage and Daniel Wong (Chapter 14, Israel)
- Devyani Agate, Deo Bhalotia, Paul Boyer, Veena Gorige, and Ken Nowak (Chapter 14, India)
- Dr. Avi Wiezel (Chapter 4, calendars)
- Dr. Edward Lockley (Chapter 3, construction failures)
- Dr. Fred Rahbar (Chapter 3, global project management and Afghanistan)
- Haitham (Henry) Halloum (Chapter 14, Jordan)
- Hassen Bas (Chapter 14, Turkey)
- Dr. Hisham (John Audi) Auode and Dr. Adel Eskander (Chapter 11, construction and planning delays)
- Holly Chan, Carol Chen, and Jing Deng (Chapter 14, China)
- Hyoung Kyun Kim (Chapters 9 and 14, global environmental issues and South Korea)
- Jin Chan Zhou (Chapter 14, Macao)
- Juan Duran (Chapter 7, dispute review boards)
- Juan Luis Alonso Mate (Chapter 14, Spain)
- June Seok Park (Chapter 9, global environmental issues)
- Matt Frechette (Chapter 14, Australia)
- Meenakshi Grandhi (Chapters 9 and 14, global environmental issues and India)
- Nancy Torrero (Chapter 3, safety issues)
- Sorason Wetchagarum and Evan Currid (Chapter 14, Thailand)
- Steven Njos (Chapter 4, global competitiveness)
- Dr. Stylianos Aniftos (Chapter 8, international standards)
- Subhransu Mukherjee (Chapters 5 and 6, privatization and international alliances)
- Svjetlana Radovic (Chapter 14, Bosnia and Herzegovina, Serbia and Montenegro, and Croatia)

- Terferi Abere, Mastewal Cherinet, Ishun Chan, and Joseph A. Smith (Chapter 14, Ethiopia)
- Dr. Vladimir L. Khazanet (Chapter 14, Russia)
- Wallant Poon, Ivan Eng, and Jack Mong (Chapter 14, Hong Kong)
- Hala El-Nagar (Chapter 11, construction delays)
- Joseph Smith (Chapter 7, global legal issues)

John McGillivray, who I am not able to thank because of his untimely death in Morocco, taught me how to live life to the fullest and contributed material on Africa for Chapter 14. Robert Hetherington who passed away prematurely was my mentor, and it was his understated faith in me that helped me to pursue an un-conventional career. I also would like to thank Soren Joseph Suver for his support, his kindness and for providing me with the opportunity to visit distant lands.

Linda Zuckerman, Wendy Hamilton, Dr. Janice Chambers, Dr. Robert Cartier, Barbara Kiblen, Lane McVae, Dr. Volkan Otugen, Rachel O'Mara, Kathy Kerns, Gayla Merryman, Dr. Rita Meyninger, Laura Thiele-Sardina, Pam Fuelling, Lynn Woodward, Dr. Panagiotis Mitropoulos, Michael Kiblen, Tim Kiblen, Dave Thomas, Gilda Pour, Dr. Rock Spencer, and my sisters Patricia Kratina and Maureen McGlathery all provided support and encouragement during the genesis of this book, and I want them to know how much their support means to me. I also apologize to them for my one-track mind during the writing of this book. I also want Larry Perry to know that I appreciate the wonderful environment that he provided where I wrote this book.

I want to thank Dr. Akthem Al-Manaseer and Dean Belle Wei at San Jose State University for providing me with a one-semester sabbatical to write this book and Professor Leslie Battersby for covering a full teaching load of classes for me why he also was working full time as a project manager.

Joseph A. Smith assisted with the material for Chapter 14 on Ethiopia, but he also conducted research and wrote a substantial portion of Chapter 7, which was way beyond the requirements for a graduate student that was also working full time in the construction industry. He came to my rescue and helped with the typing and editing of this book at the last minute, and I can never thank him enough for his timely assistance.

I also would like to thank all the people over the years who I have met throughout the world and in the United States who allowed me to ask them so many questions about their culture and for their honesty in answering those questions truthfully.

James Harper, my editor, understood the potential and importance of this book and gave me the opportunity to be the catalyst that passes along the knowledge of hundreds of people to thousands of other people throughout the world who otherwise would not have access to this information, and for that I thank him.

My parents valued education, and they always said "when" you go to college, not "if," which had a profound impact on my life. They also taught me to be learning continually, and that was the impetus for me to pursue a PhD degree. Even though they are no longer here to thank, their encouragement always follows me wherever I am living and in whatever I am trying to accomplish in life.

Chapter 1

Introduction

The struggle of the foreigner to get the words right results in the right words.—PBS news commentator, December 12, 2005

Modern telecommunication systems link the world together in an irrevocable manner, creating a global working environment, but cultural differences are still prevalent because the world is a long way from being a homogeneous society. Engineering and construction (E&C) professionals are required to work within societal and cultural constraints when they are performing work for global clients in foreign countries or working with foreign nationals within their own nations. Global variations also account for the vast differences seen in the construction techniques and processes used to construct projects throughout the world.

The twenty-first century is being shaped by global political, social, and economic events that are no longer concentrated in Western nations because Eastern nations now are moving to the forefront of global visibility. Forty percent of the working-age population is concentrated in India and China, and projections are that by 2032, three of the four largest economies in the world will be Japan, China, and India (Zakaria, 2005). Throughout the world, investors have been acquiring foreign firms, and it is becoming harder to determine which firms are domestic and which are foreign.

The definition of global competitiveness includes *firms competing for work in foreign countries,* but it also requires an analysis of the effects of global competition on domestic engineering and construction (E&C) markets. Diverting personnel and resources onto foreign projects creates more opportunities in domestic markets for native firms. Monitoring the trends and events that transpire in the global E&C arena is essential because members of domestic E&C firms are affected by price fluctuations in the cost of construction materials, the increasing cost of transporting materials, emerging innovative designs, the availability of technical personnel, foreign ownership of domestic E&C firms, and changing global economic and political climates.

In order for E&C firms to remain competitive in the global marketplace, their personnel have to be able to adapt quickly to working with people from other cultures, and they have to develop a global perspective that is incorporated into their designs, the techniques and processes used to construct a facility, and the way they manage construction projects.

The price of structural-steel members doubled during 2005, along with similar price increases for timber products. The main producers of steel now are located in Pacific Rim countries. Virgin forests in North America and other countries have been denuded, and compressed wood chips (glue lams or plystrand) are replacing wood as structural materials. The quality of construction materials is declining steadily, owing to relaxed safety requirements in the nations where construction materials are being produced or fabricated, and designs have to be adjusted to accommodate lower-quality materials.

Since firms are forced to compete for scarce resources that are produced by firms throughout the world they are influenced by global competitiveness issues, even if they are only domestic E&C firms. When domestic firms no longer produce or process the raw materials and components used in construction projects, domestic E&C firms have to compete for scarce materials in the global marketplace, which requires an understanding of the issues and challenges that surface when working with global suppliers, fabricators, and foreign E&C firms.

The competitiveness of domestic E&C firms also is affected when firms can no longer locate qualified personnel in their native countries to hire for technical positions. Increasing competition for highly qualified E&C professionals has led to firms hiring foreign nationals to fill positions because the country of origin of a worker is not as important as a firm's ability to perform the work required for the fulfillment of an engineering or construction contract.

1.1 Global Engineers and Constructors

Civil unrest, wars, terrorist activities, severe natural disasters, escalating terrorism, and the increaseing number of worldwide natural disasters affects E&C professionals as they are called on to design, build, or manage the reconstruction efforts following these destructive events. E&C professionals are required to move rapidly to disaster sites to perform damage assessments and to create reconstruction plans. Figure 1.1 contains a photo of an oil tank, located in St. Croix, U.S. Virgin Islands, that was damaged by a hurricane. This photo demonstrates the power of natural disasters, given that the oil tank was constructed of heavy-gauge steel, but it was completely crushed by hurricane-force winds. Figure 1.2 shows one stage in the slow and arduous process of repairing the oil tank.

Figure 1.3 shows a retaining-wall construction operation in Hong Kong, where the reinforcement is being lifted into place with a small crane beneath a precariously perched shantytown in the shadow of modern high-rise apartments. Figure 1.4 shows the delicate task of repairing and strengthening the Acropolis in Athens, Greece. Both these construction operations are being performed with technology that has been used on construction projects for thousands of years (except for motorization of the crane), yet there are high-technology apparatuses, such as the three-story-high universal testing machine shown in Figure 1.5, that are available to help analyze structural members used in construction projects if clients can afford them.

Much of the infrastructure in developed countries was built in the decade following World War II, the 1950s, and as it continues to deteriorate, structural fail-

Figure 1.1 Oil tank destroyed by a hurricane, St. Croix, U.S. Virgin Islands.

Figure 1.2 Oil tank repair, St Croix, U.S. Virgin Islands.

Figure 1.3 Construction of a retaining wall, Hong Kong.

Figure 1.4 Reconstruction of the Acropolis, Athens, Greece.

Figure 1.5 Three-story universal-testing machine, Seattle, Washington.

ures will become more commonplace. Major infrastructure reconstruction projects will require E&C professionals from every region of the world because no one country could possibly provide enough technical expertise to design and build all the upcoming reconstruction projects. Figures 1.6 and 1.7 show two extremes in bridge reconstruction and highway construction—when they have to be repaired or retrofitted to meet new building codes. The retrofitting process on the Golden Gate bridge in San Francisco, California, shown in Figure 1.6, was accomplished with small cranes, whereas in Figure 1.7 the roadway sections for the Eisenhower Freeway in Colorado were lifted with a massive gantry.

Social entrepreneurs provide funds to affect social change in developing countries, but even they realize that their efforts will be unsuccessful without a solid

Figure 1.6 Golden Gate Bridge retrofit project, San Francisco, California.

Figure 1.7 Gantry lifting a roadway section, Bear Creek, Colorado.

infrastructure, including transportation and sanitation systems (Hsu, 2005). A solid infrastructure is the backbone of any country, and throughout the world governments are emphasizing how important basic services are to maintaining their national stability.

When engineers are developing global designs, they need to be cognizant of the fact that suppliers, fabricators, engineers, and constructors from foreign cultures will be the ones interpreting the contracts, plans, and specifications for projects. Anyone who works for an E&C firm is now part of the global marketplace, even if they never leave their native country. *Global* is defined as "involving the earth as a whole," and *foreign* is defined as "situated outside one's own country; characteristic of, or dealing with, another country or countries" (*Webster's Dictionary*, 2005, p. 261 and p. 542). Therefore, if E&C personnel specify or order materials from foreign countries, they are involved in the global environment.

When engineers design projects that will be built in foreign countries, they make assumptions about the culture that will surround the project that could result in lengthy construction delays, increased costs, or high turnover rates in personnel. In the global environment, engineers and constructors can no longer only rely on their technical expertise to design and build projects, because now they need insight into how to solve additional nontechnical challenges related to cultural differences, language barriers, import and export restrictions, religious requirements, major economic and exchange-rate fluctuations, environmental restrictions, differing productivity rates, and contrary legal systems.

1.2 Engineering and Construction Global Ambassadors

Engineers and constructors are a unique element of society, and the influence they have globally has been evident throughout recorded history. Providing designs and construction expertise are not the only results achieved by members of the E&C industry as they are also global ambassadors. In their 1957 book, *The Earth Changers*, Neil C. Wilson and Frank J. Taylor describe "the role of America's many 'unsung' ambassadors with power shovels who are doing a great job of friendship-building for the U.S.A" (Wilson and Taylor, 1957, p. 125). Wilson and Taylor dedicated their book to "the rugged construction stiff in his hard hat and muddy boots, doing his best, and doing it well, from the Artic Circle to the Antarctic and from Suez to right around to Suez. Teaching what he knows to distant peoples, and learning from them considerably more than he teaches" (Wilson and Taylor, 1957, p. 5).

Fifty years later, as E&C professionals continue traveling around the world, to work on projects far from their homes, often risking their lives, they are still learning more from the "distant peoples" than they teach. Engineers and constructors may be the first foreigners from a certain country that the residents of the job site ever encounter, and the impression they leave behind could determine how the locals will view the citizens of these countries until their next encounter with them.

1.3 Global Risks to Engineers and Constructors

Being global ambassadors increases the visibility of E&C personnel, which exposes them to additional risks related to terrorism, kidnapping, theft, and political upheaval. Unfortunately, publicity surrounding these types of events prevents some E&C personnel from accepting overseas work assignments. Multinational firms are working to reduce the risk exposure of their employees, but individuals are also responsible for learning about how to protect themselves in foreign environments. Several of the chapters in this book address the risks associated with working overseas and provide information on precautions that could be implemented by individuals to help reduce risks.

1.4 Organization of This Book

The purpose of this book is to provide information on the issues and challenges that engineers and constructors face when working in the global marketplace. The book is organized to provide both general and specific information about the global E&C environment. General information on global issues of concern to E&C professionals is provided in the first half of the book, along with examples from different regions of the world, and the second half of the book provides details on global engineering and construction practices and information on specific regions of the world.

Chapter 2 discusses where to find information and assistance when planning to work in a foreign country. Chapter 2 also includes information on technical issues E&C personnel face when designing projects that will be built in foreign countries, such as language and translation considerations, design criteria for different cultures, technology considerations, and cultural issues that affect engineers and constructors.

Chapter 3 provides information on managing global E&C projects and the differences between managing projects in industrialized nations and managing them in developing countries. This chapter also discusses the desired characteristics of global E&C managers and personnel, how to effectively manage projects overseas, global calendar and time considerations, scheduling, safety issues, construction failures, and construction-failure investigation techniques. Appendix B contains a case study about a project that was built in Indonesia that demonstrates the concepts presented in Chapter 3.

Chapter 4 provides an analysis of global competitiveness in the E&C industry during the later part of the twentieth century and an analysis of the competitive issues that are influencing the E&C industry at the beginning of the twenty-first century. A historical analysis of competitiveness is provided in this chapter because competitiveness issues during a particular decade are the result of events that transpired in previous decades.

Chapter 5 discusses global E&C alliances because in many parts of the world foreign E&C firms cannot design and build domestic projects unless they are aligned with a domestic firm. The formation of global alliances is required for entering previously closed construction markets and for increasing the volume of work for a company. Chapter 5 focuses on multinational contractors, global contracting, the global dimension and global influences, global strategies, risks in global investments, project financing, privatization, build-own-transfer projects (BOT), and global joint ventures and partnerships.

Chapter 6 explains global financing techniques, including countertrade, cofinancing, swap financing, and other financing options to help E&C personnel understand some of the financial constraints that restrict funds for global construction projects. This chapter also includes a discussion about global payment methods that provides information on addressing global exchange rate fluctuations and government restrictions on exporting currency.

Chapter 7 guides readers through some of the legal issues and contract clauses that affect global construction contracts, and Chapter 8 provides information on international technical standards, environmental standards, the ISO 9000 series of quality standards, the ISO 14000 series of environmental standards, and how international standards are developed and implemented in the E&C industry.

Chapter 9 explains global environmental issues that affect engineering and construction projects and provides information on global treaties and protocols that are designed to reduce environmental pollution and hazardous waste generation (the global construction industry generates the largest proportion of hazardous waste of any industry).

Chapter 10 addresses global productivity issues and includes international comparisons of labor productivity rates, along with providing a discussion about management, labor, and materials in the context of how they affect productivity rates in different countries.

Chapter 11 provides insight into the types of delays that occur during both the planning and execution stages of global construction projects and provides examples of mitigation strategies that are used in the E&C industry to help prevent delays and that can be implemented to guide projects back on schedule.

Chapter 12 addresses terrorism and kidnapping issues and discusses design considerations related to acts of terrorism, as well providing information on how E&C personnel can protect themselves while working overseas. Preparing to work globally is discussed in Chapter 13, and the last chapter, Chapter 14, provides summaries of global regional differences and country-specific information on culture and customs, business methods, whether business agents are required, materials available locally, social standards, languages spoken, and holidays. Appendix A is a glossary that provides definitions of the terms used throughout this book.

1.5 Summary

Engineering and construction professionals now face additional challenges due to the interconnection of the world that they did not have to address in previous centuries, such as increasing competition for materials and technical professionals. If they are able to adjust to working in a global environment, they will maintain their competitive position; but if they cannot adapt to working with foreign engineering and construction professionals, other firms will dominate the global marketplace.

REFERENCES

Hsu, C. 2005. Social Engineering. U.S. News and World Report 139 (18).

Webster's New Universal Unabridged Dictionary, 2005. New York: New World Dictionaries/ Simon and Schuster.

Wilson and Taylor, 1957. *The Earth Changers.* Garden City, N.Y.: Doubleday and Company.

Zakaria, Fareed. 2005. India and China Rule. Newsweek. May 8.

Chapter 2

The Concept of Culture and Global Issues Important to Engineers and Constructors

2.1 Introduction

Global is a term that captures the extended consequences of actions by citizens in one country as their actions are magnified by world events. Global is defined as being "worldwide" or "involving the earth as a whole" (*Webster's Unabridged Dictionary,* 2005, p. 261). When engineering and construction (E&C) professionals work in the global arena, their objectives are the same as when they work in their native countries—to design and construct projects on time, within budget, safely, and with the highest achievable quality. The only additional obstacles in the global environment are language barriers, cultural misunderstandings, and working with personnel with varying degrees of technical education.

Engineering and construction professionals know how to achieve technical objectives, but achieving project objectives in a global environment requires more than merely technical expertise. It is during the execution stage of projects that cultural, political, environmental, religious, legal, and language barriers surface, and it is how these barriers and differences in perspective are addressed that determines the success or failure of global E&C projects.

A study of hundreds of E&C professionals conducted in Taiwan, South Korea, Japan, and Indonesia indicated that the primary reasons for not hiring foreign firms was cost of services, language differences, lack of adaptability, and cultural differences. Asian design engineers indicated that compensation and language were the most important factors, construction personnel were more concerned about cultural differences and social factors, and construction management personnel ranked technology transfer and culture as their main concerns. Technology was ranked seventh of eight major concerns, which indicates that the technical ability of a firm is not the main criteria used to select firms for global projects in Southeast Asia (Yates, 1991).

In the global arena, members of E&C firms have to cope with unique challenges in addition to trying to achieve technical goals; therefore, this chapter focuses on providing suggestions for coping with the challenges that surface on global E&C

projects. The first section of this chapter discusses what information might be available through government agencies, embassies, and nongovernmental resources. The second section explains how and where language translation issues affect designs, contracts, specifications, and plans. The third section discusses cultural issues that have an impact on design criteria, including self-reference criteria, religion, *fung shui* or *feng shui,* local requirements, cultures within cultures, and issues related to status.

The fourth section provides suggestions for adapting to foreign cultures, such as avoiding generalizations, presenting business cards, preventing the loss of face, providing recommendations for giving presentations and conducting meetings, avoiding unacceptable gestures and behavior as well as passive-aggressive behavior, and recognizing historical or heritage issues. Economics and politics are mentioned only briefly in this chapter, because both of these topics are covered in other chapters of this book.

2.2 Sources of Information

Before anyone commits to working on an overseas project, they should be familiar with any differences between their native culture and the culture where the project will be constructed so that they may incorporate *appropriate technology* (technology that is suitable to a particular culture) into their designs. In order to locate information on foreign cultures, E&C personnel should contact governmental and nongovernmental agencies that specialize in providing cultural information, and the subsections that follow provide information on some of these agencies.

2.2.1 Government Sources of Information

There are many different government agencies that publish information that would be of use to foreign E&C personnel, but this section discusses only the agencies that are directly responsible for promoting the exchange of products and services.

2.2.1.1 Government Sources in the United States

When someone is planning to work in another country or to conduct business with members of foreign nations, he or she should contact the United States and Foreign Commercial Services (US&FCS), which is part of the International Trade Administration of the Department of Commerce, because this agency provides country-specific information. Members of the US&FCS are responsible for promoting trade between United States firms and firms in foreign nations. The US&FCS has offices in countries throughout the world. The US&FSC has staff members who speak foreign languages and they can provide information on local customs and regulations, as well as names of contacts in other countries.

The US&FSC deals with international business matters and export licensing requirements for shipping construction equipment and materials overseas. The

American Association of Exporters and Importers publishes documents about current regulations related to exporting and importing products, and anything that will be exported from the United States should be checked against the Commerce Control List (CCL) to see if any of the materials or equipment (including computer hardware and software) are on the CCL of restricted products. If materials and equipment are not on the CCL, then they can be exported under a general export license (required by the president's Enhanced Proliferation Control Initiative, or EPCI). There are equivalent government agencies in every country in the world that regulate the importing and exporting of products from their nations, and they should be consulted before incorporating foreign materials into the design of a project.

2.2.1.2 Embassies

In the United States, there are 173 foreign embassies, and each one has a commercial officer or someone in a similar position who can provide guidance on working in that embassy's country and information on local business practices, customs, and legal issues. Embassy staff members also can be contacted when there are emergencies or their citizens need assistance in dealing with legal matters while they are in a foreign country.

Some governments, such as the Japanese government, provide funding to their E&C firms to help defray the costs of developing bid estimates, which is a practice that allows firms from their country to bid on more global projects. Since preparing bid estimates costs between 1 and 3 percent of the total cost of a project, having the government reimburse a firm for this cost or a portion of the cost is a competitive advantage for their native firms.

Commercial officers (attachés) usually know which firms from their native country are bidding on projects, but they may not let firms from their native country know who else is bidding on a project to avoid conflicts of interest. However, commercial officers from other embassies may provide this information. Embassy staff members are also familiar with any new legal requirements that have been implemented because of terrorist activities or political upheaval.

Embassies issue visas and provide information on visa requirements, such as how long a visa is valid, whether a work permit is required, and how long it takes to be issued a visa for their native country. Expatriate E&C personnel may have to leave a country periodically to have their work permits or visas renewed, and embassy personnel know about these requirements. Host governments may require foreigners to provide up to 100 copies of passport photos that are distributed to government officials throughout the country.

Expatriates should not surrender their passports to anyone in a foreign country unless the person seizing the passport is an authorized government official. To verify whether someone is a government official, contact embassy personnel. Customs officials may try to take a passport away from a foreigner by telling him or her that the passport photo is not correct or valid. This is a ruse used to gain possession of passports because passports from industrialized nations can be sold on the black market for thousands of (U.S.) dollars.

2.2.1.3 Ministry of Foreign Trade

In the Russian Federation, the People's Republic of China, and other eastern European countries, the Ministry of Foreign Trade should be contacted to locate the appropriate foreign trade organization that is responsible for assisting foreign E&C firms. In these countries, contracts are negotiated through foreign trade organizations (FTOs) or foreign trade corporations (FTCs), both of which are legal entities of the state that facilitate relationships between foreign and domestic organizations. The FTOs and FTCs evaluate the commercial aspects of potential imports or services, including prices, quality, quantity, delivery times, and payment terms. Every country in the world has an administrative agency or ministry that is responsible for assisting foreign firms when they are interested in conducting business in that country, and these agencies are listed on government Web sites.

2.2.2 Nongovernmental Sources of Information

Many of the major multinational E&C firms have personnel that have lived or worked in foreign countries, and they can provide information on what it is like to work in a particular country. Some E&C professionals spend their entire careers working on jobs in foreign countries. In the United States, 10 percent of the population does 90 percent of the moving, which makes it easier to locate E&C personnel who have worked on foreign construction projects. Multinational firms can be located by referring to the *Engineering News Record's* list of top U.S. and foreign engineers and contractors or similar directories that are published in other countries and on company Web sites.

Articles in professional society publications provide information on who is designing and constructing particular types of global projects. Another resource is foreign members of professional societies because they can assist with technical issues, as well as cultural issues. If foreign professional societies cannot be located on the Web, professors at domestic or foreign universities may know which professional societies are located in which particular country.

Other sources of general information about foreign countries are available on the Central Intelligence Agency (CIA) *World Factbook* Web site (www.cia.org) and in the *TIME Almanac and Information Please.* The CIA Web site and the *TIME Almanac and Information Please* provide country-specific economic information, as well as information on geography; forms of government; languages spoken; ethnicity, race, and religions; literacy rates; and populations, along with maps and other information that is updated on a yearly basis (CIA, 2005; Bruner, 2005).

World Trade Centers publish newsletters and have trade libraries that can be used by business travelers. World Trade Center staff members also provide assistance to business travelers in making hotel reservations and other travel arrangements, along with providing rooms for meetings. There are 300 World Trade Centers in over 100 countries throughout the world.

For accurate information on the vaccinations required for working in or traveling to different countries, check with the World Health Organization (WHO) or the WHO Web site (www.who.org). Malaria preventatives should be started at least one month before moving to or traveling through countries where malaria is prevalent because E&C personnel have been known to contract malarial immediately on entering a country where mosquitoes are carrying the malaria parasite. Tetanus, tuberculosis, polio, mumps, measles, and chicken pox have been eradicated in industrialized nations, but they are still a health risk in developing countries.

There are numerous Web sites on the World Wide Web that contain information on specific countries, but the accuracy of the information they provide is not verified; therefore, the Web is not a reliable source of information on foreign countries.

2.3 Language and Translation Considerations for Design

Given the precise nature of the E&C profession, engineers and constructors should carefully review the terminology they use in engineering designs, contracts, and specifications. Not all words or phrases translate into other languages with the same meaning, and some words or phrases cannot be translated at all into some foreign languages. The English language has over 700,000 words, and 40 percent of the words (280,000 words) are technical terms, many of which do not have foreign translations.

There are official languages in every country in the world, but other languages are also spoken, and Chapter 14 contains information on languages spoken in different countries. In some countries, there are more people who do not speak the official language than people who do speak it. One engineer, Franklin Fernandez, who works for a large multinational firm from the United States was assigned to be the company representative on a construction project in Kazakhstan, and when he arrived he realized he was one of only a few people working on the project who spoke English. Everyone else was speaking either Russian or Turkish, and he had to learn to translate the contract requirements into Turkish or Russian.

Foreign professional society members and university faculty are useful resources for obtaining information related to designing and constructing projects in foreign countries, and they also can provide information on what terminology standards are used in the local E&C industry. A Web site has been developed in the United States that provides Japanese translations of English engineering and construction terminology, and it is located at www.engr.sjsu.edu/mcmullin/service/bilingual/Default.htm.

One example that demonstrates the difficulties associated with translating words into other languages occurred when the words *Coca-Cola* (*Coke*) were being translated into Mandarin. The Mandarin characters for Coca-Cola translated to "bite the wax tadpole" or "beat the mare with wax," neither of which was a good representation of the soft-drink product. Different characters had to be selected that translated into "may the mouth rejoice" (Copeland and Griggs, 2001, p. 97). Another example is the name of the clown that represents the fast-food chain from the United States called McDonald's, who is called Ronald McDonald in the United States. Many Chinese have difficulty pronouncing the "R" sound, so the clown's name in China was changed to Donald McDonald.

One engineer from the United States tried to bring root-beer extract into Libya, but customs officials seized it, and the engineer was sent to jail for 6 weeks while his firm negotiated for his release. Customs officials in Libya assumed that root beer was an alcoholic beverage, and alcohol is strictly forbidden in Islamic countries, so the literal translation of *beer* caused the engineer to lose his right to work in Libya. In Southeast Asia, there is a soft drink known as ginger beer that is not an alcoholic beverage (it is a tangy version of ginger ale), but it causes problems for expatriates in Islamic Indonesia if they are seen purchasing or consuming it. Examples such as these demonstrate how essential it is to have proper translations and the importance of finding out how a term translates into other languages.

Words or phrases in contracts and specifications and other E&C project documents should be changed if they are too difficult to pronounce or if their meaning would be different in a local language. Engineering and construction professionals have had their visas revoked by foreign government officials for cultural offenses that some of them were not even aware of before they were told to leave the country.

2.3.1 Methods for Translating Information

On engineering plans, colored highlighters can be used to color-code sections of drawings and construction materials. If this technique is used, it is irrelevant whether workers know the difference between a 2 by 4 or a 2 by 6 as long as the plans indicate which material is used in each location.

Portable electronic translators can be used to translate words into other languages, but they can only translate English, German, French, Spanish, Italian, Portuguese, Russian, Chinese, Japanese, Korean, Greek, Hebrew, Dutch, Arabic, Swedish, and Norwegian, and they do not have the capability to translate technical words or phrases. Companies that sell electronic translators include Lingo Corporation (New York and Hong Kong) and Franklin Electronic Publishers (Burlington, New Jersey).

If human translators are used on E&C projects, they should be familiar with the technical terms of the trade. If they are not, a second translator also should be used so that one translator can rapidly translate conversations or documents while the other one is translating technical terms. Translators are not always familiar with local dialects, and their translations could merely be a summation of only the material they could understand from a conversation or document. Local translators do not have a vested interest in translating information properly, so it is better to use a translator from the native country of a firm. According to project manager Mark Reiser, who is the vice president of operations for a developer in the United States and who has managed a work force where over 50 percent of the workers spoke Spanish, "When over half your work force speaks another language, it is time for you [the project manager] to learn their language, even if you are working in your native country."

Written documents, such as contracts, specifications, and agreements, should be translated into a foreign language, and then the documents should be translated back into a native language by a different translator to ensure that they have the same meaning as the original documents when they are translated back into the original language. Contract negotiations normally are conducted in the language of the client, so translators always should be included in negotiating sessions.

2.3.2 Language Issues in Contracts, Specifications, and Plans

2.3.2.1 Contracts

Contracts are not evaluated consistently everywhere in the world. In Germany, even if items are not included in a contract, they still may have to be performed because contracts are subject to the *German Civil Code,* which has details on standard business practices that could be enforced on a project or that could override contract clauses. In Japan, a contract is interpreted to be statements of general business practices, not the finite details that are found in contracts in the United States. In Mexico, contracts are considered to be artistic ideals that may not be applied consistently. In Greece, contracts are interpreted to be an agreement to build future business (Copeland and Griggs, 2001). A person's word in Arab cultures is more binding than a written contract, as is the case in countries such as Afghanistan or Yemen, where people are bound more by their tribal (or clan) affiliations than by laws.

Being concise in international contracts avoids misinterpretations, and words such as the ones listed in Table 2.1 should not be used in international contracts and specifications or in any type of document that will be used for global E&C projects owing to their imprecise nature. All the terms listed in Table 2.1 were extracted from contracts and specifications. Additional information about international contracts and legal systems is provided in Chapter 7.

Using any of the terms in Table 2.1 in contract clauses or specifications makes it difficult to enforce the contract clauses that contain them in any meaningful way. Other colloquial terms, such as *canary yellow, battleship gray, acts of God, and force majeure* also cause confusion when they are used in contracts because they are regional terms that are not understood by people in foreign countries. The term *acts of God* cannot be used in Communist countries owing to its religious connotation, and other terms such as *beyond anyone's control* have to be substituted. Contracts and specifications should be reviewed carefully, and vague terms or terms that could be misconstrued by a court of law should be eliminated from legal documents before they are issued for use in the global arena.

2.3.2.2 Engineering Plans

The numbering systems used on drawings and in specifications can be misinterpreted because most of the world uses a modern version of the International Sys-

Table 2.1 Vague Terms That Are Inappropriate for E&C Contracts

A few	Convenient	Just about	Reddish
A lot	Correctly	Large	Regularly
Able	Damage	Later	Relevant
About	Decent	Light	Required
Above	Detailed	Like	Respectable
Acceptable	Different	Little	Responsible
Accordingly	Discrepancy	Local	Safe
Adequate	Earliest	Local matters	Safe distance
Agreed upon	Easiness	Long	Satisfactory
Alleviate	Efficient	Maybe	Satisfying
Almost	Elements	Medium	Scheduled
Anticipated	Encumbrance	New	Secure
Apparent	Enough	Often	Semi-
Applicable	Excessive	Old	Short
Appropriate	Expensive	Or equal	Significant
Approximate	Expressly	Overall	Similar
Area	Extraordinary	Perfectly	Simple
Around	Extreme heat	Perhaps	Slanted
Average	Fair judgment	Pipe	Small
At least	Familiar	Pleasant	Smaller
Attribute	Feasible	Plenty	Soft
Bad	Few	Possible	Some
Bargain	Firm	Practical	Sound
Better	Foreign	Practically	Standard
Big	General	Precisely	Substantial
Bigger	Good	Probably	Such
Blue	Greater	Proficient	Such as
Bright	Hard	Proximity	Support
Clean up	Harmless	Qualities of	Thing
Clearly	Heavy	Quiet	Valuable
Close	Hereunder	Quick	Value
Comparable	Improvements	Really	Very
Competent	Included	Reasonable	

tem of Units (the metric system), with the exception of the United States (although it was legalized in the United States in 1866). References are available that contain conversion factors for units of weights and measures, called conversion tables and scientific calculators and electronic translators have conversion functions.

When dimensions are listed on drawings, engineers assume that they are in meters and centimeters if the metric system is their native system of units, or they assume that they are in feet and inches if they are from the United States. If the plans do not state which system is being used, or if the inch symbol is not used, or if only the first page of the plans (the index) indicates that the metric system is being used and the pages are later separated, then gross errors are introduced into projects.

Globally, dimensions are labeled in several different ways, such as 7′6″, 7.5′, 7.5″, 7.5, 7 feet 6 inches, 7.5 feet, 7.5 inches, .75 inches, 7 meters 6 centimeters, 7.5 meters, 7.5 centimeters, .75 meters, and .75 centimeters. Sometimes the foot and inch symbols or decimal places and commas appear to be the same when blueprints are darker or lighter than the original plans or if the plans have been copied on a copier or reduced in size. If the design engineer of record is not located in the country where the project is being built, he or she may be hard to reach to clarify dimensions. The dimension system used on drawings needs to be consistent throughout the plans and specifications and explained in several different places so that 7.5 is not interpreted as 7.5 inches, 7.5 feet, 7.5 centimeters, or 7.5 meters by different people working on the same project. To avoid confusion, dimensions can be listed as 7 feet 6 inches, or 7 meters 6 centimeters, but using this system is a time-consuming process and too cumbersome on drawings. If the drawings have been reduced or reproduced on copiers, a note should be included that contains a scale that correlates with the reduced or copied versions. *Not to scale* (NTS) is a standard phrase that is used to indicate that the drawings have been altered in some manner from the original version.

The standard for lumber (2 by 4, 2 by 6, 2 by 8, 2 by 10, and so forth) also causes confusion because personnel may not understand that a 2 by 4 is actually $1\frac{3}{4}$ inches by $3\frac{3}{4}$ inches (nominal versus actual dimensions) as a result of 2 by 4s being planed to obtain a smooth surface. Yet a piece of plywood is designated as a 4 by 8, and it is actually 4 feet by 8 feet.

Symbols that are familiar to engineers in one country could be completely foreign to engineers in other countries. One way to make sure that symbols are not misinterpreted is to provide handouts of standard symbols along with the plans (charts of standard symbols are available in drafting books), and the terms used for symbols should be translated into local languages.

In many countries, people have never seen items such as overhead showers or shower stalls, bathtubs, stoves, sinks, elevators, escalators, dishwashers, garbage disposals, central heating and air-conditioning, self-flushing toilets, electronic garage door openers, revolving doors, or automatic door openers. Providing pictures of these items along with plans and specifications helps workers have a better idea of what is supposed to be built.

Two examples of symbols that are not the same in Eastern and Western cultures are toilets and bathtubs. In Western cultures, toilets are used while sitting on a seat that is approximately 18 inches high and that has a lid, or they are used in a standing position in front of the toilet with the lid on the toilet raised. Western toilets are flushed by depressing a handle on the front, top, or side of the toilet. Eastern toilets are flush to the floor, and they are used in a squatting position. They may be a porcelain bowl (or merely a hole in the ground) with "footprints," or pieces of wood, that are used to position feet. They do not have Western flushing mechanisms, so water is poured into them from a separate container. Figure 2.1 shows an Eastern-style toilet and tub. In Western cultures, toilets are drawn as an oblong circle with a rectangular box attached to one end, and in Eastern cultures,

Figure 2.1 Eastern-style tub and toilet, Borneo, Indonesia.

toilets are drawn using a circle with a small oval on two sides of the circle. In Western cultures, bathtubs are a fixture that a person either lays in or sits in. In Eastern cultures, they are a container that holds water that is ladled over the body using a pot (users stand on the floor, which contains a drain).

2.3.2.3 Specifications

Another concern related to specifications is the issue of colors, because colors are interpreted differently throughout the world. The color green represents parks and recreation in the United States, but it is associated with disease in countries with dense green jungles. Green is a favorite color in Saudi Arabia, but it is a forbidden color in parts of Indonesia. In Japan, green is used in the high-tech industry, but you do not see it used in the high-tech industry in the United States (the United States uses putty, black, and silver). For mourning, black is worn in the United States and Italy, but in Asia people wear white while they are in mourning. Other colors that signify mourning are purple in Brazil, yellow in Mexico, and dark red in the Ivory Coast. Blue is a masculine color in the United States, but red is masculine in the United Kingdom. Red is a sign of good fortune in China, and the front doors of structures could be painted red. Pink is a feminine color in the

United States, but yellow is feminine in most of the rest of the world (Copeland and Griggs, 1984). Since specifications stipulate specific colors, regional differences should be a consideration when designing projects and writing specifications.

Numbers cause similar problems because the number four is considered bad luck in Japan. In China, 4 and 14 mean death, 3 and 13 mean life, and 8 and 18 mean good fortune. If a building in China has 8 (or 18) in the address, people are more likely to buy the structure or apartments in the building. Seven is unlucky in Ghana, Kenya, and Singapore, and the number 13 is bad luck in the United States.

In some buildings in the United States, there is no elevator button for the thirteenth floor, and offices and apartments on those floors are labeled as the fourteenth floor (but there is a thirteenth floor in the structure). In the United States, the number 13 means bad luck. In China, the fourth and fourteenth floors are not labeled in structures because 4 and 14 are bad luck numbers, and their pronunciation is similar to the word *death* in Mandarin.

In the United States, some hotels and office buildings have the first floor labeled as the lobby or a plaza level, and the second floor is the mezzanine, so the third floor is labeled as the first floor, and the plans will be labeled in the same manner. If *around-the-clock engineering* or *24-hour engineering* is used to design a project, the electrical engineer might be in China, the mechanical engineer could be in India, the civil engineer may be in Lithuania, and the project might be under construction in South America. Engineers in each of these countries work on projects during their working hours, then they relay their work to engineers in a time zone that is 8 hours earlier in the day and they, in turn, relay their work to engineers in a time zone that is 8 hours earlier. If 24-hour engineering is used while a project is being designed, every engineer working on the project needs to know which standards are being used on the plans to designate levels and which numbering system is being used on the project

2.4 Design Criteria for Different Cultures

This section discusses issue areas that affect the choice of technology for engineering designs.

2.4.1 Self-Reference Criterion

When planning and designing structures for foreign countries, there are factors unique to each country that should be taken into consideration. One obstacle to incorporating cultural issues into engineering designs is called *self-reference criterion,* and it is "the habit of seeing everything in relation to what one would want or do or feel in any situation, but the problem is that those interests, tastes, and values are not relevant" (Copeland and Griggs, 1984, p. 67).

Engineers tend to create designs that are familiar to them or that are similar to the ones they learned to design in college or on the job. When they use their expertise to create designs for a foreign country, their designs could be irrelevant to the cultural needs of that country. Even within one country people have different needs depending on: whether they are rural or urban populations or live in religious sectors; what are the ethnic and tribal affiliations, historical influences, and technical capabilities; and the availability of materials.

One project in Saudi Arabia required 27 unique precast concrete elements to be incorporated into the design of a city for a population of 50,000 people. Rather than using design elements of masonry blocks, steel beams and channels, or timber planks and lumber, every structure in the city had to be built using the 27 different precast concrete shapes. Using precast concrete elements is a common construction method in Saudi Arabia, but when U.S. firms bid on the construction phase of the job, their bids were substantially higher than the bids from local contractors because of a lack of familiarity with that type of construction. Not only was the civil design unusual but the mechanical and electrical systems were designed differently because all the structures lacked rectangular rooms or false ceilings.

2.4.2 Global Design Considerations

Other global considerations that affect how structures are designed and built include:

- Unavailability of materials
- High cost of transporting materials
- Lack of access roads, railways, or docking facilities for transporting materials
- Limitations on the number of skilled laborers or government-imposed regulations on the use of large numbers of laborers
- Government restrictions on the importation of certain products or materials and the availability of replacement parts to repair facilities
- Theft of materials from job sites
- Having to build redundant systems to have spare parts immediately available
- Differing soil conditions at the site than what is shown on the plans and specifications
- Lack of appropriate equipment, skilled operators, or tools
- Complexity of plans and on-site modifications of plans

Figure 2.2 shows a brick factory that is located in China to demonstrate the limited capacity of some indigenous material suppliers in developing and newly emerging nations.

Other influences that affect designs include the availability of utilities both during construction and when the structures are inhabited after construction. Throughout the world, the supply of electricity, diesel fuel (diesel generators used in China consume larger amounts of oil than natural gas–heating systems), and

Figure 2.2 Brick factory, China.

natural gas is intermittent (or nonexistent). The technology used on a project could be too advanced to be tied into existing utility systems. One design modification that addresses this problem is to have alternative sources of energy, such as thermal storage floors, natural light, fireplaces that interface with heating systems, biomass (powered by animal waste) sources of energy, windmills, solar panels, and other "green" technologies that can be integrated into designs along with traditional electrical and mechanical systems.

Designs used in some countries are inappropriate for other countries because of cultural nuances. In many parts of the Arab world and in most Asian countries, living space is limited, so it is used as all-purpose space—for sleeping, cooking and eating, watching television (if it is available), entertaining, playing, and studying. Chairs, tables, and beds are used in Western cultures, but cushions, rugs, mats, and futons are more common in Eastern cultures because they can be used both during the day and at night. It is difficult to modify Western structures to suit the needs of Eastern cultures because Western structures have more rooms that are smaller rather than larger multipurpose rooms.

In some parts of the People's Republic of China, single dwellings have been divided up by the government to accommodate multiple families, with space allocated to each family by the government. Design requirements for Chinese structures are different from those of Western nations, to allow some measure of privacy and access to natural light, because multiple families may have to share so much of their space, including communal baths, kitchens, stairs, and other areas.

Cultural factors that should be considered when designing structures for foreign countries include:

- Family structures
- Social and business organizational structures
- Education levels and how many people are educated
- Standard of living
- Levels of materialism
- Technical orientation of the society
- Religious beliefs
- Attitudes toward women, not in terms of equal rights but whether designs need to accommodate the requirement for women not to be seen by male visitors who are not members of the immediate family
- Historical influences
- Societal structure, whether a tribal society
- Cultural and governmental influences on the concept of privacy
- Government restrictions and regulations

Culture can influence something as simple as the placement of light switches or as complex as the overall design of a structure. If cultural issues are not explored prior to the initiation of a design, changes to accommodate them later may be more expensive.

2.4.2.1 Religious Influences on Designs

Religious influences are evident in designs throughout the world, whether they are for buildings resembling cathedrals in Europe or South America or structures having large open spaces similar to mosques in Islamic countries (offices in Muslim countries may use large pillows or rugs in open spaces, and business may be conducted sitting on the floor). In Asian cultures, the Buddhist influence mandates that everything should be in harmony. Figure 2.3 shows the architecture of the Royal Palace in Bangkok, Thailand (formerly Siam), that influenced the design of structures throughout Thailand. Knowing about the exteriors and interior layouts of indigenous religious structures, government structures, and palaces provides insight into what are considered appropriate designs for structures in foreign countries.

2.4.2.2 Asian Influence on Design

Fung Shui or feng shui translates into "wind and water," and it is used in Eastern cultures to create an ideal environment to live and work in by evaluating the interaction between humans and buildings. Fung Shui is an ancient Chinese custom that examines the relationships among time, the environment, people, and structures. Fung Shui requires that all aspects of a structure be evaluated by a Fung Shui Master, who determines the date to start construction, the type of design, the number of floors, and the arrangement of rooms in the structure (even the arrangement of furniture may be evaluated). If a structure is not evaluated as a good

Figure 2.3 The Royal Palace, Bangkok, Thailand.

structure in terms of Fung Shui, it may harm the reputation of the engineering firm that designed it.

An elaborate Fung Shui setting helps to attract wealth and dispels or appeases evil spirits. A rectangular structure indicates that everything in one's life is in balance. Fung Shui customs include:

- Building so that entrances do not face west because west is the direction of the road to the end of life.
- The number 14 in Chinese is pronounced in a manner similar to the word *death,* so buildings do not have a fourteenth floor.
- Door openings should not face directly down stairs because that indicates that wealth is going down the stairs and away from the structure.
- If a door faces the end of a street, bad *chi* can enter a home. Plants, wind chimes, or mirrors can ward off bad *chi.*
- Furniture should face doors or windows not walls.

In Hong Kong, one design considered for a high-rise building was designed with the building blocking the mountains. When a Fung Shui master suggested that it would bring bad luck to people living in the building and the surrounding area, the owner had the building redesigned with a large, rectangular gap in the center of the building. Another design for a high-rise structure in Hong Kong resembled a white candle, and white candles are only used for funerals in Asia. Therefore, the owner added a swimming pool on top of the building because water "snuffs out" the fire of a candle. Therefore, the swimming pool could counteract the building's bad *chi*. Another structure in Hong Kong that has sharp edges on its exterior (it resembles a knife) is said to be cutting through the good *chi*.

2.4.2.3 Local Influences on Designs

Construction projects in developing countries or where there are overcrowded cities could include unrelated structures such as a tree farm that has a dentist office or a power plant that includes a water treatment facility because it may be the only means of having structures built that supply basic services. In remote locations, contractors usually are responsible for providing support structures before building primary structures, and support structures might include housing and/or dormitories, utility and road systems, sanitation facilities, sports facilities, medical facilities, a commissary, a recreation center, religious buildings, and mess halls (cafeterias that serve food to workers). In remote regions of the world, while community structures are being constructed, engineers and constructors might be required to live in temporary housing that could include any of the following:

- Trailers
- Mobile homes
- Tents
- Grass huts
- Mud houses
- Corrugated-steel houses
- Woven-straw houses
- Simple wooden houses (built on 4-foot-high footings to allow the monsoon rains to flow under them)

Global engineers and constructors at one time or another probably will work on designs that seem inefficient, but the main purpose of the designs may be to employ large numbers of people or to use local materials and processes. In developing countries, if a facility is too technically advanced, local personnel will not be able to operate the facility once foreign E&C personnel have left the country. Having technology transfer requirements on projects encourages native workers to learn how to operate facilities, but if the institutional memory is neglected, facilities are shut down temporarily or are abandoned because subsequent generations no longer know how to operate them.

Engineers might be required to provide elaborate designs in situations where a simple design would suffice because clients might assume that they are being cheated if a design is too simple (this occurs when clients have unlimited funds and they want the best that money can buy, such as is the case in Dubai in the United Arab Emirates). Examining existing structures within a country provides insight into what types of designs are appropriate for each culture.

If a design is unfamiliar, native workers need to be supplied with the equivalent of shop drawings (engineering designs that have elaborate details) or a small model so that they can visualize what the structure is suppose to look like before they perform the work. In some cultures, workers have never seen structures similar to the ones they are trying to construct for foreign firms.

2.5 Cultural Issues That Affect Engineers and Constructors

Cultural issues that are specific to certain regions of the world are discussed in Chapter 14. Therefore, this section only discusses general cultural issues that affect E&C personnel on global projects.

2.5.1 Cultures within Cultures

Global E&C personnel need to be aware of *cultures within cultures,* because people in one region of a country may be quite different from people in other regions. Published information about a particular country may only contain generalizations about the country, its people, and its culture. Relying on only published information can result in cultural insensitivity, so it is better if a person spends time observing local people before attempting to interact with them.

If foreigners try to speak a few words of the local language, it is always appreciated, and native speakers will help them pronounce words and phrases correctly. Attempting to communicate in the language of a host country demonstrates that you are interested in the local culture. Knowing anything about a local culture, the history of a country, its geography, or locally famous people also demonstrates interest in a country to its native citizens.

When someone is working with foreigners, it is more important to have an understanding of *why they are the way they are* than to know vast amounts of data about their native country. Clues to behavior can be found in religious beliefs, history, social customs, politics, and interpersonal interactions.

2.5.1.1 Censorship

Citizens of Communist countries may not have any idea how their governments operate because of censorship. In order to find out how citizens really view their country, read local graffiti. In Palestine, one graffiti artist stated that "Graffiti should comfort the disturbed and disturb the comfortable" (BBC News, September 16, 2005). Graffiti may be the only way for oppressed people to express their frustration or anger without fear of reprisals. In China, *diazabo* are illegal bulletin boards where people post political statements, poetry, complaints, or other expressions of discontent in locations where there is a great deal of pedestrian or automobile traffic. In China, the *diazabo* and illegal magazines are the main source of nongovernment information. Posters are posted quickly onto walls while a crowd surrounds the wall to protect the author from being arrested by government officials.

It is useful for global E&C personnel to know that Eastern cultures put great value on economic, social, and cultural rights and Western cultures put value on civil and political rights. Freedoms that people are used to in Western cultures are completely foreign to people in Eastern cultures. Therefore, citizens of Eastern countries are not motivated by the same things that motivate people in Western cultures. In some Western cultures, people are defined by what they do. In Eastern cultures, identity comes from a person's religion, family, village, or clan (tribe). In Asian cultures, it is important for a person to sell himself or herself first (establish personal integrity), then sell a company, and then sell a product or service. If the last step is done first, clients are not interested in the product or service. When operating in other cultures, it is not polite to always talk about how things are done in another culture, because the focus should be on the native country of coworkers, business associates, clients, suppliers, and subcontractors.

2.5.2 Issues Related to Status

Status (a person's position in society relative to other people) is still significant in many cultures, and if young personnel are sent to work on projects in countries where there is a rigid class system, they will not be afforded the respect they are used to receiving in their native country. Many cultures still revere their elders, and seniority improves one's status and position within a firm. In these cultures, business associates will only deal with the oldest person on a team, even if a younger person is in charge of the project or office.

Since titles are interpreted differently throughout the world, titles on business cards are not a good indication of a person's status. Cues that can be used to determine the status of a person within his or her society include:

- How someone sits in his or her chairs (whether someone is slouching or sitting up straight). In China, South Korea, or Japan, every person in a meeting will be sitting up straight with their hands in their laps or at their sides and never on a table.
- How someone speaks and in what tone.
- Where people stand when they are in a group.
- Who enters and exits a room first and who sits in what position at a table. (In Western cultures, the person in charge sits at the head of the table, and in Asian cultures, the person in charge sits at the middle position of the table.)
- Who is the first person to start eating during a business meal or the first person to offer food to guests.
- Who is the first person to start speaking in a business meeting.
- Who is not carrying anything. In Eastern cultures, the higher a person's status, the less likely he or she will be carrying his or her own briefcase, computer, or documents.
- The size of someone's home (managers have larger homes or apartments) and the type of car he or she owns.
- Whether someone has a driver for his or her vehicle or drives his or her own vehicle.

- The quality of clothing. (People wear business suits throughout the world, but there are subtle differences in the quality of material, the cut of the suit, the fabric used in shirts and ties, and the quality and shine of shoes.)

Office size and location are not always an indicator of status. A large office or an office with windows indicates higher status in Western cultures. In Eastern cultures, having an office that is not shared may indicate lower status, because most managers sit among their subordinates.

The higher a person's status is within a company, the more likely executives rather than their assistants will greet them. An unforgivable mistake is to assume that the person greeting someone is a driver or an assistant and to treat him or her with less respect than an executive.

There are caste systems (a social ranking system) in many parts of the world, and people who live where there are caste systems who are from a higher social class will not take instructions from nor work for people with a lower social rank. Individuals with higher status will not perform work that is considered to be beneath them or that is normally performed by members of a lower caste.

In Eastern cultures, if thumbnails or one thumbnail is considerably longer than other fingernails, it could be a sign that a person is not from the working class (he or she does not use his or her hands to make a living). Other ways of determining where someone fits in a caste system are to inquire as to where they were educated, whether they come from a large or a small city, and what their parents do or did for a living. However, if personal inquiries are considered rude in the local culture in a business setting then asking these types of questions is considered rude.

In most of the world, business attire advertises someone's rank within an organization. Business attire could be a suit, but it also can be other forms of native dress that are considered to be business attire for a particular country. Examples include batik shirts in Indonesia (shirts with patterns created by hand drawn wax symbols), Mao Zedong clothing in China, Nehru jackets or vests in India (no collar with a small opening at the neck), and embroidered shirts in Mexico. Suits are always acceptable attire for business in any country until someone can determine what forms of local dress are also acceptable for work. For women, it is important to respect the local culture, and if you are in an Islamic country, arms, legs, and possibly the head and face should be covered in public. In some Islamic countries, if a woman is wearing slacks (pants) in public without a chador (veiled-over garments), she could be mistaken for a prostitute. Shorts are rarely seen on adults outside the United States, and business personnel should not wear jeans to work nor outside of work to social engagements with coworkers.

2.6 Suggestions for Adapting to Foreign Cultures

This section provides broad examples of some of the types of cultural issues that cause problems for E&C personnel when they are living in unfamiliar cultures.

2.6.1 Avoiding Generalizations

Avoid generalizations about regions of the world. For example, Japanese, Chinese, and Koreans have unique heritages, and their customs, languages, and attitudes are quite different. If a country was invaded and controlled by another government in the past, its citizens still may be resentful toward their conquerors (such as how some of the Chinese and South Koreans feel about the Japanese). An example of an inaccurate generalization is the notion that Japanese people are always really calm. In reality, some Japanese companies provide padded rooms where employees use baseball bats to hit posters of executives, or employees are required to go through training where they receive "badges of shame" (physical badges that they must wear during training programs) if they are unable to attain desired goals. Some Japanese training programs require potential executives to scream a song while they are standing in front of a busy subway station entrance.

2.6.2 Business Cards

Small gestures such as a slight bow while giving someone a business card in China, Japan, or South Korea, with two hands, and taking a business card from someone else with reverence (rather than putting it in a pocket without looking at it) are an essential part of conducting global business. Some type of comment is always appropriate when a business card is handed to someone because it demonstrates that the business card is being read by the recipient. Business cards should be handed personally to everyone present at a meeting. Business cards should be translated into the local language, and the translation should be printed on the back of the business cards because most people understand languages better when they are written rather than spoken by foreigners (everyone's accent sounds strange somewhere in the world).

2.6.3 Losing Face

In every culture in the world, people do not like to be publicly ridiculed, and in Asian cultures, public ridicule is called *losing face*. When people are concerned about not losing face (being embarrassed), they may write everything down that they plan to say at meetings, and they will read it, or they will memorize everything they will say during meetings. If someone loses face at work, he or she might not return to work, and if the embarrassment is severe, people have been known to commit suicide to atone for the embarrassment to their firm, their family, or their country. In some cultures it is rude to interrupt people while they are talking (Asia), but in other cultures if no one interrupts, a speaker will continue talking until a meeting or gathering is over.

In feudal cultures, insults are not easily forgotten, nor forgiven, and they are passed down from generation to generation. A personal insult in an Arab culture could be interpreted as an insult to an entire family. Nothing may be said when an insult takes place, but something could be done years later for revenge.

People whose native language is not being spoken need time to translate what is being said into their language, time to think about their answer, and then time to translate their answer back into the first language. A common practice is for people to raise their voice when a foreigner does not respond to a question quickly or if they look perplexed after they are asked a question. Instead of repeating the question in a louder voice, the question should be asked in a different manner.

2.6.4 Recommendations for Presentations and Meetings

When engineers or constructors give presentations, clients are judging them on their technical expertise and in relation to local cultural standards. Major differences between Eastern and Western cultures surface during E&C proposal presentations. In some Western cultures (such as the United States), engineers and constructors use charts, graphs, and flipcharts, and they write on them during the presentations. In Eastern cultures, this is interpreted as a sign that someone is not prepared to give their presentation, because they have to add things during the presentation. Well-rehearsed, polished presentations should be the only type of presentation given in Eastern cultures. In Asian cultures, a presenter should never sit on the edge of a table during the presentation because members of the audience do not understand why someone would disrespect a table or other object that is not intended to be sat on (especially if they eat on the same table).

Meetings are conducted differently in every country, and they vary from autocratic lectures to free-form discussions. To find out what is acceptable, observe local meetings before trying to conduct a meeting or participate in them to learn about what is expected of meeting chairs and team members. Watching meetings provides insight into whether interrupting someone is acceptable or whether someone should wait until a speaker is finished and they are recognized by the chair of the meeting before they speak in a meeting.

When engineers and constructors are presenting proposals, they should not provide all the technical information required for a project because clients could duplicate technical processes if they have enough information.

Dress appropriately for the culture where presentations or meetings are being held, and pay attention to details when imitating other forms of dress, such as the length of pants, how long suit jackets are, whether jackets match pants, if shoes are polished, whether shoes have leather or rubber soles, what fabric is used for clothing, how ties are tied, and other small details, because it is these details that distinguish leaders from their subordinates.

Traffic congestion is prevalent throughout the world, and meeting times are affected not by the distance people must travel but also by the travel time required to go that distance. In many parts of the world, people use public transportation or bicycles, and they are too embarrassed to tell this to a foreigner when they are late for meetings. When someone arrives at a meeting late they should quietly apologize for being late because no other explanation is necessary, as it wastes the time of the people attending meetings.

2.6.5 Gestures and Behavior

There are some gestures and behaviors that are unacceptable in certain regions of the world, and this section provides examples of a few unacceptable gestures. Never touch anyone that is from an Islamic culture with the left hand because the left hand is used for cleaning up in the restroom and not for touching other people or eating. In Islamic cultures, if a dog is touched, the hands have to be ritually washed seven times. Shaking an index finger in someone's face and giving the thumbs up sign are obscene gestures in some Islamic countries.

Women should never be approached or touched in Islamic cultures, because it could result in severe punishment to the women. In Saudi Arabia, if a woman from a traditional family dates a foreigner, she could be punished by imprisonment or killed in front of her family for shaming them.

Shouting and vehement refusals are part of the Islamic business culture, and they are not signals that a business meeting is over, merely that more bargaining needs to take place. Giving in too early during a negotiation could be interpreted as a sign of weakness. When working in an Islamic country, it is wise to hire a local driver. If a foreigner accidentally kills someone in a car accident, the penalties are severe, and it is best to leave the area immediately (sometimes leaving the country is advisable).

In Western cultures, people point with their index finger, but in Eastern and Islamic cultures, pointing with the index finger is rude. Even pointing with a foot (when legs are crossed) is rude, and tables have cloths around the front of them to hide the feet of people sitting there so that their feet are not pointing at the people sitting across from them. If it is absolutely necessary to point at something, point with a thumb.

Hugging business associates is acceptable behavior in countries in the Middle East, South America, or France, but people from other countries shake hands in business situations. The length of time someone holds onto someone else's hand when they shake it also varies throughout the world. People from Arab cultures might hold onto the other person's hand while they are shaking it, and they may pump the hand repeatedly. In the United States, handshakes are abrupt. Asian handshakes are gentle so as to keep from offending someone. Some people shake so aggressively that they leave the other person with sore fingers.

There are situations that are acceptable in some cultures but not in others. For example, people in Asian cultures may not use handkerchiefs in front of other people, and sometimes they blow the contents of their nose onto the ground (they consider handkerchiefs to be disgusting). This behavior is offensive to people in industrialized nations. In the United States, Italy, and China, people talk loudly in public places or on their cell phones. Checking e-mail, or voice-mail messages while talking with someone personally is rude in most cultures. People might ask permission to speak if they are from India, or they may interrupt and start talking without being recognized, as is done in New York City. In Arab and Asian cultures, beverages or food are always offered to guests, and it is rude to refuse the offer.

2.6.6 Passive-Aggressive Behavior

In countries ruled by dictators or benign dictators (where there are elections but citizens have no choice but to vote for the person or party currently in office), people realize they have little power, so they may resort to passive-aggressive behavior when they receive a bad evaluation or they are feeling out of control. In construction, passive-aggressive behavior is manifested by workers slowing down their work, missing deadlines, pretending not to know how to do something, becoming argumentative, pretending not to understand instructions, and doing work incorrectly on purpose.

Another reason for passive-aggressive behavior is when there are large disparities in compensation between management and workers, such as ostentatious housing or cars for management and spartan accommodations for workers. Knowing the source of passive-aggressive behavior helps a manager to figure out how to deal with it. If people have no outlet for their aggression, or if they have no control over their lives, it becomes more difficult to manage them.

2.6.7 Historical Heritage Issues

In countries that were colonized by European countries, foreigners still could be referred to as *sir* (even if the person is female). The title of sir is used for all levels of foreign personnel, so it is not an indication of who is in charge. In some parts of Africa, Asia, and Southeast Asia, there are people who have never seen Europeans (or people of European descent), so it is a novelty to them when someone has blue or green eyes, freckles, or red or blond hair. Their curiosity could lead them to trying to rub freckles off of someone's skin because they think that freckles are dots put on the skin with ink, or they may touch or pet someone's hair.

2.7 Economics

Many of the issues related to economics are discussed in other chapters under areas such as competitiveness, global alliances, financing of projects, and currency fluctuations. Basic economic information, such as gross domestic product (GDP) per capita for every country in the world, is published by the World Bank, the *TIME Almanac,* and the Central Intelligence Agency *World Factbook* Web site (World Bank, 2005; Bruner, 2005; CIA, 2005).

2.8 Politics

In a study conducted in Southeast Asia, E&C personnel from Japan, Taiwan, South Korea, and Indonesia indicated that it was not important for foreign E&C personnel to know about politics in the project country unless it was something that would directly affect their work (Yates, 1991). Knowing what political system is in power (and what it means) is only important to E&C personnel if there are signs of an imminent coup or the nationalization of private facilities. Politics might influence who can be hired to work on E&C projects, and the firing of personnel could be affected by government regulations. Governments set requirements for safety and environmental standards, and these are discussed in Chapters 8 and 9.

2.9 Summary

This chapter discussed cultural issues that influence global E&C personnel. The first section provided information on government and nongovernmental sources of information on securing work in foreign countries. Language and translation issues that affect the design of structures were included in the second section, along with how language affects contracts and specifications.

The third section presented information on design criteria for different cultures, and the fourth section explained technology considerations that affect the design of foreign projects. General cultural issues that affect engineering and construction personnel were presented in the fifth section, along with suggestions on how to adapt to foreign cultures. Other subjects discussed in this chapter included global issues related to status and education, presentations and meetings, gestures and behavior, passive-aggressive behavior, and economics and politics.

DISCUSSION QUESTIONS

1. Discuss which global issue areas E&C personnel should investigate before working in a foreign country.
2. Discuss which of the sources listed in this chapter could provide information on foreign legal requirements, cultural nuances, bid requirements, economic data, and visa requirements.
3. Discuss who should be contacted before materials and supplies are incorporated into global designs.
4. Explain why proper translations are essential for the engineering plans, contracts, and specifications that are used to build global projects. Provide two examples of technical E&C terms that cannot be translated into other languages.
5. Explain why electronic translators are not useful for E&C projects.
6. List 10 vague terms that are not included in Table 2.1 that should not be used in global contracts and specifications.
7. Discuss what would be the most effective way to list dimensions on engineering plans that are used for global construction projects.

8. Discuss why having an understanding of what colors and numbers mean in each culture is important to E&C personnel.
9. Discuss another method for communicating what symbols mean on engineering plans besides the one mentioned in this chapter.
10. Discuss four ways that E&C personnel could move beyond their self-reference criterion when they are designing and building construction projects.
11. What are the five most important global design considerations, and why are they the most important ones?
12. Explain why structures in Western cultures look so different from structures in Eastern cultures.
13. What are the five most important cultural factors, and why are they the most important ones?
14. How does *Feng Shui* influence the design of structures in Eastern cultures?
15. What numbers should be avoided if possible when designing global projects, and why should these numbers be excluded from global projects?
16. Explain why some global construction projects employ more people than are really necessary to build projects and why more elaborate structures could be required than are necessary.
17. How do cultures within cultures affect the design and construction of global engineering and construction projects?
18. Why is it essential that E&C personnel avoid generalizations about foreign cultures when they are working with personnel from foreign nations?
19. What is the proper way to distribute and receive business cards in Eastern cultures?
20. How could the concept of losing face affect the behavior of project management team members that are managing workers who are from Eastern cultures?
21. Why is it important to know how to conduct presentations and meetings in foreign cultures?
22. What gestures and behaviors should never be used in Islamic countries or while working with people of the Islamic faith?
23. Why do project management team members need to understand passive-aggressive behavior and why do people use that behavior when they are working on construction projects?
24. How should someone react, or what should be said, when a native worker either pets their hair or tries to rub freckles off their skin?
25. When are politics an issue for E&C personnel on global construction projects?

REFERENCES

CIA. 2006. *Central Intelligence Agency World Fact Book.* Washington, D.C.: Central Intelligence Agency, United States Government.

Copeland, L., and Griggs, L. 2001. *Going International.* New York: Random House.

Bruner, B., ed. 2005. *TIME Almanac and Information Please.* Needham, Ma.: Pearson Education.

Webster's New University Unabridged Dictionary. 2005. New York: New World Dictionaries/Simon and Schuster.

World Bank. 2005. *World Development Report.* Washington, D.C.: World Bank.

Yates, J. K. 1991. "International Training Survey of U.S. Engineers and Constructors, American Society of Civic Engineers," *Journal of Professional Issues in Engineering Education and Practice* 117 (1): 27–47.

Chapter 3

Managing Global Engineering and Construction Projects

3.1 Introduction

Engineers and constructors have to balance tradeoffs between technical and cultural objectives when they are involved in managing projects in the global arena. Global engineering and construction (E&C) environments require a framework that can be used as a guide for successful management of E&C projects. The development of an implementation program for managing projects requires an evaluation and ranking of projects with respect to project objectives and the implementation of appropriate management techniques and document-retrieval systems.

To compensate for cultural variations throughout the world, a three-dimensional model that addresses traditional, technical, and global dimensions is provided in this chapter. Projects can be evaluated, ranked, and plotted on the three-dimensional model, and it can be used as a guide for determining management strategies and implementing document-retrieval systems. The degree of sophistication required for a management information system (MIS) can be assessed based on the location of a project in the three-dimensional model.

During the past few decades, a common perception was that as developing countries industrialized, they would adopt models and MIS prevalent in developed countries, such as those used to manage E&C projects. However, traditional institutions in developing countries and in the newly industrialized countries (NICs) have proven to be remarkably resilient, persistent, and enduring, and many of these institutions have not yielded as countries have modernized their infrastructure. This reluctance to modernize illustrates how it is difficult simply to translate industrialized technology to developing countries without investigating whether what is being transferred is appropriate technology and appropriate management techniques.

The objectives for projects in developing countries and NICs differ greatly from the objectives of companies in developed nations (i.e., profit). In order to address these differences properly, the MIS used to verify the achievement of objectives needs to be adapted to supply appropriate input to management and decision processes.

The three-dimensional graphic model included in this chapter can be used as a guide by project managers to help them determine project objectives. The model also can be used by multinational corporation (MNC) executives dealing with project planning and initiation, project managers responsible for project development and implementation, and host-country project administrators who are involved with the latest technical and managerial methods. The three-dimensional model is based on the following:

1. Projects that rank high on the traditional dimension are labor-intensive and require high levels of supervision and in-depth planning.
2. Projects that rank high on the technical dimension are capital-intensive and require active participation of multinational companies.
3. Projects that rank high on the global dimension are human-intensive and require appropriate technology adaptable to local conditions.

For global project-management techniques to be successful, project managers should have a thorough understanding of their work environment and plan accordingly. The factors that should be considered in global environments include (Kerzner, 2003):

- Appropriate technology
- Economics
- Government and politics
- Operational difficulties
- Society and culture
- Technical skills

3.2 Management Functions and Project Objectives

The success or failure of projects in national or global arenas is influenced by the quality of project management and project-management team members. One aspect unique to developing countries and some of the NICs is that underlying project objectives are influenced by governments and clients. Identifying the objectives that need to be achieved on projects is crucial, and during the identification of project objectives, it becomes apparent that not all project objectives are compatible. In both developed and developing countries, project-management personnel perform traditional management functions, such as planning, organizing, staffing, directing, controlling, and coordinating. These six management functions are as important in developing countries as they are in industrialized nations. Project management personnel need to reconcile and integrate all the objectives identified for a project into a set of overall project objectives that will ensure that all parties responsible for project implementation have a similar understanding of the objectives.

3.2.1 Underlying Project-Management Objectives in the Global Arena

When project-management teams execute the scope of work for a project properly, its technical requirements are achieved, but the cultural dimension is equally important because a project cannot be accomplished without the goodwill, commitment, and direct guidance of political leaders and the full support of all the social groups within a country that are interested in promoting economic growth and social progress. Therefore, successful project management is a tradeoff between technical and sociocultural objectives.

As with overall objectives, subobjectives in developing countries are not always compatible with some of the primary objectives set for projects. The primary objective of developing countries (as stated in most national development plans) is to raise the standard of living for citizens by expanding output and providing basic services and facilities for educational and cultural activities. The subobjectives of a host country might be to provide jobs and welfare; promote the development of rural areas; preserve national resources, cultural values, and tribal values; promote training and the use of local labor; increase the use of local materials; and promote traditional and appropriate technology. Since many of these subobjectives are not mutually compatible, it might be necessary to weigh one subobjective against another subobjective or several others.

For comparison, three countries that demonstrate differing objectives and subobjectives are Saudi Arabia, Singapore, and India. In Saudi Arabia, concrete construction, especially prefabricated concrete construction, is used for most types of construction because of the following conditions: substantial development in the local building materials industry that led to self-sufficiency for most construction materials, highly competitive global E&C firms within the nation that influenced and improved the efficiency of the local construction industry, the capability of the local construction industry to provide precast components and other accessories, a high level of quality control enforced by the Ministry of Housing Affairs to reduce the long-term maintenance costs of structures, and government emphasis on using traditional ideas in the design of structures.

In Singapore, the small physical size of the country led to the use of multistory construction. The government was committed to providing low-cost housing to the largely low-income population without sacrificing a decent living environment, which required not only the delivery of physical structures but also a support infrastructure. The ability to achieve this objective was enhanced by the long-term integrated planning of the government, as evidenced by the master plan of the government. Singapore, like Saudi Arabia, experienced rapid economic and industrial growth in a short period of time that resulted in the sudden expansion of the E&C industry. But Singapore was faced with a shortage of construction laborers because of rapid industrial growth and the policy of the government of phasing out all foreign workers. Productivity rates were falling, which also played an important role in the Housing and Development Board of Singapore exploring innovative E&C technologies. The stable government of Singapore played a significant role in its ability to implement a long-range plan for construction.

India did not have the same characteristics that Saudi Arabia and Singapore did when it was first exploring ways to expand its construction industry to accommodate rapid industrial growth. One of the main differences is the vast diversity that exists in India that prevented its government from establishing guidelines for mandatory construction methods such as were used in Saudi Arabia. The presence of diverse technical personnel and a stable infrastructure was superseded by a shortage of decent housing and widespread poverty. Therefore, members of the E&C industry in India were concerned mainly with reducing the cost of structures by using innovative methods such as precast concrete for foundations rather than cast-in-place concrete. Not only does using precast concrete reduce costs, but it also reduces construction time. India has a 3- to 4-month wet monsoon season, when it is difficult to cast concrete in place, but concrete can be precast under controlled conditions off-site and then put in place; thus allowing other trades to continue working instead of waiting until the concrete foundation is placed and cured (or waiting for the end of the wet monsoon season).

The examples of Saudi Arabia, Singapore, and India demonstrate that although countries may have similar per-capita incomes or other similar economic factors, each country should be investigated to determine the components of its political, economic, technologic, and sociologic objectives before determining appropriate construction methods, technology, and management techniques.

The matrix in Figure 3.1 displays the relationship between the six management functions mentioned previously, sample underlying objectives, and related subobjectives for a project in a developing country derived from a sample project in the Middle East. Although the objectives and subobjectives differ from country to country, there are some overlapping objectives. Even within countries, objectives are not always compatible, and in some cases, there are direct conflicts. Project managers need to realize that conflicting objectives are inevitable and that procedures or techniques should be developed for analyzing and accommodating existing conflicts.

3.3 Three-Dimensional Project-Objective Model (POM)

One technique for quantifying and analyzing conflicting objectives is the process of evaluating, ranking, and plotting projects on a three-dimensional matrix. Figure 3.2 was developed after studying a series of multinational projects to help the parties responsible for decision making integrate separate objectives into one set of project objectives.

This figure displays the relationship between project complexity and project dimensions. Project complexity relates to the degree of project sophistication, which varies from highly complex to highly conventional, as indicated on the x axis (project complexity). The project dimension, as shown on the y axis, varies

	MANAGEMENT FUNCTION					
PROJECT OBJECTIVES	PLANNING	ORGANIZING	STAFFING	DIRECTING	CONTROLLING	COORDINATING
INTERNATIONAL PROJECT MANAGEMENT UNDERLYING OBJECTIVES:						
MULTI-NATIONAL CONSTRUCTION COMPANY'S OBJECTIVES:						
PRIMARY OBJECTIVE: P R O F I T	P	P	P	P	P	P
SUB-OBJECTIVES: COMPLETE JOB WITHIN BUDGET AND TIM	P	S	S	S	P	S
HIGH QUALITY OF WORKMANSHIP/PERFORMANCE	P		P	P	P	S
SAFE/SATISFACTORY WORKING CONDITIONS	S		S	P		P
DELEGATION OF AUTHORITY		P		P		
MOTIVATION OF HUMAN RESOURCES		P	S	P		
PROMOTE TEAMWORK		P		P		
P = PRIMARY OBJECTIVE S = SECONDARY OBJECTIVE						
HOST COUNTRY'S OBJECTIVES:						
PRIMARY OBJECTIVE: RAISE THE LEVEL OF LIVING OF ALL THE PEOPLE THROUGH EXPANDED OUTPUT AND SERVICES FOR EDUCATION, HEALTH, AND CULTURAL ACTIVITIES.	P	P	P	P	P	P
SUB-OBJECTIVES: PROVIDE JOBS/WELFARE	P	S		S	S	P
PROMOTE DEVELOPMENT OF RURAL AREAS	P	S		S	S	S
PRESERVE NATURAL RESOURCES	P	P	S	S	S	S
PRESERVE CULTURAL/TRIBAL VALUES	P	S	P	S	S	S
PROMOTE USE OF LOCAL LABOR	S	P	P	S		S
PROMOTE TRAINING OF LOCAL PERSONNEL	P				S	S
INCREASE UTILIZATION OF LOCAL MATERIALS	P			S	S	P
PROMOTE USE OF TRADITIONAL TECHNOLOGY	P	S	P	S	P	P

Figure 3.1 The relationship between the six management functions and project objectives.

in degree from highly technical projects (equipment-oriented) to highly human resources–oriented projects. Therefore, as one moves from point A to B on the figure there is a greater need for global companies to possess expertise related to sophisticated and automated MIS. Projects in the direction of point A require industrialized technology or global company services, but they also require more emphasis on supervisory personnel and management consultants. However, management methods, along with the technology employed, should be adapted to local conditions.

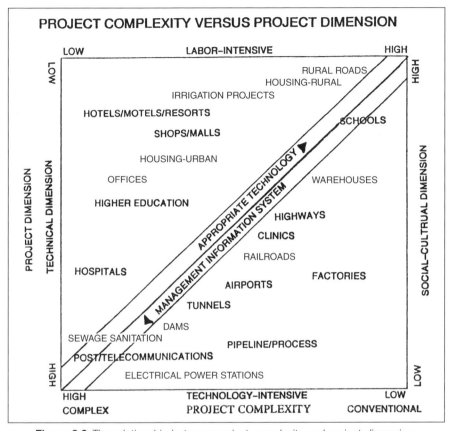

Figure 3.2 The relationship between project complexity and project dimension.

In the global market for conventional (infrastructure-oriented, labor-intensive) construction projects, firms from NICs such as China, South Korea, and Turkey are building these types of projects. Global firms from industrialized nations are focusing on large, complex (technology-intensive) projects, the area of the E&C industry that matches their competitive strengths. However, other countries, such as Japan and Italy, are able to effectively manage and construct these types of projects.

While the model shown in Figure 3.2 is useful for development planning, it cannot be used to address all possible scenarios. The model addresses situations where a project may be complex and yet highly labor-intensive. Moreover, although the model should take into account three dimensions—technical, traditional, and global—it only covers two dimensions. The third dimension (global), which is not included, deals with the interaction of traditional and technical dimensions, and it should be goal-oriented, human-intensive, and sensitive to local conditions.

In order to represent all three dimensions accurately in the model, Figure 3.2 was reanalyzed and a triangular model called the *project-objective model* (POM) was developed that addresses all three dimensions. The POM is shown in Figure 3.3. This figure illustrates that projects with highly human resource–intensive rankings require more cross-cultural training, acceptance of local society, and use of local materials. The third dimension relates to the methods and techniques used for construction that should be adapted to local conditions as appropriate technology, and this concept is valid for both domestic and global projects. Projects with a highly labor-intensive ranking require greater supervision, more intensive planning, and careful organization of work in order to provide a system for motivating workers to increase productivity. Projects with highly technology-intensive rankings, such as refineries or power plants, require advanced technology and more sophisticated MIS.

3.3.1 Ranking of Projects

One of the fundamental questions that should be addressed in the planning and management of global projects is how simple or complex a project is with respect to technical, economic, political, social, and other issues. The evaluation and ranking of projects with respect to project objectives helps to implement successful project management programs. For example, road construction in a rural area

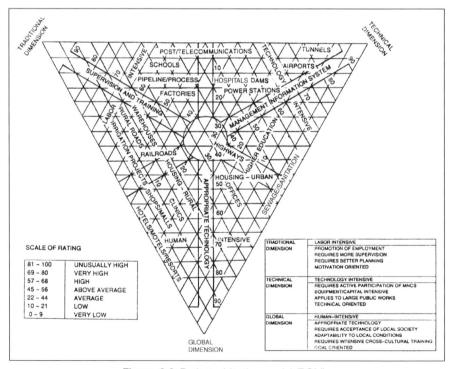

Figure 3.3 Project-objective model (POM).

requires highly labor-intensive programs, whereas petrochemical complexes need highly capital-intensive programs. The ranking of projects with respect to project dimensions varies from country to country, and most sources identify a project as either labor or capital intensive.

3.3.2 Development Activities and Construction Work

Engineering and construction projects are important to economic policies of developing nations because they contribute to economic growth and improvement of the standard of living. Basic infrastructure such as utilities, sewage systems, roads, irrigation canals, flood-control systems, and crop-storage facilities are important elements for economic development, and many developing countries devote a large share of their annual public investment to these types of E&C projects.

3.3.3 Project Complexity versus Project Dimension

Too often technology that is selected for use on a project is technology that engineers and constructors have used on previous projects without adequately examining alternative technologies. Since developing countries usually set modernization as a goal, they often look for or think they require the latest technology no matter how far removed that technology is from traditional techniques and behavior patterns of the present economy. One alternative is to reject modern technology and rely instead on upgrading indigenous technology. However, the real issue is not whether to choose between importing technology or creating it at home but, rather, which elements to import and which to obtain from domestic sources.

Development represents a deliberate effort to produce change, but efforts to produce change do not mean that genuine development occurs. Change can be effective only if traditional and modern management concepts are integrated and absorbed into the construction of projects. The level of integration and the tradeoff between modern and traditional management techniques vary from project to project and from country to country.

3.3.4 Project Complexity

Complex projects are relatively more capital and technology intensive and usually require specialized equipment, logistical support, and management skills. Although the types of projects ranked as complex vary from country to country, projects such as petrochemical processing plants, large power plants, sophisticated pipelines, and large-scale infrastructure projects that are built under severe time constraints generally fall into this category. Monitoring such projects requires sophisticated and automated MIS.

On the other hand, conventional labor-intensive projects use standard equipment and support services. Conventional projects include roads, airports, basic water and sewage pipelines, irrigation systems, bridges (when they span short distances), and single-level structures. These types of projects require less sophisticated MIS than complex projects.

3.3.5 Project Dimension

Human resource–intensive projects should be motivated by social and cultural factors. For example, one influencing factor is the need to provide jobs for low-income citizens who otherwise would be unemployed if there was no project. A project will be more labor intensive the higher the social and cultural dimension. Labor-intensive projects, by their nature, require less sophisticated MIS and more supervisory staff than those that are equipment intensive. However, they also require more intensive planning.

The scarcity of skilled supervisors constrains the feasibility of human resource–intensive projects in some countries. In addition, a shortage of engineers and management experts with the ability to organize and supervise projects of this type creates serious problems. These types of problems were overcome in countries such as Kenya and the Honduras by integrating human resource–intensive programs into the development programs sponsored by government ministries (Baum, 1985).

3.4 Hierarchy of Objectives

The objectives of each entity in an engineering or construction organization can be linked to a network of objectives and subobjectives. The objectives, at each operating level, contribute to those at the next-higher operating unit level. Each subobjective may have a *hidden subobjective* that hinders the progress of the project. However, subobjectives must be compatible with overall project objectives, and they need to be communicated effectively to employees from different cultures.

For example, suppose that a foreman wants to increase productivity and reduce absenteeism (or employ family members), whereas the area superintendent wants to optimize resources (or obtain resources for his own home). The objective of their project manager may be to build the project under budget and on time (to get back to his native country more quickly), and the specific goal of the ministry of the host country may be to increase the electric capacity of the country to 1500 MW by the year 2010 (and create construction jobs for 2000 people). Each of these plans is related to the overall organizational plan, as depicted in Figure 3.1.

Subobjectives are a vital link to establishing *proper* communication between project managers, project team members, and workers, and global projects can be directed and controlled using the *management-by-objectives* (MBO) approach. Management by objectives is a systems approach for aligning project goals with organizational goals and the goals of other subunits, as well as individual goals, and it allows employees at all levels to provide input into setting objectives and achieving deadlines (Kerzner, 2003).

Management by objectives is a useful technique when managing global projects because on global projects the project managers and project team members could be from different countries, and therefore communicating objectives, and what is required for the accomplishment of objectives, becomes a primary concern. Figure 3.4 shows how the strategy of managing by objectives is incorporated into the different phases of a project.

3.5 Determining Objectives

Determining project objectives requires an examination of the organization or project in a global context. Analytical questions should be raised at this point that include questions such as the following:

1. What are the development plans of the country?
2. How does this project conform to the development objectives of the country?
3. What is appropriate technology for this project?
4. Is there an adequate supply of raw materials and skilled labor? If not, where can they be obtained, what is the cost to transport them, and how long will it take to have them delivered to the site?
5. Are there any environmental concerns?
6. How much political risk is involved in the project?
7. What is the source of funding for the project, and is it a reliable source?
8. Who are all the decision-making bodies, and how much influence will they have on the project?
9. Has there been input from the local authorities or tribal leaders?

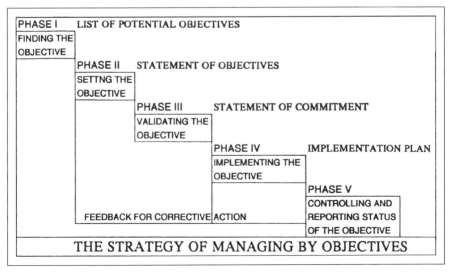

Figure 3.4 Strategy of managing by objectives (MBO).

10. Are their any sociocultural issues that will affect the project?
11. Will there be an adequate number of qualified supervisors for the project, and if not, are there funds available to train supervisors?
12. Will there be any special safety issues for the location of the project such as the possibility of kidnapping of personnel or harm to personnel?
13. Will the project be affected by government technology requirements such as having duplicate positions for expatriate personnel and native personnel?
14. Will the support staff be expatriate or native personnel?
15. Should local personnel be hired through temporary (temp) agencies to avoid government requirements for paying compensation to employees when their employment is terminated by their employer?
16. Are there any national or local laws or regulations that will negatively affect the project?
17. Does the project have the support of the local community, and if not, what can be done to increase local support?
18. Will the project be affected by any cultural or religious requirements, such as approval of the local *feng-shui* or *fung-shiu* master in Asian countries or religious celebrations for ground-breaking ceremonies in India?
19. If local E&C personnel are used on a project will they have equivalent degrees?
20. What types of cultural offenses will force expatriate E&C personnel to be repatriated back to their native countries?

By addressing these types of questions, a list of potential objectives can be developed for projects. Explicit attention to project objectives should include efforts to ensure that all the parties who will be involved with executing the project share a common view of the objectives and the strategies for meeting them. Figure 3.5 contains a sample hierarchy of objectives in a work-breakdown-structure format.

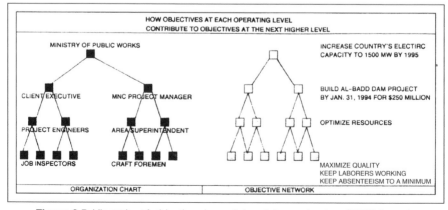

Figure 3.5 Hierarchy of objectives in work-breakdown-structure (WBS) format.

3.5.1 Setting Objectives

All team members should be involved in the setting of objectives that have a direct impact on their work and the monitoring of the accomplishment of objectives because this improves team performance. Setting objectives requires management-team members to issue a formal statement of objectives. Because the whole organization is accountable for the implementation of objectives, the formal statement should be in writing, properly communicated, supported by top management, and interlocked with other groups. The objective statement should identify development objectives, specific objectives of the project, and any additional objectives that will be achieved as a result of the project, such as integrating a particular group of the population into the development process.

3.5.2 Validating Objectives

Formal objective statements should be confirmed and verified to ensure that the objectives are reasonable, measurable, and achievable within the stated period of time. One important planning tool for linking objectives in a time framework with activities and resources is the *work-breakdown structure* (WBS). To create a WBS, a project is divided into smaller tasks that aid in organizing, defining, and displaying the project. Then progress on each task is summarized at higher and higher levels. Tasks are the base of a pyramid, and as the pyramid gets narrower, it represents summary levels until the apex of the pyramid is achieved, which represents the overall project summary. The progressive summary levels could be activity numbers, area numbers, building numbers, and the total overall project.

Actual cost expenditures, physical percent complete, earned and actual labor hours, cost and schedule performance indices, and cost and schedule variances can be calculated at any WBS level to verify project objectives and subobjectives. Work-breakdown structures are used to validate management coordination, time and schedule, total risk and the impact of decision making, organization charts, the cost-breakdown structure, and construction methods. During the validation phase, the statement of objectives is translated into a statement of commitment to ensure that resources, facilities, materials, methods, personnel, and management are available in order to reach a desired goal.

3.5.3 Implementing Objectives

On global projects, project managers should have the entire project team participating in the implementation of objectives and the development of an implementation plan that should serve at least the following three purposes:

1. Establish a *clear* understanding between project managers and project-team members as to what they must do in order to achieve objectives.

2. Help those within a team communicate more effectively when their input is required for a decision (this is especially important if different languages are spoken by team members or if team members are from a culture where they are not allowed to contribute to decisions because decisions from managers are final).

3. Help managers to assess progress toward the selected goals by comparing accomplishments with estimates.

3.5.4 Controlling and Reporting the Status of Objectives

The controlling and reporting phases are used to highlight deviations from baseline plans and highlight deviations for corrective action. Project managers and members of project teams jointly appraise performance through an iterative process of continual evaluation and iterative planning with the aid of a MIS and scheduling programs such as Primavera Project Planner (P3), Microsoft Project, or other project management software that provides critical-path method (CPM) schedules, cost control, project documentation, and other reporting functions. Appraising performance is more difficult if a MIS is not being used on a project because of limited access to technology, as is seen in the case of developing countries or other countries where the use of technology in restricted by the government. In these situations, standardized manual methods for tracking performance should be implemented at the start of a project.

Control systems are a tool that helps managers to determine whether objectives and subobjectives are being achieved, and if they are not achieved, project team members have to determine why (delay analysis) and provide mitigation strategies. Delays and mitigation strategies on global projects are discussed in Chapter 11.

3.6 Managing Projects in Foreign Countries

Even though many of the project-management issues in foreign countries are the same as in domestic countries, other challenges may arise that are location specific, and managers need to be equipped to deal with these challenges. In paternalistic cultures, managers are considered to be "father figures" who help with work and personnel problems. Socializing after work hours with workers may be mandatory, as is seen in some of the Asian cultures, and managers should be able to balance work and their personal lives.

Project managers who work in remote locations cope with issues that do not affect domestic projects, such as making sure that food is available for personnel, that there is adequate housing for personnel, that workers have recreational facilities, and that there is some type of entertainment for personnel. Project managers also have to watch for signs of culture shock, which are discussed in Chapter 13.

Project managers also should address cultural issues and help to integrate workers from different cultures into projects. Motivating people from different cultures is challenging because they are not all motivated by the same things. If people are from cultures where the availability of products is limited, they probably are not motivated by money because there is nothing to buy in their native countries. People from societies that barter for goods may continue bartering for their wages and other benefits. In some cultures it is difficult to instill a sense of urgency if people are not used to moving quickly when there is high heat and humidity or complex bureaucratic systems.

Leadership styles vary throughout the world and techniques employed in one country may be ineffective or inappropriate in others. Participative-leadership styles do not work well in cultures where people are used to dictatorial leadership. South American cultures have paternalistic leaders that show concern for workers and their families. Japanese workers are used to being involved in consensus-style management, Chinese workers live under a dictatorship that controls almost everything in their lives. Western Europeans leaders obtain their positions by virtue of their knowledge, and workers from the United States are used to visionary, or charismatic, leaders. In tribal societies, workers only take orders from tribal leaders, and their leaders inherit the right to be leaders. Techniques such as managers helping to perform work may be viewed with suspicion by workers in some parts of the world, because they do not understand why it is necessary for someone at a management level to be doing their work.

Issues of inequality are understood by people from every culture, and avoiding situations where there is inequality, or perceived inequalities, helps to maintain a productive work force. Maintaining high levels of productivity is important on construction projects, and Chapter 10 addresses global productivity issues.

Project managers also have to deal with personnel that have different levels of education. In some countries having an engineering degree qualifies someone to be a technician, and in other countries an engineering degree indicates that someone can perform advanced scientific analysis.

Project managers have to interface more with members of government agencies in some countries where there are strict regulations and government participation requirements for projects. Knowing about local politics is useful in these situations because it helps to prevent mistakes that cannot be rectified by going through proper channels or agencies. Also, knowing about *hidden* systems for accomplishing tasks in countries sometimes can expedite problem solving. Hiring local business agents is one means for finding out about hidden systems.

3.7 Characteristics of Global Project Managers

There are several attributes or characteristics that are essential for global project-management candidates, and they include (Lucas, 1986):

- Professional competency
- A clear sense of priorities

- The ability to develop and implement short and long term plans
- Foresight
- Communication and language skills, one cannot communicate effectively with people who speak different languages if they cannot already communicate effectively in their own language
- State-of-the-art knowledge of equipment, materials, and engineering
- The ability to lead or follow someone else's lead
- The ability to make decisions independently and quickly
- Knowing how to use responsibility and authority effectively
- The ability to solve conflicts both interpersonal and cultural
- Respect for other cultures

The last item is essential for global E&C personnel, because one of the main reasons expatriate personnel are released early from assignments overseas is their inability to operate within different cultures (cultural insensitivity). Sometimes visas of expatriate personnel are revoked by the government of a host nation, and they are required to exit a host country within 24 hours if cultural infractions are severe.

Additional assets that are important for global project-management personnel include (Copeland and Griggs, 2001, p. 209):

- Self-reliant
- Self-motivated
- Full of creative solutions to problems where others might be stuck
- Diplomacy and tact
- Ability to take on sole responsibility
- Able to get results with limited resources
- Social and personal rapport
- Acceptance by native population
- Emotional stability
- Technical competence
- Tolerate ambiguity
- Sincere respect for and interest in others
- Curiosity and intellectual curiosity
- Breath of knowledge and character
- Geocentric
- See opportunities not constraints
- Think in world terms
- Adaptable
- Patient
- Acceptance of failure without despair

Traditional E&C college-level educational programs concentrate on preparing technically competent engineers, constructors, and project managers, but few educational programs emphasize the importance of understanding cultural issues.

Introducing imported skills and management techniques always raises the possibility of importing new values and behaviors, which may be entirely foreign to and inconsistent with the values and culture of a country. Therefore, cultural training should be part of the technical education process to ensure the survival of local social values and the strength of a country. Global topics that are useful for engineers and project managers to learn about include courses in language, world cultures, global business strategies, political science, world history, and global management, finance, and economics. Two other areas that are useful when working overseas are cultural anthropology and archaeology because they provide a historical and ancient perspective on the development of other cultures.

Other subjects are "social and business etiquette, history and folklore, current affairs, relations between your native country and the foreign country in which you are working, cultural values, geography, sources of pride, artists, musicians, novelists, sports, great achievements of the culture, religion and its role in their way of life, and the political structure and who is in charge" (Copeland and Griggs, 2001, p. 216).

One anomaly that is created by educational systems occurs when engineers and constructors leave developing countries to study in industrialized countries, where educational systems emphasize technology that is appropriate to highly developed countries or Western cultures. One example of this problem is seen when engineering students take courses in structural-steel design, concrete design, and computer-aided design in the West, and then they return to a developing country such as Afghanistan, where the majority of structures are made out of mud and bricks, and where few computers and limited Internet connections are available. This is the reason that foreign E&C students try to remain in Western countries where they received their education, because there may be little or no opportunity for them to apply their knowledge in their native countries.

Another problem created by educational systems is that some of the Western-educated engineers and constructors end up working in key government posts in their native countries where they influence development plans and apply their Western education without emphasizing indigenous social and cultural aspects of designs. This could result in projects where there is no local expertise for competent operation of projects, or parts and services are not available for maintaining projects. In some situations, projects may be abandoned because the native population reverts to their traditional way of doing what the project tried to automate for them.

To compensate for a lack of personnel with a global perspective, projects may have different project managers during each project phase. This allows project managers with different expertise to be assigned to a project when a particular expertise is required during a certain phase of a project. Although this type of a process sacrifices institutional memory, it takes advantage of other aspects, such as technical and managerial expertise and cultural familiarity. It also demonstrates how difficult it is to hire project managers that have appropriate backgrounds in all of the areas of expertise required for global engineering and construction.

3.8 Calendars and Time Considerations

This section provides information on calendars that are used throughout the world, along with an explanation on how time is interpreted in different societies.

3.8.1 Calendars and Holidays

Software scheduling programs have embedded calendars that usually are based on the Gregorian calendar (Gregorian calendars have 12 months of 30 or 31 days, and an extra day is added at the end of February every four years). If scheduling software is used for construction projects in countries that do not use the Gregorian calendar, then local calendars have to be input manually into the computer software. Countries that use the Gregorian calendar include Belgium, China, Denmark, Egypt, Germany, Great Britain, Greece, Holland, Italy, Japan, Norway, Poland, Portugal, Russia, Spain, Switzerland, and Turkey (Bruner, 2005).

Other calendars that are used throughout the world include (Bruner, 2005):

- Lunar—19-year period, with 7 of the years having 13 months with $29\frac{1}{2}$ days.
- Chinese—12 months with 29 or 30 days, with an extra month added at regular intervals.
- Egyptian—12 months with 30 days each month, and a week is added at the end of each year as a festival week.
- Hindu—Indian national—The year starts on March 22, except leap years, when it starts on March 21; there are 5 months with 31 days and 7 months with 30 days.
- Islamic—12 months with either 29 or 30 days, and the year begins on the first day of Muharram; 2005 is equivalent to the year 1425–1426, because the Islamic calendar started when Muhammad emigrated from Mecca to Medina.
- Jewish—A lunar-solar calendar that gets the months from the moon and the years from the sun; there are 12 lunar months and 11 days each year and a leap month every 3 years; holy days begin in the evening; the day starts when three stars are visible in the sky; and the calendar year begins in September or October on Rosh Hashanah; 2005 corresponds to the Jewish year 5765–5766.
- Julian/Roman—12 months with 30 or 31 days, with every fourth year 366 days.

Chapter 14 provides information on holidays observed throughout the world that have to be incorporated manually into project-scheduling programs.

One religious holiday that is observed by over a billion people worldwide is *Ramadan*. Ramadan lasts for approximately a month, and it is on different dates each year, depending on lunar observations. Ramadan ends with *Eid al-Fitr*, which is the festival of sacrifice that includes the ritual slaughtering of animals. Muslims are required to make a pilgrimage to the holy city of Mecca in Saudi Arabia at least one time in their life during the *hajj* period. During Ramadan Muslims cannot eat or drink during daylight hours, which affects construction projects, because workers may be in a weakened condition from lack of food and water. In some Islamic countries, work is either suspended during Ramadan, or work hours are reduced to 4 hours per day.

Another problem during Ramadan is that non-Islamic travelers could be stranded in a country until the end of Ramadan if airlines or railways are overloaded with people making the pilgrimage to Mecca. Major airline and railway hubs in non-Islamic countries may also be affected during Ramadan.

3.8.2 Time Issues and Workweek Differences

Military time is stipulated as the standard time format by the International Organization for Standardization (ISO). Military time uses a 24-hour clock; therefore, after 12 noon, the hours are listed as 13, 14, 15, and so forth until midnight, which is hour 24. International dates are year, month, day of the month (ISO, 2006).

In Islamic countries, weekends are observed on Thursdays and Fridays, and the workweek starts on Saturday, so expatriates only have Mondays, Tuesdays, and Wednesdays to contact their home office if weekends are observed on Saturdays and Sundays in their native countries (or vice versa). In nations where Catholicism, Christianity, Protestantism, and Orthodox Christian are the main religions, people usually do not work on Sundays because church services are conducted on Sundays. In the Jewish religion, the Sabbath (the day of worship) is from sundown on Friday until sundown on Saturday, and if people are orthodox, they will not operate mechanical devices during the Sabbath.

Another issue that adversely affects project schedules is the concept of *rubber time,* which loosely translates to *it will get done when it gets done.* Enforcing deadlines is challenging in cultures that operate on rubber time, especially if there are also religious beliefs that include fatalism, where rewards are obtained in the next life, or everything is in God's or Allah's hands, so why hurry in this life. People arrive for meetings hours late even if they are only in another part of a building. When workers are not meeting deadlines, it could be due to issues such as rubber time, religious beliefs, clan or tribe loyalties, or it could be a matter of providing workers with written deadlines instead of verbal deadlines. In tribal cultures and countries with large government bureaucracies, the family members of workers, their community (the local tribe or clan), their religion, and their dealings with government bureaucracies all take precedence over work activities. Allowing workers time off to deal with issues related to these areas helps to keep workers from being distracted at work.

To meet deadlines on global projects, rewards could be tied to specific deadlines if that is what local workers are used to in their culture. If workers do not personally benefit from completing tasks on schedule, they do not understand why meeting deadlines is essential. Offering monetary rewards when deadlines are met may not work in every culture because there is nothing to buy in Communist countries (state-controlled societies) with the extra money. Foreign products are more of an incentive to people to meet deadlines. Time off, a party, special foods, or small gifts could be used to celebrate project milestones to demonstrate the benefits of completing a task on time.

In some Asian or Southeast Asian cultures, questions should not be asked in a manner that requires a direct yes or no answer because workers always answer questions with a yes response to please the boss. In place of yes or no questions, ask broad, hypothetical questions. In Japan, there are 16 ways of saying no without saying no and displeasing someone. In India, people could be shaking their head from left to right to acknowledge the speaker instead of meaning no. In South Korea, when someone says "Yes," it may only mean that they hear what the other person is saying and not an indication of an affirmative answer to a question.

3.9 Technology Considerations

This section discusses global issues related to the use of technology in different parts of the world.

3.9.1 Appropriate Technology

Determining what is appropriate technology for a particular region of the world is a time-consuming task that engineers have to undertake in order to design global structures. What is appropriate in one country may not be appropriate for other countries because of:

- Climate conditions. (One example is when structures are built on raised footings in regions of the world where there are wet monsoons and another is buildings that contain numerous windows to create cross-ventilation in situations where there is no mechanical air-conditioning).
- Unavailability of materials
- Frequent natural disasters
- Cultural taboos
- Construction laborers lack of familiarity with processes or materials
- Lack of hand tools or heavy equipment
- Religious restrictions
- Limited education of workers
- Lack of training in the skilled trades
- Government restrictions or requirements
- Social standards
- Fear of the unknown by local citizens

3.9.2 Computers and Electronic Equipment

In developing countries, the use of computers to generate designs (AUTOCAD) is not as prevalent as it is in industrialized countries because labor costs are relatively inexpensive; therefore, engineers, architects, and draftsman hand draw their designs. The use of computers is also limited in some regions of the world because of limited finances or the unavailability of reliable utility sources. Pirated (unlicensed) of computers, monitors, software, flash drives, videos, compact disks, digital videos, and cameras have made technology more affordable but possibly less reliable. Unlicensed systems do not come with technical support (or warranties), so they may last for only a short period of time, and their warranties could be phony. Sometimes the place of business where computers are purchased moves around to avoid angry customers.

A reliance on computers could work against engineers and constructors if they are assigned to remote job sites where computers are not available or do not function properly. Spare parts for computers may not be available locally, or electronic devices may not work properly in adverse weather conditions (excessive humidity or heat) or with intermittent power. If any of these conditions are present, engineers have to rely on manual methods or calculators to perform calculations and to create designs.

Language is also a barrier to the use of technology. As of 2005, computer systems could only translate text into 16 languages, and 6500 languages are spoken throughout the world, with more than one language spoken in each country. For example, people in India speak 14 languages. Many Europeans are bilingual or trilingual (they can speak two or three languages); therefore, Europeans are able to use computers if they are familiar with one of the languages available; but in other regions of the world, people usually speak only their native language (*TIME*, 2005).

Other issues that affect the use of technology globally include the availability of fabricators and raw materials (or poor-quality materials), limitations on the availability of transportation systems, and limited or no access to communication systems and supplies (availability of the Internet, wireless technologies, faxes, phones, cell phones, satellite phones, copiers, paper for copiers, toner, and printer cartridges). Also, government restrictions on the exportation of certain technologies to specific countries, proprietary processes, and confidentiality agreements also restrict the use of technology.

3.9.3 Substitutions for Technology

Global engineers learn to be resourceful in order to provide adequate substitutions for materials or processes when governments limit the exporting or importing of materials and supplies that are called for in specifications. Import and export permits could be denied by government officials, or the items received could be defective or damaged in transit. One U.S. firm was fined $300,000 for exporting computers to a developing country that is on the Commerce Control List (CCL) published by the U.S. government.

Materials may arrive in countries on schedule, but for various reasons, customs officials may not release them from port facilities. Project team members may have to design alternate strategies for working around materials not released by customs officials. Engineers and constructors also could be detained when entering a country to work, even though they have all the appropriate paperwork that was issued to them in their native countries.

The required technology level of a design could be too advanced for the skill level of the local construction work force. If the skill level of workers is not known when a structure is being designed, it creates an added burden for site engineers when plans have to be redesigned at job sites to accommodate the skill level of workers or when workers have to be imported from other countries to perform the work (both of which substantially increase the cost of a project). In one developing country, welders had to be brought 13,000 miles from the United States at a cost of $50,000 to $100,000 per person for transportation, extra housing, benefits, food, and R&R (rest and relaxation) costs (not including salaries). In some countries, such as Indonesia and China, citizens cannot relocate without "transmigration" permits or residency permits, which may or may not be approved by their native government officials.

Engineers also need to investigate what equipment and materials are available locally and what electrical system they are designed for before they are incorporated into designs. Electrical systems are different throughout the world, with 110 volts and 60 hertz used in the United States, and 220 to 280 volts and 50 hertz used in other parts of the world. Voltage transformers can adjust the voltage, but outlets require one of 14 different plug adapters because the shapes of outlets vary throughout the world.

3.9.4 Climate Considerations in the Choice of Technology

Textbooks on construction equipment and methods address the issue of the degradation of power at high altitudes by including formulas for calculating the reductions in power. But there are no formulas for calculating the reduced efficiency or life of construction equipment due to the salting of roads in areas of heavy snowfall or the effects of sandstorms on engines, which renders them useless.

Climate considerations increase in regions with extreme weather conditions, and one method for determining the adverse effects on construction equipment is to investigate the used construction equipment market in the specific region where construction equipment will be used on a project. If the local equipment is rusted, especially the undercarriage, it is a sign that the roads are salted during the winter months, and if there are signs of pitting in the engine, or on the exterior surface, it is caused either by sandstorms or large pieces of hail generated during thunderstorms.

The selection of materials for a structure also should take into consideration climatologic conditions. It is not appropriate to use paper-clad drywall (sheetrock) in climates where high humidity causes mold or where there are monsoon rains or other causes of flooding, such as hurricanes. Rather than using cardboard covered sheetrock, a mold-resistant gypsum-board product can be substituted, such as the one manufactured using fiberglass cloth (manufactured by Densglass). Paper promotes mold growth, whereas the fiberglass alternative does not grow mold. Furthermore, steel rusts more quickly in climates where there are prolonged periods of high humidity and in damp climates such as areas where there is fog or monsoon rains. Timber is problematic in areas where there are termite infestations; therefore, masonry, stone, and steel are more appropriate in termite regions. If termites are active, support beams that are located close to the ground should be steel beams rather than wood beams. Freezing temperatures require auxiliary methods for maintaining heat while concrete is curing, such as covering the concrete with straw or burlap or using heaters. Hot temperatures require water to be sprinkled on concrete to keep it moist while it is curing so that it does not crack from curing too quickly.

3.9.5 Human Technology

In some cultures, engineers or constructors may not admit that they do not understand a technology in order to "save face," to avoid embarrassment. A person could be called an engineer, but that is no assurance that he or she has completed a 4- or 5-year accredited college-level engineering program. A person could be called an engineer by virtue of his or her seniority or experience or after he or she has completed a 2-year engineering or technology program. Only a small percentage of countries require professional licensing for engineers and constructors, and in countries that do not issue licenses, some engineering professionals will obtain licenses in countries that issue them.

The professional registration process is not uniform throughout the world. In order to take the professional licensing examination some countries require graduation from a nationally accredited college or university, a 2- to 5-year internship, and a passing score on two 8-hour written examinations (the United States), whereas other countries merely require a set number of years of work experience, such as 5 years in Yemen, after graduating from college.

The status of a person is normally determined by socioeconomic factors and his or her heritage or tribal affiliations, but exceptions may be made if someone earns a college degree. In Islamic countries, someone of a lower status is able to supervise or teach people with a higher status if they have more education than their employees or students. Degrees should be listed on résumés, and supervisors should be told about them verbally in case they are not provided with copies of résumés.

3.10 Estimating

Global estimating issues are discussed in Chapter 10.

3.11 Scheduling

Depending on the education level of E&C personnel, computer-generated critical-path method (CPM) construction schedules may have to be translated manually into hand-drawn bar charts because not all E&C personnel are familiar with CPM scheduling techniques. Critical path method schedules still may be used on projects, but computer printouts should be converted into hand-drawn bar charts before they are distributed for updating by personnel. People will not admit that they do not understand the scheduling methods used on projects; therefore, inaccurate information ends up being incorporated into schedules. The ability to keep a project on schedule is challenging on global projects due to influences that are beyond the control of project-team members. Such influences may include:

- Government regulations
- Owner interference
- Material delays
- Design modifications
- Weather conditions
- Language barriers
- Cultural misunderstandings
- Technology failures
- The quality of labor
- Religious restrictions
- Tribal interference
- Festivals
- Labor slowdowns in countries where workers are not allowed to strike
- Unions and strikes
- Bureaucratic limitations
- Slow decision-making processes
- Material substitutions
- Rationing of personnel
- Inadequate support staff
- Philosophical differences

On global projects, some engineering personnel lack the requisite technical expertise, some do know how to interpret drawings, and others may miss indicators of potential breakdowns. Every member of a project team should be included in meetings to avoid communication problems and misunderstandings.

3.12 Permits and Codes

The processes required to obtain building and other permits should be investigated before a firm agrees to design a structure in a foreign country, because it might be impossible to secure permits without bribes or government connections in certain countries. Information that is gathered from government sources is not always a reliable measure of how the permitting process really functions or of how long it will take to obtain a permit within a particular region of a country. In some nations, such as the United States, state and local governments can enact legislation that is more stringent than federal government requirements. Permits are issued by municipal governments rather than by state or federal governments. In the United States, the *Uniform Building Code* is used throughout the country, except in regions where there are freeze/thaw conditions, where local building codes set additional requirements for burying pipes. If the ground freezes, pipes have to be buried at a greater depth (usually 8 feet) than they are buried in other regions of the United States where the ground does not freeze (3 feet).

Another issue that can cause problems on global projects is that E&C personnel may be using older versions of national or international standards, specifications, or publications, such as steel and timber manuals, uniform building codes, National Institute of Standards and Testing standards, American National Standards Institute standards, and the International Organization for Standardization (ISO) standards. To avoid confusion, engineers should cite in the design specifications which edition of a text or set of standards applies to a particular project.

3.13 Construction Safety Issues

Safety is a major issue on global construction projects because not all governments enforce safety standards or inspect projects. Several factors that contribute to the high number of accidents in the global-construction industry are misunderstandings related to who is responsibile for safety issues, poor management, and blatant neglect since some governments do not sanction firms if their workers are injured while working at job sites.

Construction foremen, project superintendents, and project managers play an important role in defining safety performance on global projects, and they should provide orientation sessions for workers, organize the work environment safely, and effectively manage time pressures.

3.13.1 Project-Level Safety Programs

Safety is more manageable at the project level because a safety program is easier to implement if it is specific and concise and supervisors play a decisive role in the program. Control can be established during company meetings, job meetings, or special safety meetings (weekly or biweekly) with supervisors, project managers, construction managers, foremen, and workers. Another form of safety control is effective communication between workers and their supervisors, including translating safety materials into native languages and regular inspections to help determine whether work is being performed properly.

3.13.2 Safety in Contracts

Construction contracts apportion responsibility and authority for safety on a project, and defective or unclear clauses in contracts can lead to confusion about who is responsible for safety during construction. Contracts should clearly specify who is the authority responsible for safety and include hazard-prevention requirements for each project.

3.13.3 Construction Equipment

Construction equipment should be monitored routinely, and old or defective equipment should be repaired and inspected periodically because new equipment is not always available in some regions of the world. Workers should be properly trained to operate modern or highly technical equipment and safeguards always should be incorporated into the design of machinery. Moreover, there should be no provisions for disengaging safety features, which is a primary cause of injuries on global-construction projects.

Instructions for using equipment and tools should be translated into local languages and explained to workers by other native workers in their own language, because illiteracy contributes to written instructions not being understood by workers. The following should be performed routinely on projects (MacCullum, 1990, p. 19):

- Periodic inspections (maintenance)
- Replacing parts before they wear out (repairs)
- Training operators on how to use equipment properly (having qualified operators)
- Employee screening (tests)
- Translating and explaining
- Site and operating rules

3.13.4 Lack of Management Control

Safety managers (or the entity who is responsible for safety on projects) should be able to communicate effectively in the local language and to execute all measures related to safety. Safety should be integrated into all levels of management, and the authority in charge must be able to comply with safety requirements. When the entity in charge is not well defined, there is a lack of organization and control, and consequently, the application of safety measures is jeopardized by outside influences.

Project team members can help to prevent accidents by designing proper signalization such as fencing, flags, and signage; by the organization of work; by identifying hazardous substances; by stipulating appropriate construction procedures; and by coordinating jobs. Project-team members should also enforce the use of protective clothing and use it themselves to set a good example while they are working at job sites.

3.13.5 Safety Issues

Another global-design consideration is providing redundant safety systems during the construction and operation of facilities. Without rigid government safety standards, safety considerations are not a high priority during the construction of projects; therefore, safety inspectors who work for the firm that is constructing a project should perform inspections to maintain safe working conditions at job sites.

Governments do not always provide compensation to the family members of workers who are killed or injured while working at job sites. If compensation is provided, it may only be a small amount. In some industrialized countries, such as the United States, insurance companies have actuarial charts that list the value of each body part. Workers who have their eyes injured receive higher compensation than workers who injure an index finger, and thumbs are worth more than the other four digits.

Safety issues that affect global-construction projects include:

- Using regional forms of scaffolding, such as bamboo scaffolding that is tied together with nylon or jute rope, which is used throughout Asia. Figure 3.6 shows a photo of a building surrounded by bamboo scaffolding that was constructed in China.

Figure 3.6 Bamboo scaffolding, China.

- Workers not wearing shoes, because it is easier for them to climb bamboo scaffolding when they are not wearing shoes, or workers may not own shoes or boots, or they are more comfortable going barefoot in hot climates.
- Workers borrowing shoes or work boots that do not fit properly.
- No work gloves, hard hats, or safety goggles. Figure 3.7 shows a Balinese (Indonesian) construction site where workers are not wearing hard hats, shoes, goggles, or gloves.
- Workers not using, or improperly using, fall-protection harnesses when working on the upper floors of structures.
- No shoring used in trenches.
- Overloading during construction; access to a crane may be limited to one or two days because of the high cost of cranes.
- When cranes are being used, there may not be any cages to move materials safely.
- Intentional injuries to opposing clan or tribe members.
- Workers who carry daggers in their belts at all times.
- Workers who do not have enough to eat, and thus their judgment or strength is impaired by malnutrition.
- Tools that are too large for the hands of local workers.
- Cultural taboos on handling certain items with the left hand.

Figure 3.7 Balinese job site, Bali, Indonesia.

3.13.6 Use of Illicit and Legal Drugs

The consumption of drugs and alcohol, as well as other substances, can alter proper performance and result in lost productivity, poor decisions, absenteeism, turnover, and involvement in accidents (Altayeb, 1992). The use of drugs by construction workers is a problem everywhere in the world, and the only thing that differs is which drugs are the problem. One example is in Indonesia, where construction workers smoke *clove* cigarettes during their breaks. Clove is a spice that comes from an East Indian tree that is used as a sedative in holistic medicine. Thus, if workers return from cigarette breaks in a lethargic condition, the cigarettes they are smoking may be causing their lethargy.

In Yemen, workers take *qat* breaks. Qat is a local leaf that is chewed and left in the mouth. The qat leaves contain amphetamines, which causes workers to talk and move more quickly. Qat impairs judgment, which is dangerous when heavy machinery or power tools are being operated by anyone who is chewing qat. In the Fiji Islands, there are *kava* ceremonies, where people drink from a communal kava bowl. Kava affects people differently, some become lethargic and others become manic.

3.13.7 Drug Testing

Drug ingestion, smoking, drinking alcohol, and consuming other mind-altering substances all contribute to construction accidents. *Random* and *for-cause drug testing* helps to reduce the number of accidents. However, at remote job locations, it could be impossible to perform drug tests because of a lack of health officials to draw blood or analyze urine specimens. Religious or cultural beliefs also may prevent workers from taking drug tests.

3.13.8 Securing Sites

Securing job sites is a challenge faced by E&C personnel throughout the world. Materials that are not available locally usually are the materials that are stolen from construction job sites. Bid estimates should include funds to cover the cost of security and to compensate for materials stolen during construction. Sometimes an extra 30 or 40 percent of actual material costs is added to bid estimates to compensate for the theft of materials. If local personnel are used to secure job sites, additional problems occur because of their loyalty to their clan, tribe, family, or community, not to the owner of the project. Having a warehouse and inventory-control system in place can reduce theft but not totally eliminate it on job sites.

3.13.9 Design of Structures

Construction-worker safety usually is the responsibility of project managers or construction contractors, but safety performance also can be dictated by the design of structures. The goal of design firms is to create a product, and there may be little emphasis put on worker safety during the design stage of projects. It is difficult, but not impossible, to develop designs that incorporate safety measures that may be implemented during the construction stage that are appropriate for foreign environments and the skill levels of native workers.

3.13.10 Health Problems

Health problems in the construction industry are gaining in importance throughout the world, as is demonstrated by the implementation of the Kyoto Protocol Treaty in 55 countries (see Chapter 9) and other global treaties that restrict the generation of pollution and hazardous wastes. Hard hats, safety goggles, and respirators are some of the safety devices that can be used to protect workers. The most common health hazards at construction sites are heat, dust, aspiration of mineral particles or asbestos, noise, dermatitis, burns, and toxic chemicals.

Job sites in tropical locations may use mobile foggers to spray insecticide to kill mosquitoes that carry the malaria parasite. Pesticides that are banned because they are carcinogenic (cause cancer) in many industrialized countries (such as DDT) are still used to kill mosquitoes and insects in developing countries. Therefore, workers and management personnel will all be exposed to carcinogenic materials.

Occupational diseases are also a serious problem in construction, and direct costs and claims increase if measures are not taken to remove hazards—such as asbestos, which is still legal to use in developing countries—before construction begins and to substitute less toxic materials. Job-site personnel should stay current on new developments related to safety, and they should implement methods that have been proven to reduce health hazards.

An analysis of contributing factors helps to determine how risks can be minimized and/or prevented at job sites. Accidents occur repeatedly when the same mistakes are made; therefore, the principal causes of accidents at a particular site should be identified, and methods should be developed to help avoid them. Developing prevention programs and improving construction practices help to reduce the incidence of construction accidents.

3.13.11 Injuries

The most common causes of construction injuries are overexertion, falls, and being struck by and against objects, along with being rubbed against or abraded, caught in between something, or struck by something (Culver et al., 1993). Strains and sprains are the most frequently occurring type of injury, followed by cuts, lacerations, punctures, dislocations, fractures, contusions, and bruises. Methods for reducing these types of injuries include providing proper signalization, increased awareness, and improved communication.

One of the most hazardous jobs in the global E&C industry is excavation. Causes of accidents related to excavations include (Rekus, 1992, p. 29):

- Accepted engineering practice
- Soil type and conditions
- Recent excavations
- Weather conditions (freezing and thawing)
- Surcharged loads
- Vibration and other types of movement
- Incompetent workers

General requirements that should be followed during excavation include (Rekus, 1992, p. 34):

- *Surface encumbrances:* must be removed or supported if they present a hazard.
- *Proper identification of utilities:* localize and identify utility lines before digging.
- *Access and egress:* must be provided by ladders, ramps, or steps on trenches deeper than 4 feet.
- *Exposure to traffic:* reflectorized warning vests must be used when excavation is taking place near traffic.
- *Overhead loads:* employees should be aware of loading and unloading of equipment.
- *Emergency response:* necessary equipment for emergencies such as lifelines, stretchers, first-aid kits, and harnesses must be available.
- *Hazardous atmospheres:* excavations dug near chemical plants, sewer lines, or landfills may allow dangerous liquids, gases, and vapors to seep through the soil.

Other safety precautions that can be done during excavations include (Rekus, 1992, p. 37):

- *Sloping:* the sides of an excavation should be adequately sloped. It is the simplest method of cave-in control and prevents loose material from rolling back in. The angle of slope depends on the soil type, loads, and distress.
- *Shoring:* it has been proven that most excavation related deaths involve trenches whose walls were not shored. Shoring presses against sidewalls and prevents cave-ins. Shoring can be expensive because of the extra materials involved, which is one reason why it is not used.
- *Shielding:* this protects employees from cave-ins by providing a sheltered space. It is a device shaped like a box that consists of two steel plates separated by structural members, and it is open at the bottom and top. The steel plates face the trench's sidewalls. Work is performed between these two plates, and as it progresses, the device is dragged along the bottom of the trench by other equipment.

3.13.12 Site-Safety Programs

A global site-safety program should accomplish the following (Terrero and Yates, 1997, p. 28):

- *Proper identification of authority.* This assigns responsibility for safety and health to the appropriate entity, either the owner, the construction manager, or whoever is indicated in the contract.
- *Contract assignment of a safety coordinator.* One person or various people, depending on how large the project, should be assigned to coordinate, inspect, and monitor project safety.
- *Description and identification of hazardous operations.* A plan should describe all hazardous operations that are going to take place throughout the project, including items such as the following:
  - Demolition
  - Excavation and trenching
  - Working on elevated platforms
  - Crane operations
  - Lifting materials to floors
  - Possible bursting of formwork while concrete is being placed in it
  - Lifting heavy materials
  - Welding operations
  - Using electric hand tools
  - Equipment operation
  - Unknown location of gas lines during trenching
  - Installation of defective materials
- *Adequate equipment for emergency procedures and general rules.* A record of inspections and test results should be kept, along with a list of measures to be taken in case of "what if." Procedures and rules should be established that are followed at all times especially during emergencies. These procedures should cover storage of materials, removal of toxic materials, and constructive procedures.

3.13.13 Characteristics of a Safety Plans

The following characteristics should be considered when trying to establish a global safety plan or program:

- The plan should be focused on areas where accidents occur frequently. The goal should be to implement safe work practices and eliminate workplace hazards (hazards should be prioritized).
- Employees should be familiar with the rules and procedures. A program cannot be implemented if employees do not know how to work safely; therefore, the processes included should be easy to understand and explained in native languages.
- Deficiencies should be identified at the start of projects.
- The safety level of projects should be measured, and there should be a way to measure quality control on existing safety measures.
- Safety procedures should be revised and updated periodically in order to keep them up to date with project activities.

- Project safety is accomplished through distribution and implementation of authority and responsibility and adequate descriptions of what is expected of all participants in a project.
- A safety program should establish clear and attainable objectives and state how objectives will be met on the project.
- Motivation and recognition are key factors in a successful program. Employees should be motivated and encouraged to promote and practice safety issues. Employees also can help in identifying critical factors to help establish safety rules and procedures.

The following subsections provide guidelines and methods for implementing safety measures and for motivating workers (as well as minimizing the risks associated with accidents). The suggestions are categorized by who should be responsible for their implementation.

3.13.14 Responsibilities of Supervisors

Supervisors are the most effective promoters of safe working practices, because they usually control the way work is performed, and they exert the greatest influence on employee attitudes. Supervisors can increase safety if they (Hislop, 1990, p. 24):

- Implement methods for measuring safety levels on projects
- Control hazards by the application of specific technical controls
- Educate and inform workers about the reporting and handling of injuries
- Classify employees by job duty whenever applicable
- Establish a safety-incentive program and encourage its use through an award system
- Properly maintain records and injury reports
- Appoint a safety officer and have a written safety program
- Deny benefits to employees injured while impaired by drugs or alcohol
- Focus attention on the basic elements when losses have been or are being experienced
- Establish clear objectives and the purpose of safety programs

3.13.15 Foremen Responsibilities

The foreman should be responsible for:

- Handling workers equally (not showing any preference for one worker over another, which is especially important when workers are from different cultures);
- Keeping unnecessary pressure off workers;
- Observing the performance of inexperienced workers to see if someone needs further orientation and/or training;

- Introducing new workers to other workers and providing them with time to become familiar with safety rules and procedures before working on their own.

A critical problem in global construction is the implementation of adequate safety programs. Since safety is gaining in importance, safety programs have to be managed properly by project personnel. Safety management should be implemented during every phase of a project, and it should comply with all local laws and codes.

3.14 Construction Failures and Investigation Techniques

Efforts to reduce construction failures by studying their causes have led to a meaningful reduction in their occurrence. Trying to reduce construction failures is a continuous process, and government safety organizations throughout the world are dedicated to this goal. In the 2000s, catastrophic failures are occurring throughout the world. For example, a parking garage under construction in Las Vegas, Nevada, collapsed because of improperly cured concrete in 2005, and a roof failures of an indoor waterslide in Russia and an ice skating rink in the Bavarian Alps killed or injured hundreds of people.

Technical causes of construction failures are those that are actual physical proximate causes. For example, improper compaction of soil could lead to excessive settlement of a foundation. Procedural causes are related to human errors, and they include things such as communication problems or shortcomings in the design and construction process that cause physical failures to occur on construction projects. An example of this would be a contractor placing reinforcing steel lower in a slab than is shown on the plans. Another example is when a testing laboratory makes an error in checking soil compaction (Lockley, 1998).

3.14.1 Failure Definitions

Design and construction failures are defined as "a human act; [an] omission of occurrence or performance; [a] lack of success; nonperformance; [an] insufficiency; loss of strength; and [a] cessation of proper functioning or performance" (Kaminetzky, 1991, p. 20). *Structural failures* occur when there is "a reduction of the capability of a structural system, or component, to such a degree that it cannot safely serve its intended purpose" (Janney, 1986, p. 1). A *construction failure* is a failure that occurs during construction, and such failures are "considered to be either a collapse, or distress, of a structural system to such a degree that it cannot safely serve its intended purpose" (Janney, 1986, p. 1). *Forensic engineering* is defined as "the application of the art and science of engineering in matters that are in, or possibly related to, the jurisprudence system, inclusive of alternative dispute resolution" (Specter, 1993, p. 1).

3.14.2 Failure Categories and Causes

Construction failures may be classified into three categories: safety, functional, and ancillary. Causes of failures fit into five general areas of deficiency (Thornton, 1985, p. 14):

- Design
- Construction
- Material
- Administrative
- Maintenance

A large majority of structural failures can be attributed to human errors, such as the following (Levy and Salvadori, 1992, p. 264):

- Knowledge is not currently available and thus the error is unavoidable.
- Delayed communication of acquired knowledge.
- Ignorance of recently acquired knowledge.
- Misunderstanding of accepted knowledge.
- Outright ignorance.
- Incorrect procedures.

A study was conducted that analyzed government (the Occupational Health and Safety Administration in the United States) safety records from a 10-year time period, and it determined that the three main causes of construction failures are (1) formwork failures and collapses, (2) inadequate temporary bracing, and (3) overloading and/or impact during construction. Other causes of construction failures that were provided in a survey of global E&C industry professionals include (Lockley and Yates, 2002):

- Failure to have a qualified person in charge
- Designs not reflecting the actual construction loads and field conditions
- Construction sequences not consistent with design considerations
- Improper definition of responsibility
- Financial pressure to complete a project early
- Incomplete connections—installing a few bolts and intending to complete the bolting process later
- Failure to use the materials specified or prefabricated elements being damaged during their handling and erection
- Unauthorized modifications to the designs specified in the contract document
- Supporting members damaged by other contractors as they are installing their work (i.e., duct work and plumbing)

- Poor communication, failure to follow design plans, failure to follow recommended industry practices, and carelessness
- Inadequacy of a system during erection
- Lack of common sense, including working while intoxicated or under the influence of drugs or using improper safety equipment
- Poor communication between designers and constructors
- No consideration for soil conditions
- Incorrect crane operations
- Not thinking fast enough
- Working too fast
- Incompetent supervisors
- Nature, gross design errors, terrorism, or contractor negligence
- Inadequate original designs that are not known to the contractor
- Insufficient or improper checking of the shop drawings
- Decisions by those with insufficient knowledge or education
- Lack of proper inspection
- Unreasonable schedules
- Inadequate training and education
- Unknown or erroneous geotechnical information

3.14.3 Methods for Reducing the Incidence of Construction Failures

When investigating structural failures, after trying to determine what happened and why it happened, an additional task is to suggest methods for preventing similar failures. It is impossible to totally eliminate structural failures, but the safety record of the construction industry can be improved by reducing the overall number of structural failures.

Factors that could help to prevent structural accidents from happening include the following (Petroski, 1985, p. 209):

- Good communication and organization in the construction industry
- The inspection of construction by structural engineers
- Increasing the general quality of design
- Improving the structural-connection design details and shop drawings
- Proper selection of architects and engineers
- Timely dissemination of technical data

A survey was conducted that included hundreds of global E&C industry professionals to determine the best methods for reducing construction failures, and the following methods were identified (Lockley and Yates, 2002):

- Design and detailing of critical connections by the engineer of record
- Design and supervision of construction of temporary structures by a professional engineer

- A clear definition of responsibility among the engineer, the fabricator, and the contractor
- Constructability reviews during the design stage
- Full-time inspection of construction by structural engineers
- Education and training of construction teams
- A comprehensive quality assurance and quality-control plan
- Structural redundancy in the design to avoid progressive collapse
- Peer review of the structural design and details by an independent professional

Other methods that were suggested by E&C industry professionals for reducing the incidence of construction failures include the following:

- Design engineers should review all shop drawings, shoring, and formwork design.
- Construction personnel should be certified before they design and build temporary structures.
- Realistic construction schedules should be used (this requires proper education and the enlightenment of clients and owners).
- Structural engineers should review temporary bracing and shoring designs, details, and construction processes and submit forms to public agencies indicating that such a review has been conducted by them.
- Superintendents that work for contractors should be educated about temporary bracing and stability issues.
- Part-time job-site visitation should be done by all design principals, not just structural engineers.
- Full-time inspection should be done by an independent construction professional.
- All construction activities should be reviewed prior to the performance of the work by professional engineers with construction experience. This includes the selection and positioning of cranes, concrete pumps, and truck movements.
- Site-specific safety plans should be provided that address issues that are highlighted by the engineer of record (including any dead loads during construction).
- Contractors and subcontractors should be hired based on a prequalification system for quality, safety, and liability.
- In-depth inspections (rather than cursory walk-throughs) should be done by local authorities.
- Stronger supervision on the part of the contractor should be provided to avoid shortcuts (methods for performing work that are quicker and less reliable) by workers.

3.14.4 Investigating Failures

The following list provides steps that can be followed during a forensic investigation (Lockley and Yates, 2002):

1. Select a principal investigator who may act independently in the investigation or as the manager of a staff of engineers and technicians.
2. Assemble (if needed) a staff of in-house engineers and technicians and/or outside specialists.
3. Investigate the scene of the failure as quickly as possible.
4. Conduct an overall visual examination of the failure site.
5. Generate as many hypotheses of causes of failure as possible.
6. Record visual information using photographs, video cameras, sketches, or drawings.
7. Collect samples for field and laboratory testing.
8. Conduct field tests or arrange for laboratory testing.
9. Conduct eyewitness interviews.
10. Review documents relating to the design and construction of the facility.
11. Review the structural design, and perform an analysis of the structure.
12. Analyze the data collected, and draw conclusions.
13. Prepare and submit a written report to the sponsor of the investigation.

3.14.5 Checklist for Investigating and Documenting Structural Failures

The following checklist contains comprehensive steps and procedures that can be followed during a failure investigation. The checklist provides guidelines for helping to plan, prepare, and execute a successful failure investigation (Lockley and Yates, 2002).

[ ] Select a principal investigator to conduct and manage the investigation.
[ ] Assign in-house personnel to assist in the investigation.
[ ] Retain outside specialists as consultants if necessary.
[ ] Assemble an inspection kit to probe, measure, sample, and record the wreckage.
[ ] Conduct an overall site examination to evaluate the scope and nature of the failure.
[ ] Generate as many hypotheses of causes of failure as can be developed based on past records of similar failures.
[ ] Establish a coordinate system for defining the location of fallen debris.
[ ] Record the position and orientation of the debris after the collapse.
[ ] Protect the evidence so that it can be documented properly.
[ ] Document the site of the failure using photographs, videotapes, sketches, or drawings.
[ ] Use a pocket recorder to record thoughts and impressions as they occur.
[ ] Create written or taped records of eyewitness accounts to the failure.
[ ] Collect samples and tag them for field and laboratory testing.
[ ] Complete field tests and document the results.
[ ] Arranged for all necessary laboratory testing.

[ ] Review the contract documents, inspection reports, project schedules, project correspondence, weather records, and any other related documents that reflect the history and life of the structure.

[ ] Examine all violations of safety procedures that may have caused the failure.

[ ] Review the structural design and either perform or arrange for an independent structural analysis to determine what behavior would have had to be anticipated for the structure.

[ ] Examine and record deviations from the design.

[ ] Determine whether there were any unconventional uses of materials or unusual construction erection techniques.

[ ] Determine whether a curtailment (even partial) of on-site construction by the design professionals occurred during construction.

[ ] Review weather bureau records to log the weather during the period being investigated.

[ ] Make a final analysis of the data and arrive at any conclusions.

[ ] Prepare a written report summarizing the findings and recommendations.

This checklist is provided to aid investigators in planning construction failure investigations, collecting and recording field and test data, generating a failure theory, and analyzing and drawing conclusions related to failures.

3.15 Case Study: Indonesia

Applying industrialized technology and techniques to rural construction projects in developing countries is difficult because of the unique nature of construction in developing countries. Appendix B provides a case study that discusses the types of problems that are encountered in developing countries in the areas of labor recruitment, scheduling, cost control, safety, training, retention of employees, and management practices. The information provided in the case study is from a project that was built in a rural area of East Kalimantan, Indonesia, on the island of Borneo. Borneo is the third largest island in the world. The area surrounding the project is one of the largest jungle areas in the world.

3.16 Summary

The success or failure of domestic or global engineering and construction projects is influenced by the quality of project-management team members. The function of a project-management team is to plan, organize, staff, direct, control, and coordinate. To be successful, project-management team members should effectively apply these functions to achieve underlying technical as well as sociocultural objectives. Therefore, successful project management in the global arena is a tradeoff between these two objectives, a concept that is called a*ppropriate project management.*

The evaluation and ranking of projects with respect to project objectives is essential for a successful project-management implementation program. Just as traditional management techniques are applied differently in developing countries, having roots in the unique cultures of the countries, so too should appropriate project-management techniques.

Although it would be unrealistic to expect global companies to sacrifice their own benefits and profits for the sake of the sociocultural concerns of a host country, increasing global competition indicates that multinational E&C companies cannot afford to ignore these concerns. Hopefully, this chapter has helped to raise awareness of these issues as they relate to managing projects in the global arena. The three-dimensional model presented in this chapter, which is called the Project-Objective Model (POM), can be used by managers when they are identifying project objectives, subobjectives, and appropriate technologies for use on global construction projects.

This chapter also discussed the desired attributes of global project management personnel and provided information on topics that could be investigated before working on global construction projects. Issues that affect the choice of appropriate technology were included in this chapter, along with information on foreign calendars, global time, and scheduling issues. The last two sections discussed safety concerns and construction failures that occur on global construction projects. A failure-investigation checklist was provided along with suggestions for preventing construction failures.

DISCUSSION QUESTIONS

1. Why is it important to adapt management information systems to the cultural environment of projects?
2. Why is it harder to develop project objectives for developing countries than for developed countries?
3. Discuss what are some of the underlying project-management objectives for developing countries.
4. Discuss why the construction industry in Saudi Arabia, Singapore, and India developed differently.
5. Where would a project that is an indoor waterslide park in Russia be ranked on the global, technical, and traditional dimensions on Figure 3.3?
6. Construct a management functions versus project objectives model similar to the one shown in Figure 3.1 and a hierarchy of objectives similar to the one shown in Figure 3.5 for a project in the eastern European country of Bosnia and Herzegovina using the information provided in Chapter 14.
7. What issues do project-management team members have to address when working at remote job sites in developing countries?
8. Discuss the 10 essential characteristics of a global project manager and why these are the most essential characteristics.

9. Explain appropriate technology and what it means in terms of using it in different regions of the world.

10. Explain how to address the issue of a global project-management team composed of members who use different calendars.

11. Discuss how to achieve 5 working days a week on a project that is being built 13 time zones ahead of the time zone where project management is being conducted (an example would be an Islamic country in the Middle East and a Christian country in North or South America).

12. Discuss how to implement a manual schedule-monitoring system in countries where computers are not available.

13. Discuss how to locate local materials in Russia.

14. Develop a spreadsheet that contains a quantifiable system to track the use of materials versus the climatologic conditions under which they will be used that could be used on projects throughout the world.

15. Discuss what type of work should be assigned to an engineer who graduated from an engineering program that stressed memorization over analytical reasoning when he or she is part of a project-management team.

16. Discuss which of the global scheduling influences that are beyond the control of project-management team members would affect the scheduling of a project in Indonesia.

17. Provide a discussion on how to explain the safety issues that affect construction projects included in this chapter to a worker in Indonesia who has never worked on a construction site.

18. Discuss how a supervisor should deal with the issue of workers using legal or illegal drugs that impair their judgment while they are working at construction job sites.

19. If all the workers at a construction job site are from the same tribe or clan, discuss who should be assigned to secure the job site and why they should be assigned this task.

20. If it is illegal for expatriates to bribe local government officials because of anticorruption laws in their native countries, explain how they should secure permits and licenses in countries where bribes have to be paid to obtain permits.

21. Discuss what types of measures engineers could incorporate into engineering designs that would help to reduce construction accidents.

22. Discuss what type of position someone should have before he or she is appointed to implement a safety program that contains the elements discussed in this chapter at a global construction–job site and why he or she should be appointed to this position.

23. Explain which of the causes of construction failures should be addressed first and why, if only six of them can be addressed on a construction project.

24. Discuss which project-management team members should be assigned to investigate construction failures and why they should be the ones to investigate failures.

25. For the case study in Appendix B, discuss how construction workers could be located if there were no newspapers to advertise positions and no access to the Internet (computers).

26. For the case study in Appendix B, discuss how work could progress on a project, if it takes 6 to 26 weeks to train craft workers, if workers are constantly quitting.

27. For the case study in Appendix B, discuss how technology that was used on the liquefied natural gas plant could be transferred to the canal project.

28. If a construction project is operating on rubber time (and workers adhere to *asal bapak senang*) and the workers are paid a daily rate, such as was the situation in the case study in Appendix B, how could project-management team members ensure that tasks are completed on time?

29. Since engineers may not be the ones directing construction in cultures with rigid class systems, as was the situation in the case study in Appendix B, where foreman performed this task, explain how an engineer can ensure that quality requirements are being met during construction.
30. If workers are from cultures where they do not compete for promotions (see Appendix B), how could a project manager determine who should be promoted on construction projects?

REFERENCES

Altayeb, S. 1992. Efficacy of drug testing programs implemented by contractors. *Journal of Construction Engineering and Management* 118 (4): 780–789.

Baum, W. C. 1985. *Investing in Development, Lessons of World Bank Experience.* Washington, D.C.: Oxford University Press

Bell, Glen. 1984. *Failure information needs in civil engineering.* In *Reducing Failures of Engineered Facilities.* New York: American Society of Civil Engineering.

Copeland, L., and L. Griggs. 2001. *Going International.* New York: Random House.

Culver, C., Marshall, M., and C. Connolly. 1993. Analysis of construction accidents: The workers' compensation database. *Professional Safety* 38: 22–27.

Hislop, R. D. 1990. A construction safety program. *Professional Safety* 35 (1): 17–20.

ISO, 2006, International Organization for Standardization. www.iso.org.

Janney, Jack R. 1986. *Guide to Investigations of Structural Failures.* New York: American Society of Civil Engineers.

Kerzner, Harold. 2003. *Project Management: A System Approach to Planning, Scheduling, and Controlling.* New York: Van Nostrand Reinhold.

Kaminetzky, Dov. 1991. *Design and Construction Failures: Lessons from Forensic Investigations.* New York: McGraw-Hill.

Levy, Matthys, and Mario Salvadori. 1992. *Why Buildings Fall Down.* New York: W. W. Norton and Company.

Lockley, E. 1998. A forensic engineering investigation into construction failures and their documentation. PhD dissertation, Polytechnic University, Brooklyn, New York.

Lockley, E., and J. K. Yates. January–February 2002. Documenting and analyzing construction failures. *Journal of Construction Engineering and Management* 128 (1): 8–17.

Lucas, C. 1986. International Construction Business Management: A Guide for Architects, Engineers, and Contractors, McGraw-Hill Publisher, New York, New York.

MacCullum, D. V. 1990. Time for change in construction safety. *Professional Safety* 35: 17–20.

Petroski, Henry. 1985. Learning from failures. *Journal of Construction Engineering and Management* 55 (2): 36–39.

Rekus, J. F. 1992. Safety in the trenches. *Occupational Health and Safety* 61 (1) 26–37.

Specter, Martin. 1993. Internal document. National Academy of Forensic Engineers. Hawthorne, N.Y.

Terrero, N., and J. K. Yates. 1997. Construction industry safety measures. *Cost Engineering* 39 (2): 23–31.

Thornton, Charles. 1985. Failure statistics categorized by cause and generic class. *Reducing Failures of Engineered Facilities.* New York: American Society of Civil Engineers.

Bruner, B., ed. *TIME Almanac and Information Please.* 2005. Needham, Mass.: Pearson Education Company

Chapter 4

Global Competitiveness in the Engineering and Construction Industry

4.1 Introduction

World War II provided opportunities for engineering and construction (E&C) professionals from industrialized nations to work globally, and during the 1970s, the global reach of E&C firms from developed countries began to peak. By the 1980s, new projects and acquisitions were limited by declining growth rates in developing countries, fluctuating oil prices, and the fiercely competitive nature of E&C companies from newly industrialized countries (NICs).

In order to obtain information on global competitiveness, executives from 57 global E&C firms (including owner organizations, constructors, construction management firms, architecture firms, engineering firms, and developers) and government agencies were interviewed for three or four hours, and the transcripts of their interviews were evaluated to obtain data on the global operations of E&C firms. Any of the quotes in this chapter that do not include citations are from these interviews. In order to maintain the confidentiality of those who participated in the interviews, their names and the names of the firms or agencies for whom they work are not included in this chapter.

Access to E&C industry executives for the interviews was facilitated by members of the Construction Industry Institute (CII), and the CII provided funding for the research investigation. The CII is located in Austin, Texas, in the United States, and it is made up of E&C firms and owners who fund and participate in research projects with academics. The results of CII research are available in publications that can be purchased on the CII Web site, www.construction-institute.org.

This chapter discusses historical patterns that have affected the competitiveness of E&C firms, along with current trends influencing competition in the global E&C industry, including the forces that affected competition in the 1980s and the 1990s, the forces that will affect the E&C industry in the 2000s, the changing nature of competition, declining competitiveness, obstacles to competitiveness, and future strategies.

4.2 The Definition of Competitiveness

In the beginning of the twenty-first century, the nature of competition in the E&C industry changed dramatically as the industry moved away from *adversarial com-*

petition toward *cooperation.* If members of the E&C industry realize that "the global economy is not a zero-sum game (i.e., when someone wins, someone else does not have to lose), it is easier for them to soften their adversarial competitive stance. Adversarial relationships are less effective when partnering with firms from cultures that encourage cooperation rather than competition (Naisbitt and Aberdene, 1990, p. 39).

One definition of *global competitiveness* is "simply the ability of domestic producers to sell sufficient goods and services outside the country to pay for domestic imports. Emphasis should be on increasing national wealth through productivity gains" (Hudson Institute, 1987, p. 13). Another definition is "the capacity of a country to add value to the world economy and thereby gain a higher standard of living in the future without going into deeper debt" (Reich, 1990, p. 59).

4.3 Competitiveness Issues

Competitiveness issues are related to the declining relative ability of some E&C firms to achieve global contract awards. It is essential for members of the E&C industry to understand that "with a growing number of competent firms, and increasingly homogeneous technologies, the pattern is one of convergence in competitiveness, and price competition in a shrinking overall market has been very intense. Many causes lie as much in *improvements elsewhere as in problems in any one country*" (Naisbitt and Aberdene, 1990, p. 35).

The E&C industry is populated with competent firms from all over the world, and there are now many competent firms that are capable of performing the same work as previous world leaders. There is more competition for projects that are labor-intensive or projects that do not require high levels of technology. Firms from Turkey, Brazil, China, Spain, India, and South Korea are now competing for major projects against firms from the United States, Japan, Great Britain, France, Germany, and Italy (volumes of construction work and revenues are discussed in Chapter 5). Some design and engineering firms in Japan, South Korea, and Europe are ahead in technology innovation; therefore, they have a competitive advantage over E&C firms from other nations.

4.3.1 Cooperative Strategies

Global partnerships should benefit all of the parties involved in a project because they allow E&C firms to gain access to opportunities within the native countries of their partners, which were closed to their firms previously due to government restrictions (such as China). If firms have some type of comparative advantage, it helps them market their services to potential partners.

Foreign firms have no particular advantage over domestic firms, because domestic firms with some sense of foresight can duplicate any advantage that a foreign firm develops. If firms lose out to foreign competitors it is not always because the competition is superior in a particular area, but rather it is because foreign firms are better at the application process in that area. Stagnation results from lower productivity, higher wages, and fluctuating exchange rates, but if a company produces a superior product or is able to manage projects more efficiently and effectively, it can maintain a competitive advantage over other firms.

With a shift toward cooperation and away from adversarial relationships, E&C firms are able to concentrate more on their clients and on becoming product-based so as to provide products tailored to well-defined markets as firms eliminate functional divisions. One strategy that is already being used is permanent project teams. Team members move as a team from project to project (as is done in some firms in the United Kingdom), and the team concept satisfies the needs of clients who want single-point responsibility for the coordination of their projects. Every project team member could be from a different country, and project teams often resemble a mini-United Nations because of hiring practices that ensure qualified experts in different technical fields.

The fragmented structure of the E&C industry is being replaced by an integrated arrangement that is composed of clients who have become experts, independent firms combining to form consortia, integrated large firms, and major conglomerates. Strategic alliances are formed to "combine skills synergistically, gain market access, and ensure proper skills and input at the right stage of a project" (Morris, 1989, p. 5).

Successful strategic alliances include traditional joint ventures, but they also may include technology transfer and partnering to obtain licensing agreements. It is easier to market a specific product or provide an array of services rather than products produced by separate firms. This is why major conglomerates are so successful in the global arena. There has been a reduction in the overall number of E&C firms competing in the global marketplace due to the formation of large conglomerates able to provide all of the E&C services required by owners.

Performing E&C work in foreign countries requires up-front capital to mobilize for overseas projects, including the purchasing and transporting of equipment and materials and the recruiting of qualified personnel who are willing to relocate or the hiring of qualified local professionals. Access to capital is one of the major factors that limits the number of firms that are able to perform work globally. With the recent mergers of smaller firms with larger firms, there are fewer firms that are able to bid on or negotiate for global projects.

Developing joint ventures, consortia, or alliances helps to bring together the major elements required for successful projects. A firm from one nation may provide management expertise and technology, whereas a firm from another country may provide financing and guarantees, and another firm from a third country may provide supplies, materials, and equipment, and laborers are hired from the local population.

Providing only E&C management services reduces risks in politically unstable environments because there are no risks associated with having to abandon equipment or materials. In 1979, some firms experienced losses in the millions of dollars when heavy construction equipment being used for the petrochemical industry was abandoned during the Iranian revolution, and this led to firms moving more to-ward providing only management services in politically unstable countries.

Another advantage of adopting cooperative strategies is that some foreign gov-ernments provide startup funds to bid on projects. In Japan, up to 75 percent of the bid-estimate cost is paid for by the government. In other countries, such as South Korea, military personnel perform overseas construction work while they are on active duty, and they are paid military not free-market wages, which in-creases the competitive advantage of South Korean engineering and construction firms.

If E&C firms are considering global collaborations, they should be aware that not only are there cultural differences between firms in Western societies versus Eastern societies but also differences within each of these types of societies, espe-cially in the area of professional ethics. In the United States, there are 40-hour workweeks, managers are not paid overtime, vacations are for 2 to 4 weeks, and there are fewer than 10 paid holidays. French employees do not work overtime, and they receive at least 6 weeks of vacation plus many holidays off throughout the year. Some businesses in France completely shut down during August and early September. In Eastern cultures, workweeks exceed 40 hours, firms are paternalistic and managers father figures, and workers are not rewarded as individuals, only as part of a team.

Collaborations may be formed between an Asian firm and a Western firm more readily than between firms from different Asian countries. South Korea has a long history of domination by either China or Japan, which makes members of South Korean firms more willing to collaborate with people from countries outside the Asian market. In 2005, Chinese citizens publicly protested the use of textbooks in elementary schools that glossed over the atrocities that were done to Chinese cit-izens by the Japanese during World War II, which demonstrates that resentment against the Japanese still exists in China.

The next two sections provide an analysis of the competitiveness of E&C firms during the latter half of the twentieth century to provide the reader with a basis for understanding competitiveness in the twenty-first century.

The nature of competitiveness in the E&C industry is influenced by the follow-ing five forces (Porter, 1990):

- The threat of new entrants
- The threat of substitute products or services
- The bargaining power of suppliers
- The bargaining power of buyers
- Rivalries among existing competitors

The following sections explore these five forces in relation to the E&C industry during different time frames since competitive issues are long-range planning is-sues. The information provided helps to explain the competitive position of E&C firms in the 2000s that have resulted from the events that took place and the strategies that were implemented during the last two decades of the twentieth century.

4.4 Competitive Forces in the 1980s

This section discusses the competitive forces that affected the E&C industry during the 1980s.

4.4.1 The Threat of New Entrants

Between 1966 and 1971, firms incorporated in the United States performed nearly 70 percent of the foreign construction volume (Bah Neo, 1976). By 1980, the U.S. share of global construction contracts had dropped to 60 percent; by 1985, it was 35 percent; and by the early 2000s, it was only 19 percent (*Engineering News Record,* various issues between 1970 and 2005).

One major force exerted on the global E&C industry during the 1980s was the emergence of a new generation of competitors. The United States, Great Britain, France, Japan, Italy, and West Germany were the leaders in performing global construction prior to the 1980s, but they were joined by aggressive new firms from Brazil, China, South Korea, Turkey, and other newly industrializing countries (NICs) in the 1990s.

4.4.2 The Threat of Substitute Products or Services

During the 1980s, few innovative substitute products, except for mining, petrochemical, and tunneling technologies, were introduced that provided E&C firms from any one nation with a competitive advantage. However, what changed was how innovative individual firms were in *applying* technology.

4.4.3 The Bargaining Power of Buyers and Suppliers

In the 1980s, the bargaining power of owners intensified the level of competition between E&C firms. Both domestic and international owners became more knowledgeable in terms of E&C contract negotiations, including E&C equity-position-contract requirements and implementing the *incentive* approach, whereby owners dictate the requirements for E&C firms to secure repeat business through various contract stipulations.

As the number of financially capable global E&C firms increased during the 1980s, the interest of owners in cost-plus contracts declined because of the increasing profit margins or fees for these contracts. The advent of increasing competition prompted owners to seek greater equity-risk sharing by E&C firms. This allowed financially stable global firms to expand into new markets, and it was a bonus to firms that received financing or subsidies from their governments.

Suppliers also increased their bargaining power by looking for owners or firms that would enter into strategic alliances or long-term commitments (4 to 40 years) with their firms in an effort to stabilize their production levels and lower the amplitude of supply and demand swings. Strategic alliances are designed "to guarantee a stable or controlled rate of production at a fair profit margin for the supplier, to guarantee a constant supply of a product at a known unit price, and to help all alliance members to develop specific products, solve problems, or provide unique services for the owner" (Tuman, 1989, p. 28).

Alliances require a high level of trust between parties. Therefore, countries other than the United States, such as Japan, whose business culture is known for high levels of commitment and trustworthiness, flourished during the 1980s. The competitive business environment of U.S. firms, coupled with the strong government antitrust policy concerning horizontal mergers, alliances, and collusive behavior, restricted the formation of strategic alliances in the United States during the 1980s.

4.4.4 Rivalry among Existing Competitors

Marketing strategies and client satisfaction became more pertinent to the E&C industry as competition increased in the 1980s. Total quality management (TQM) programs, constructability programs, just-in-time inventory management, and full-service E&C organizations were instrumental to the industry. Domestic competition in the United States increased to unprecedented levels as large firms merged with other large firms to form E&C megacorporations. Competition is not necessarily detrimental to domestic firms because innovation takes place when competitive pressures create an environment that supports innovation (Porter, 1990).

4.4.5 Management and Organizational Changes in the E&C Industry in the 1980s

Engineering and construction firms underwent management and organizational changes during the 1980s. Members of E&C firms that were interviewed provided different reasons as to why firms were having competitive difficulties in the 1980s, including increasing foreign competition, labor problems, unexpected decreases in interest rates, and a tightening of the E&C market, but bad management decisions also contributed to failures. In addition, mergers also influenced competitiveness, because foreign firms were penetrating the U.S. market by merging with domestic firms.

Another organizational strategy that was implemented during the 1980s was the phasing out or elimination of several layers of middle management, which flattens the hierarchy of organizations. Twenty-nine of the people interviewed from global E&C firms concluded that the largest adjustment that their organizations underwent during the 1980s was the downsizing or elimination of middle-management layers. The two primary competitive advantages of a flatter organization are flexibility to move into new markets quickly and increased efficiency to adapt quickly to changing market conditions.

4.5 Changes in the E&C Industry Structure in the 1990s

Changes in industry structure, or discontinuities, create new methods of competition. Typical causes of innovation include (Porter, 1990, pp. 45–46):

- New technologies
- New or shifting buying needs
- Emergence of new industry segments
- Shifting input costs or availabilities
- Changes in government regulations

In this section, each of these sources of innovation are discussed in relation to how they affected the E&C industry in the 1990s.

4.5.1 New Technologies

In the last decade of the twentieth century, the global marketplace was more lucrative for design firms than it was for construction firms because of the type of technology transfer industrialized countries were supplying to developing and newly industrialized nations. Technology was transferred to developing countries at a faster rate than skilled construction workers, such as surveyors or heavy equipment operators, which helped engineering firms from industrialized countries to maintain their competitive advantage. Even though technology has been exported to other countries, industrialized nations are still the leaders in computer-aided design and drafting (CADD) and in the design of petroleum refineries, power stations, chemical plants, and other large industrial facilities.

During the 1980s, firms and owners implemented new measures to foster innovation and productivity improvements, including the following:

- In the United States, the Construction Industry Institute (CII) was formed by members of the E&C industry. This institute was established to foster, enrich, and improve relationships among construction team members, including owners, contractors, designers, and vendors, so that they could learn to work together to improve the competitive positioning of U.S. companies and to develop and finance E&C research projects.
- Strategic alliances were formed that allow construction team members to work together by sharing resources to improve productivity.
- Owners began sharing more risks with E&C firms through contract negotiations. Along with shared risks come (1) an increase in quality products from the E&C firms for the owner, (2) an increased understanding of how important a cooperative relationship between owners and E&C firms is to producing

a quality product, and (3) a realization by owners that the lowest-cost project does not necessarily translate into the best project for the owner.
- The increased popularity and implementation of constructability programs contributed to improved relationships and higher productivity by members of the E&C industry.
- Productivity improvements methods were developed and implemented that helped increase productivity at construction sites.

4.5.2 Shifting Buying Needs

Two other forces driving competition in the 1980s were "declining economic growth rates, both domestically and in developing countries, coupled with the collapse of oil prices" (U.S. Congress, 1987, p. 119). Overall, the debt of developing nations increased during the 1980s at rates similar to postwar periods, and this translated into fewer large international construction projects such as airports, dams, and petrochemical plants in the 1990s. The buying patterns of domestic owners also changed during this time period. The first decade of the twenty-first century is seeing global economic patterns similar to those that took place during the 1970s when oil prices were rising at unprecedented rates, there was widespread global political instability, the cost of manual labor increased substantially, and the cost of real estate increased rapidly in industrialized countries.

4.5.3 New Industry Segments

During the 1990s, the most prevalent new business segment in countries such as Japan, the United States, and South Korea was the development and manufacture of high-technology items such as computer software, machinery, and electronics. However, few E&C contractors obtained a clear advantage over other competitors in these areas. An executive from a leading U.S. E&C firm stated that "the U.S. construction industry's strongest tie to high-technology developments during the 1990s was the building of the facilities to house the high technologies and not the high technologies themselves." This is related to the fact that the construction industry is primarily a service industry.

The E&C industry did contribute new innovations during the 1990s, and two of these were identified by one E&C executive as advances in the power industry and in waste management services. Infrastructure facilities such as roads, waste-water and water-treatment facilities, and potable water were privatized in many countries, but the power sector experienced the greatest success with privatization. The combination of a growing concern for the environment and the congestion of landfills opened up a new market segment for E&C firms in the privatization of waste-management services.

4.5.4 Shifting Input Costs and Availabilities

Increasing input costs influenced the competitive position of firms in the E&C industry during the 1990s. Increases in input costs, which were caused by the high value of some currencies relative to other currencies in the global marketplace, translated into stagnated growth and an influx of less expensive foreign E&C labor into industrialized countries. Also, real interest rates during the 1980s were higher in the United States than in most other advanced nations, which slowed domestic investment in long-term capital investments such as construction, and the effects of this were evident during the 1990s.

Another changing input was labor costs, which negatively affected the United States and other developed nations, because they had high dollar-per-man-hour costs. Therefore, since the technological competency of developing and newly industrialized nations grew at a rate faster than it did in industrialized countries, the appeal of and worldwide demand for services from industrialized countries diminished to levels well below those of previous decades. However, in the first part of the 1990s, some of the firms from developed countries increased their competitive position by applying technology and capitalizing on lower labor costs in other countries.

4.5.5 Government Regulations

Another major force that affects the E&C industry is the reluctance of some governments to enact legislation that would allow E&C firms to compete financially with foreign competitors. Before the 1980s, the technical and management expertise of firms from developed countries established them as worldwide leaders in market developments. However, one phenomenon that has long-term implications on the future competitive positioning of E&C firms is the emergence of firms from developing and newly industrialized countries and the willingness of their governments to adopt policies that increase their competitiveness, such as providing funding to pursue projects in foreign countries.

Many of the nations that were rebuilt at the end of World War II by means of the U.S. Trade and Development Program, Export-Import Bank loans, World Bank financing, and capital improvements by the Overseas Private Investment Corporation now have native firms that are competing as opponents in the global construction marketplace with subsidies from their native governments. Due to anticorruption legislation, such as the U.S. Federal Foreign Corrupt Practices Act (FCPA), which requires U.S. businesses to conform to U.S. business standards on all international projects, firms from industrialized nations are not able to compete on an equal basis with firms that are not subject to these types of government regulations.

4.6 Forces Driving Competition in the Twenty-First Century

This section contains information on the forces that are driving competition in the twenty-first century.

4.6.1 The Threat of New Entrants

It is not possible to determine which countries will dominate the global E&C industry in the first part of the twenty-first century because the dynamics of world political and economic structures are unpredictable and always changing, depending on which regimes are in power. During 2004 and 2005, eleven countries experienced regime changes. However, it is possible to explore the types of transitions or adjustments that members of the E&C industry will have to undergo in order to anticipate new and emerging competitors.

In the twenty-first century, leading E&C competitors will not necessarily be the same ones that emerged in the 1980s and 1990s, as they may include firms from other countries that want to increase the standard of living within their countries such as China, India, Indonesia, and eastern European countries. Global engineering and construction projects are a common method for rapidly increasing economic expansion.

With the increasing complexity of corporate structures it is possible for E&C professionals to work for a company that they assume is a domestic firm when, in reality, foreign investors may hold a controlling interest in the firm, or a firm owned by a domestic company may perform all of its work outside the borders of the native country of the firm. It is becoming increasingly difficult to distinguish exactly which firms are truly foreign firms; therefore, the relative amount of global E&C competition has declined in real terms.

In countries previously viewed as only a source of inexpensive labor, potential consumers and construction markets are being identified. The formation of the European Union is affecting the centralized trading power of Europe and providing several new market niches for foreign firms. China and India are also emerging as world power brokers with consumer markets that have over a billion people in each country.

4.6.2 The Threat of Substitute Products, Services, or Strategies

There are two types of substitute products that have the greatest potential for improving the price-performance tradeoff in the E&C industry during the 2000s, and they are (1) the electronic transfer of data via the *application* of higher technologies, such as wireless computers, telecommunications networks, nanotechnology, AUTOCAD, the bar coding of inventory, voice-activated computer interfaces, digital video technology, and satellite cell phones, and (2) the supply and marketing of privately owned facilities to those currently dominated by the public sector, such as power-generation facilities, roadways, waste services, and prisons.

Three substitute services or strategies growing in popularity among owners are (1) creative contract developments on the part of E&C firms, (2) the supply and marketing of privately owned facilities to those currently dominated by the public sector, and (3) *partnering*, which is the formation of strategic alliances. Design/build contracts increased in popularity during the latter part of the 1990s, and they are expected to increase at even greater rates in the power, processing, and petrochemical sectors in the twenty-first century.

Another current trend is the turnkey, lump-sum, or date-certain contract, with an increasing equity position taken by E&C firms. The proven ability of some firms to provide technical and managerial expertise to develop a project, such as cogeneration power facilities, from inception to operation with an attractive profit margin when they are completed, has created interest by owners in this type of contract.

Partnering relationships and the formation of strategic alliances have increased, and firms are forming alliances with major firms outside their borders. Since a large number of countries in the world have cultures where cooperation and relationships are two of the primary values, partnering will continue to be used more frequently on E&C projects. *Partnering* is defined as "a long-term commitment between two or more organizations for the purpose of achieving specific business objectives by maximizing the effectiveness of each participant's resources" (Baker, 1990, pp. 7–12). Partnering is a relationship wherein (Baker, 1990):

- Everyone seeks a win-win solution.
- Value is placed on long-term relationships.
- Trust and openness are valued.
- An environment to make a profit exists.
- Everyone is encouraged to openly address problems.
- Everyone understands that no one benefits from the exploitation of another partner.
- Innovation is encouraged.
- Each partner is aware of the other's needs, concerns, and objectives, and each partner is interested in helping the partner achieve its objectives.

In the twentieth century, most of the strategic partnerships were between companies from the same country; but during the twenty-first century, there is an increasing incidence of strategic partnering between companies from different countries and vastly different cultures.

4.6.3 The Bargaining Power of Buyers and Suppliers

The advent and continual development of strategic alliances will help to alleviate the adversarial relationships that are commonly associated with traditional lump-sum or negotiated contracts between owners, architects, and E&C firms in the private sector. The bargaining power of suppliers poses a problem in the area of *teaming,* or horizontal integration, of foreign suppliers and E&C firms. Some countries, such as the United States, have laws that, when coupled with the legal and trading policies of other countries, do not allow the teaming of foreign suppliers and E&C firms to obtain a monopoly or unfair competitive advantage over other domestic firms. At the same time, many E&C firms have had trouble penetrating protectionist markets. Therefore, as new markets emerge, the foreign E&C firm that is already *teamed* with a local supplier has an advantage over firms just entering a market.

4.6.4 Rivalry among Existing Competitors

While strategic alliances provide improvements to owner and E&C firm relationships in the private sector, the public sector does not experience these improvements. An executive from a global E&C firm that has actively pursued government contracts for the past five decades expressed his views on government contractual relationships:

> The contracting process is far too political in the public sector. It requires too many regulations, requirements, impediments; the cash flow is poor; the agencies that are getting the contracts are weak; and it is not going to get any better in the next 10 years. The bottom line is that there are too many risks involved, and yet people are bidding contracts time after time and going out of business.

Worldwide political events have changed the dynamics involved in government defense spending, and defense spending diverts funds away from public projects. The downsizing and closure of military installations affects both domestic and foreign E&C markets if domestic governments are also closing bases in foreign countries.

The downside to military and infrastructure projects is the long lead times required to enact federal budgeting policies and the tightening of federal budgets. The time required to design and build these types of projects, approximately 3 to 7 years, and the financing impediments imposed by government legislation allow only those E&C firms with high levels of capital and the ability to transfer their workloads to other specialty sectors a competitive advantage in the government sector.

Midsized E&C firms that are highly integrated horizontally and geographically and that have a limited supply of capital will find the rivalry among their competitors to be extremely intense. There are three main reasons for this scenario:

- Owners are continuing to require higher levels of equity risk by E&C firms, which requires a stronger capital or financing base.
- Limited capital restricts organizations from making the investments required to become technologically efficient in *all* sectors.
- Diversification, coupled with limited capital, restricts the ability of an organization to become specialized (vertically integrated). This is especially important when marketing a firm for joint-venture partnership relationships.

4.7 Changes in the E&C Industry in the Twenty-First Century

This section addresses the issues that will affect engineering and construction (E&C) industry competition in the twenty-first century.

4.7.1 Technology

Engineering and construction technology and its adaptation require large amounts of capital and high levels of scientific advancement. Implementing technology and increasing innovation require more than merely investing capital in current technologies such as computers, computer-aided design development (CADD), optical-recognition devices, digital cameras, digital video, bar-coding hardware, and robotics, as it also requires the training of personnel required to operate these technologies. Therefore, it is important to define the difference between construction technology and the *management* of technology. *Construction technology* is "the combination of resources, processes, and conditions that produce a constructed product" (Tatum, 1988, p. 344). The *management of technology* is "defining, directing, and delivering a real solution to a customer defined need by" (Manley, 1989, p. 11):

- Capturing the definition of a customer's need
- Selecting the necessary (and appropriate) technologies
- Converting technology to product definition
- Targeting construction technology and process
- Delivering a customer-defined solution

The benefit of using construction technology lies in its management, innovation, and implementation. However, owners have to provide an atmosphere that stimulates technology innovation and cooperation.

4.7.1.1 The Effects of Education on Technology

The content of engineering programs is another area that affects global competition because most countries do not have an accreditation process for E&C programs similar to the accreditation process used in the United States, which is the Accreditation Board for Engineering and Technology (ABET). The competitiveness of E&C firms is affected by the *quality* of their E&C employees. Quality requires more than having employees who merely possess a university degree, because experience is vital in the E&C industry. There is a trend toward eliminating mandatory internships in E&C educational programs as a requirement for graduation, so the industrial experience level of college graduates is declining in the 2000s.

Another situation that is making it harder for firms to hire qualified technical personnel is changes in the processing of visas during the 2000s. Visas are not being issued in time for foreign students to enroll in graduate programs during the term of their acceptance to universities. Graduate programs in the United States have experienced declines in enrollments by 30 or 40 percent in the first part of the 2000s. Projects could experience delays while firms are obtaining visas for foreign technical personnel. Foreign nationals are sometimes allowed to work in the country where they receive their college degree as expatriates on temporary work permits, called *practical training* visas, immediately after graduation from college, but the amount of time is usually restricted to 1 or 2 years.

If foreign nationals are turned down for a visa in the United States, the appeals process takes 1 to 2 years (or more). Therefore, members of firms that hire foreign nationals need to be cognizant of the fact that foreign nationals cannot remain in the country during the appeals process because they are considered to be "out of status," and they could be deported at anytime by the U.S. government.

Before bidding on or negotiating for a project in another country, members of E&C firms should investigate the formal (and informal) visa-acquisition process of the host country to determine if members of the firm will meet local visa requirements. Information on the visa-application process can be obtained through appropriate government agencies or from the embassy of the country where the work will be performed for a project.

4.7.2 Stimulating Innovation in the Public Sector

One force restricting innovation in the public sector is current contracting methods. Some countries, such as the United States, have laws that do not allow government agencies and E&C firms the freedom to share risks in contracts. Therefore, E&C firms absorb most of the risk, and this limits technological advances. However, some government agencies have developed programs designed to improve productivity through the combined efforts of the government and the private sector.

Engineering and construction firms innovate from within their own organizations, and this creates *firm against firm* competition, which is essential to increasing the overall competitiveness of firms. The more intense the internal level of competition, the better prepared a company is to compete in the global marketplace. Unfortunately, in Eastern cultures, when personnel are too competitive within a firm, it is detrimental to the firm.

In cultures where relationships and cooperative strategies are valued above all else, firms are fiercely competitive externally, but they are always building relationships internally—and Japan, South Korea, and other Asian cultures are examples of cultures where internal harmony is valued by companies. In these cultures, work efforts are rewarded based on long-term cooperative relationships of the *team* rather than of individuals. Employees are embarrassed to receive personal praise or compliments because they are used to being complimented only as part of a *group*. This is in contrast to the United States, where individuals compete against each other within firms, individual praise and rewards are common practices, and work is focused on achieving short-term goals.

A large part of the research that is conducted in the E&C industry is conducted within firms, and the results, as well as the products of the research, are kept within the organization as *proprietary processes* to help increase the competitiveness of a specific firm not the global competitiveness of the entire industry.

4.7.3 Stimulating Innovation in the Private Sector

Members of the E&C industry will spend the 2000s learning to apply and efficiently use currently available technologies in innovative ways and in convincing clients to use new technologies on construction projects. Engineering and construction professionals are discovering that some of the technologies that are used effectively in industrialized nations are rendered useless in developing countries. A primary example of this is cell phones that work well (or sometimes not so well) in industrialized countries, where there are thousands of repeaters, but that do not work efficiently (or at all) in regions of the world where there are a limited number of repeaters. Satellite phones offer an alternative technology that is more reliable for communicating in remote regions of the world. Another example is construction equipment that does not operate during sand storms, which are common in Saudi Arabia, or in extremely cold climates, as is the case in Siberia, some locations in China, and the interior regions of the United States and Canada.

Technology advances do not always guarantee that a firm will secure work globally if the technology is not being used appropriately for a particular region of the world. Spare parts and supplies for electronic devices are not always available, and this forces engineers to perform calculations manually or on a slide ruler or an abacus (which is a wooden frame that contains several series of beads that are used to add or multiply numbers).

4.7.4 Shifting Buying Needs

Two key issues related to shifting buying needs are service and innovation. Although these issues have been stressed repeatedly in the past, they became even more important during the twenty-first century. An increasing demand for service will influence the future buying needs of owners in the private sector, but it is difficult to determine what constitutes *service* (Porter, 1990). According to one of the interviewees, there is a philosophy throughout the industry that the job of an E&C firm is to first make a profit and then to satisfy clients so that the firm can make more profit. Short-term planning ideologies should be reexamined if firms want to remain competitive in the global marketplace and in countries where projects are part of building long-term relationships, such as in Greece and Japan.

One aspect that will separate successful E&C firms in the public sector is innovation, because government contracts are awarded through a competitive bid process in many nations. Therefore, innovation is the only thing that distinguishes one firm from another firm. Many governments have started spending more money on infrastructure rehabilitation because of declining infrastructures. In many countries the infrastructure was constructed shortly after World War II (during the 1950s), and concrete only lasts 50 years.

4.7.5 New Industry Segments and Issues

During the 1990s, and early 2000s, several countries experienced regime changes, some of which were peaceful, such as when pressure was exerted by thousands of people protesting in the streets against the current governments in Lebanon and the Ukraine in the early 2000s, and others that resulted from violence in places such as Bosnia and Herzegovina, Serbia, Afghanistan, and Iraq, where private citizens were killed by military action or by acts of terrorism during regime changes.

Terrorism and kidnappings have always been a concern for E&C industry professionals, and these threats will continue to plague members of E&C firms because no one can predict where they will surface. Competition is limited in some regions of the world because of terrorism and kidnapping, and some executives of E&C firms have decided that pursuing projects in these regions are not worth the potential risks. In 2004, there were 651 major terrorism incidents worldwide that resulted in 9321 casualties of which only 1 percent, or 103 people, were U.S. citizens (*U.S. News and World Report,* May 9, 2005). Major global E&C firms are adopting new methods for protecting their employees in politically unstable regions of the world by hiring private security firms and training employees to protect themselves. Chapter 13 provides additional information on the subject of terrorism and kidnapping.

4.7.6 Government Regulations

Government regulations constantly change, so it is difficult to provide timely information on regulations and laws. Contacting embassies is the most efficient method for locating information on current laws and regulations that affect E&C firms and their employees. The effects and severity of government regulations are discussed in this section.

Engineering and construction firms that are heavily concentrated in public, lump-sum contract projects will be affected the most by changing government regulations. In the United States, the federal government has deregulated services that were previously only performed by its agencies. The 1990s demonstrated that private industry could build and operate facilities more efficiently than governments at the local, state, and federal levels (e.g., cogeneration power plants, roadways, and waste-management facilities). The increased productivity performance records in privatized facilities support the trend toward privatization of public facilities.

When governments are faced with decreasing revenues, they turn to privatization as a means of providing basic services to their citizens. In politically unstable countries, there is always a risk that the government will nationalize private facilities. Some E&C firms self-insure by including contingency factors in contracts that represent 40 to 50 percent of the total contract value to compensate for political and economic risks.

4.7.6.1 Forms of Government

Members of E&C firms need to be prepared for constantly changing regulations and laws as government officials frequently move in and out of major offices in politically unstable countries. Engineering and construction professionals should have a basic understanding of political systems because there are different forms of government throughout the world, including democratic regimes, dictatorships, republics, republics with a unicameral legislature (one legislative body), unicameral parliaments, Islamic emirates, Communist countries, parliamentary republics, federal republics, constitutional sultanates, monarchies, constitutional monarchies, military dictatorships, or other forms of government. Table 4.1 provides a list of the different forms of government used throughout the world (Bruner, 2005). The potential for facilities *nationalization* (government seizure of private facilities) increases in unstable countries or in countries that are controlled by one person, such as monarchies or dictatorships. In republics, the chances of an unforeseen nationalization of facilities are reduced because it requires an act of congress or parliament to nationalize facilities.

The sector that is the most affected by rigorous changes in government regulations, both in the number of regulations initiated and in the fines imposed, is the environmental and hazardous waste sector. The Kyoto Protocol Treaty, which is an amendment to the United Nations Framework Convention on Climate Change, was ratified in February 2005, and it sets binding targets for 2012 for the 55 countries that have ratified the protocol for the reduction of greenhouse gas emissions of at least 5 percent from 1990 levels (a list of countries that have emissions targets under the treaty is included in Chapter 9). The Kyoto Protocol Treaty will affect the E&C industry in terms of how facilities are designed and constructed to reduce emissions and hazardous wastes.

4.8 The Changing Nature of E&C Industry Competition

Several forces are influencing competition in the twenty-first century, and this section discusses each of the forces.

4.8.1 Obstacles to Competitiveness

The global E&C market is influenced by financing restrictions, integrated world markets, an oversupply of production capacity, a decline of labor union wage-setting power, the deregulation of many industries, and privatization. Members of the E&C industry should monitor financial markets because of their influence on

Table 4.1 Forms of Government

Government Type	Examples
Administrative region	Hong Kong, Taiwan, Serbia and Montenegro
Constitutional democracy	Sierra Leone, Tuvalu
Constitutional parliamentary democracy	United Kingdom, Northern Ireland, Scotland, and Wales
Multiparty democracy	Australia, Eritrea, Malawi, Nepal, Taiwan, and Uganda
Democratic state	Poland
British commonwealth	Bahamas, Mauritius
Democratic republic	Moldovia
Republic	Albania, Angola, Argentina, Armenia, Austria, Azerbaijan, Bangladesh, Belarus, Benin, Bolivia, Botswana, Burundi, Cameroon, Cape Verde, Chad, Chile, Colombia, Congo, Côte d'Ivoire, Cyprus, Dominica, Dominican Republic, Ecuador, Egypt, El Salvador, Finland, France, Gabon, Gambia, Georgia, Guinea, Guyana, Haiti, Honduras, Hungary, Indonesia, Ireland, Israel, Italy, Kazakhstan, Kenya, Kiribati, South Korea, Lebanon, Liberia, Macedonia, Maldives, Namibia, Nauru, Nicaragua, Palau, Panama, Paraguay, Peru, Philippines, Portugal, Romania, Rwanda, San Marino, São Tomé and Principe, South Africa, Sri Lanka, Suriname, Switzerland, Tajikistan, Tanzania, Togo, Turkmenistan, Uruguay, Vanuatu, and Zambia
Authoritarian republic	Uzbekistan
In flux	Palestine State and Western Sahara
Papal law	Vatican City
Federal republic	India, Mexico, United States, Venezuela, and Yugoslavia
Constitutional government in free association with the U.S.	Marshall Islands and Micronesia
Multiparty republic	Madagascar, Mozambique, and Nigeria
Constitutional republic	Iceland, Kyrgyzstan, Russia, and the Ukraine
Republic parliamentary	Greece
Republic with a unicameral legislature	Djibouti (an Islamic emirate)
One-party republic in flux	Iraq
Communist state	Laos, Vietnam, the People's Republic of China, and Cuba
Communist dictatorship	North Korea
Parliamentary republic	Algeria

Table 4.1 (*Continued*)

Government Type	Examples
Parliamentary democracy	Bosnia and Herzegovina, Czech Republic, Estonia, Germany, Ghana, Grenada, Jamaica, Latvia, Lithuania, New Zealand, Papua New Guinea, St. Lucia, Senegal (socialist leanings), Slovakia, Slovenia, Turkey, and Zimbabwe
Federal republic	Ethiopia (a dictatorship)
Socialist multiparty state	Seychelles
Constitutional sultanate	Brunei (a monarchy), Andorra, Antigua and Barbuda, Bahrain, Barbados, Belgium, and Belize
Federation of provinces	Canada
Federation	United Arab Emirates–Abu Dhabi
Constitutional monarchy	Bhutan, Cambodia, Kingdom of Denmark, Japan, Kuwait, Lesotho, Liechtenstein, Luxembourg, Malaysia, Monaco, Morocco, The Netherlands, Norway, St. Kitts and Nevis, St. Vincent and Grenadines, Samoa, Saudi Arabia (based on Sharia-Islamic law), Sweden, Thailand, Tonga and Tunisia
Monarchy	Swaziland
Parliamentary monarchy	Solomon Islands, Spain
Traditional monarchy	Qatar
Absolute monarchy	Oman
Independent sovereign republic	Mongolia
Constitutional hereditary monarchy	Jordan
Military government	Burkina Faso, Mauritania, Myanmar (formerly Burma), Niger, Pakistan, Sudan, and Syria
President with a military counsel	Equatorial Guinea
Military dictatorship	Fiji, Libya
Overseas departments of France	French Polynesia, French Guinea, Guadeloupe, Martinique, Réunion, New Caledonia and Dependencies, Southern Antarctic Lands, Wallis and Futuna Islands
Territorial collectives of France	Saint Pierre, Miquelion, and Mayotte

the interest rates that are charged for construction projects. The interest-rate situation can be illustrated in the following way: "An interest advantage of one-half of a percentage point is equivalent to a 5 percent edge in price. If European and Japanese firms offer financial packages with 2 percentage points lower than the United States, it would result in a 20 percent advantage in price" (Goldstein, 1983, p. 21). Steps have been taken to eliminate this type of situation, including the Organization for Economic Development (OECD) setting export bank rates in developed countries. A major drawback of the OECD approach is that its rate consensus is not an enforceable treaty, and it is up to individual firms not to cheat in light of their moral obligations because the OECD does not have enforcement powers over the treaty (Goldstein, 1983).

Given the difficulties caused by financing shortages or interest-rate discrepancies, some firms are forced into joint ventures with firms from other countries merely to provide competitive financing packages. This type of joint venture can be counterproductive because it takes control away from firms when it comes to specifying domestic products to be used on joint-venture projects, and this also may impact other domestic industries.

Foreign governments may offer financial assistance to their national construction firms that includes terms that are not acceptable in other countries. Some European and Asian governments provide mixed financing packages in which the government makes from 50 to 100 percent loans at low- or zero-interest rates. Foreign governments may not require loan payments to start immediately when a project is completed, or they may extend loan payments out beyond normal repayment schedules. In effect, what these financial methods do is create a situation where taxpayers are subsidizing loans.

Some governments in European nations offer more subtle forms of aid to their domestic construction industry in the form of contractor's inflation insurance and foreign-exchange insurance, both of which could be duplicated in other countries. Some governments assist their domestic construction firms by making up the difference between what the contractor offers the client as a price and the market price.

If firms try to use advanced technology to create a competitive advantage, it could be in direct conflict with the long-term economic goals of developing nations. Many of these nations have set goals for increasing national employment that are contrary to using technology as a competitive advantage.

Another problem for global E&C firms is the practice of foreign clients or governments demanding participation in projects by local firms. The requirement for local participation could be as high as 70 percent, such as is the situation in China. If local firms are not familiar with a foreign E&C firm, it is difficult to form a joint venture and be able to maintain the quality that firms are used to maintaining on their own projects. In China, foreign firms cannot work on E&C projects without a joint-venture partner. Even with a joint-venture partner, foreign firms might be limited to performing only a certain percentage of the work. Governments could require that foreign contractors apply and receive licenses before performing work in their countries, and licenses could have certain restrictions on the type of work that can be performed within the country.

Included among the obstacles to future growth are technology and the nature of services, but another major obstacle is the fear of change: "In most service institutions there is an equivalent of the tenured faculty, a powerful group with no incentive to change. In the absence of a market to force action, this stand-pat instinct usually prevails" (Hudson Institute, 1987, p. 69). One impetus for change is the introduction of foreign firms and foreign E&C professionals into domestic markets. In addition to geographic preferences, global clients are hiring individuals from foreign countries when they are searching for technically competent personnel or proprietary processes.

Engineering and construction firms are consolidating and moving into the areas of operation and maintenance, renovation, infrastructure rehabilitation, hazardous waste management, and technology innovation. In the global marketplace, E&C firms increasingly are undertaking *turnkey projects,* where they create markets in which they propose a new project, provide or arrange for project financing, design the project, build the project, and participate in its operation and maintenance.

4.8.2 Nationalism Issues

Concern for protecting native E&C markets is increasing worldwide for a number of reasons, including the following:

- Stagnant domestic economies
- Regional political conflicts
- Rising unemployment rates (over 10 percent in Germany and 44 percent in Bosnia and Herzegovina in 2005)
- Reductions in the employment of technical personnel by governments due to a lack of funds
- Difficulties associated with adjusting to membership in the European Union, as is evidenced by the rejection of the European Union Constitution by Belgium in 2005
- A resurgence of religious fundamentalism
- Technical professionals returning to their native countries after receiving degrees from industrialized nations due to increases in the standard of living and services in their native countries
- A decrease in the number of engineering and technical degrees awarded in industrialized countries and an increase in levels of engineering and technical degrees awarded in newly industrialized countries (NICs)
- A lack of cultural exchange precipitated by a reduction in the number of people moving or traveling overseas
- Threats of terrorism
- Negative cultural and ethnic stereotypes
- Globally unbalanced trade deficits
- Threats of regional wars
- Decreasing government funding for companies to pursue projects offshore
- The ease of using telecommunication systems to acquire expertise
- Misunderstandings about foreign cultures and religions
- The closing of borders to foreign nationals and reciprocal responses
- The increasing disparity in incomes between countries. Some foreign nationals are too expensive to hire compared to local personnel
- Seeing the long-term effects of the inappropriate technology used in the 1970s and 1980s

- The desire of governments to mine their own national resources rather than to allow foreigners to mine them
- People concentrated in areas where basic services already have been provided for citizens
- The success of technology-transfer programs from industrialized countries to developing countries
- A worldwide shortage of construction materials and the increasing cost of construction materials
- Major fluctuations in oil prices that influence the cost of construction materials manufactured using petroleum products

If government officials start enacting protectionist legislation there will be a major decline in global projects.

4.8.3 External Influences on Competition

Trade deficits are calculated differently throughout the world, and this fosters misunderstandings. One example is in the United States, where the trade deficit does not include *services* in total exports, only goods and tangible items; therefore, consulting fees, which represent billions of dollars of exports, are not included in export totals, and this makes it appear that the United States is importing a substantial amount more than it exports. There are also companies that incorporate offshore, which makes it impossible to estimate trade deficits accurately.

Another concern is the issue of control. If E&C firms become truly global in the way they are owned, managed, and operated, governments no longer will be able to influence their operations, which could lead to global monopolies. Self-regulation is the preferred method of operation because it increases options for collaborative strategies.

Competitiveness is influenced by the ever-changing focus of clients and owners as they seek different buying factors, including (in order of importance) the following (Transmar Enterprises, 1989, p. 20–24):

- Quality of key project personnel
- Project-management capabilities
- Quality of project controls
- Previous experience with similar projects
- Responsiveness and flexibility
- Detailed engineering capability
- Quality of senior management
- Construction capability
- Ability to handle engineering in one office
- Evaluated cost of services

Since firms usually operate under domestic laws, government officials need to have a global perspective when enacting legislation that might improve their domestic corporate climates but that is detrimental to firms that operate globally.

Some executives prefer to explore potential strategies for avoiding competition, and they could implement the following noncompetitive strategies (Hasegawa, 1988, p. 192):

- Enter into less competitive markets; provide protection against the anticipated entry of competitors by selecting a market with natural barriers.
- Building barriers, such as management barriers, or patents, equipment capacities, technological levels, distribution channels, or regulations.
- Cooperation strategies: bring benefits to companies by eliminating excessive competition, reducing overlapping costs, teaming to develop common technologies, and joint efforts to promote market growth. Three forms of cooperation are:
  - *Vertical cooperation:* a small and a large company form an equal partnership for conducting research and development (R&D), procurement of machinery and materials, production, and sales. Large companies commercialize products using the technologies of the small ventures through vertical cooperation.
  - *Horizontal cooperation:* two large companies, or two small companies, or a large and a small company combine their technologies and experience to commercialize products or services.
  - *Paternalistic cooperation:* a large company provides the small venture development business with funds, goods, or facilities in exchange for the joint development of the ventures' business technologies into salable products.

It is important for members of E&C firms to realize that *once you go global, you must do business in environments in which some companies—the local ones— are more equal than others.* There are always new ways to compete against local firms, but it requires knowing as much as possible about local conditions, cultures, and politics.

4.9 Summary

This chapter discussed the forces that influenced the competitiveness of the E&C industry in the 1980s and the 1990s, such as the threat of new entrants, the threat of substitute products or services, the bargaining power of suppliers, the bargaining power of buyers, and rivalries among existing competitors. Previous changes in the structure of the E&C industry were discussed in relation to new technologies, new or shifting buying needs, the emergence of new industry segments, shifting input costs or availabilities, and changes in government regulations. This chapter also discussed the forces that are driving competition in the twenty-first century. An analysis of the changes in the E&C industry in the twenty-first century and the changing nature of global E&C industry competition was provided, along with discussions on obstacles to competitiveness and future strategies to remain competitive.

It is the aggressiveness and level of commitment that individual members of the E&C industry make that determines the competitive positioning of E&C firms in the global arena. Many noteworthy ventures already have been undertaken by once rival adversaries in the E&C industry to improve their global competitive positioning, such as the formation of strategic alliances among owners, contractors, and design firms; joint commitments between government agencies and members of the private sector to increase the construction technology base of their nations; the pooling of resources and efforts by contractors and labor unions or labor representatives to improve craftsmen training and skills; and the development of in-house training facilities by owners and E&C firms. If these types of steps are implemented across the entire E&C community in a country, it will have a significant impact on the global competitiveness of the E&C industry in that country.

If only one phrase is used to describe the effects of change on the E&C industry during the 1980s and 1990s, it would be *innovative, proactive thinking.* During the 1980s and 1990s, for the first decade since the pre–World War II era, managers from industrialized countries found themselves faced with foreign and domestic forces that caused their firms to lose ground as leading providers of design and construction services to the rest of the world. Leaders of the E&C industry came to realize that the *their native country's way* did not for the first time mean the *global way.* However, some managers challenged by the *native way is the only way paradigm* did transform their organizations and move them toward the forefront of domestic and global E&C arenas at the close of the twentieth century.

DISCUSSION QUESTIONS

1. Explain why the global E&C industry is moving away from *adversarial competition.*
2. This chapter cites newly emerging nations that appeared during the latter part of the twentieth century, but which nations will be newly emerging nations in the beginning of the twenty-first century, and why will these nations be industrializing in the first part of the twenty-first century?
3. Which members of construction project teams would benefit from strategic alliances, and why would they benefit from them?
4. Why are citizens of Asian countries more willing to form strategic alliances with firms from Western nations rather than some Eastern nations?
5. Why is the application of technology more important to E&C firms than new innovative substitute products?
6. Why are long-term commitments for strategic alliances 4 to 40 years in Eastern cultures versus 1 to 5 years in Western cultures?
7. What events in the 1980s irrevocably changed the nature of the E&C industry in the 1990s and beyond?
8. How did the introduction of high technology in the 1990s affect the global E&C industry?
9. What would be required to convert the citizens of China and India into global consumers?
10. What needs to exist for global partnerships to benefit all the parties in a partnership?
11. How do the antimonopoly laws in some countries prevent E&C firms from teaming with their suppliers?
12. How does government spending on the military affect firms in the E&C industry?

13. What events are altering the quality of technical personnel who work in the E&C industry?
14. How can E&C firms be compensated for the risks of nationalization when they are building projects in politically unstable nations?
15. Why should E&C industry professionals be familiar with the different forms of government listed in Table 4.1?
16. What are some of the techniques used by governments to help their native E&C firms be more competitive in the global marketplace?
17. Discuss 10 reasons for protecting native E&C markets that have a direct affect on foreign E&C firms.
18. Why are trade deficits an inaccurate measure of the global viability of a nation?
19. If an engineering firm does not meet all the buying factors that are the focus of clients, what other factors could it use to market its services to clients?
20. Explain how some E&C firms could avoid global competition.

REFERENCES

Bah Neo, R. 1976. *International Construction Contracting.* New York: Bowker Publishing Company.

Baker, S. T. 1990. Partnering: Contracting for the future. *Journal of the American Association of Cost Engineers* 32 (4): 1–12.

Engineering News Record. 1965–2003. New York: McGraw Hill Publishing Company.

Forbes, 1990. Today's Leaders Look to Tomorrow. 121 (7): 30–158.

Goldstein, Gray. 1983. Are U.S. firms handicapped in winning foreign jobs? *Journal of Construction Engineering and Management* October 109 (3): 41–45.

Hasegawa, Funi, and Shimizu Group FS. 1988. *Built by Japan: Competitive Strategies of the Japanese Construction Industry.* New York: John Wiley and Sons.

Hudson Institute. 1987. *Workforce 2000: Work and Workers for the 21st Century.* Washington, D.C.: U.S. Department of Labor.

Manley, W. F. 1989. Productivity improvement through a planned change process. *Engineering Management Journal of the American Society of Civil Engineers* 1 (1): 11–18.

Morris, C. R. 1989. The coming global boom, *Atlantic Monthly* 264 (4): 51–72.

Naisbitt, J. and P. Aberdene. 1990. *Megatrends 2000: Ten New Directions for the 1990s.* New York: Morrow.

National Research Council. 1988. *Building for Tomorrow: Global Enterprise and the U.S. Construction Industry.* Washington, D.C.: National Academy Press.

Porter, M. 1990. *The Competitive Advantage of Nations.* New York: Free Press.

Reich, R. B. 1990. Who is us? *Harvard Business Review* 30 (1): 53–64.

Bruner, B., ed. 2005. *TIME Almanac and Information Please.* Needham, Mass.: Pearson Education.

Tatum, C. B. 1988. Technology and competitive advantage in civil engineering. *Journal of Professional Issues in Engineering Education and Practice* 114 (3): 256–264.

Transmar Enterprises, Ltd. 1989. *The Transmar Study of the Engineering Construction Industry.* Houston, Tex.: North America Publishers.

Tuman, J. R. 1989. Information technology, organizational culture, engineering management for a global world. *American Society of Civil Engineers Engineering Management Journal* 1 (2): 24–53.

United States Congress Office of Technology Assessment. 1987. *Technology and the Future of the U.S. Construction Industry.* Washington, D.C.: Government Printing Office.

Yates, J. K., Subhransu, Mukherjee, and Njos Steven. 1991. *Anatomy of Construction Industry Competitiveness in the Year 2000.* Source Document No. 64. Austin, Tex.: Construction Industry Institute.

Chapter 5

Global Engineering and Construction Alliances

5.1 Introduction

During the past century, members of engineering and construction (E&C) firms have earned the reputation for supplying top-quality methods, management, and technology in the global marketplace, and they hope to maintain this reputation during the twenty-first century. Engineering and construction professionals are usually willing to adopt new and innovative methods for penetrating foreign markets, and during the next decade, they will explore and implement a variety of new business strategies to remain competitive in the global arena. Members of some E&C firms participate in global alliances to remain competitive or to penetrate regulated markets.

This chapter provides insight on the types of business strategies that are being implemented or may be implemented in the future by the executives of E&C firms. Global dimensions and global strategic management are discussed to provide information on who operates in the global arena. In addition, the risks that influence global investments and the strategies for dealing with risks are also presented in this chapter. Multinational contracting is discussed, to highlight the areas that differ when operating globally, along with project financing, privatization, build-own-transfer, joint ventures, and partnerships, as all of these financial techniques are used in the global marketplace.

5.2 Multinational Contractors

During the latter part of the twentieth century, the United States E&C industry dominated the world market. The rebuilding of Europe and Japan after World War II provided U.S. firms with the opportunity to make inroads into the global construction market that were not available in the first half of the twentieth century. More than half the revenue generated by U.S. E&C firms during the 1980s in construction came from global projects (Morgan, 1988). However, the 1980s brought forth a new era in the global-construction scenario—with firms from several countries becoming competitive in the global marketplace. This situation arose as a result of a number of related factors (Committee for Economic Development, 1990):

- Lagging U.S. productivity
- Rising global competitiveness
- Deteriorating global economic conditions
- Falling oil prices
- Lack of U.S. government subsidies in acquiring global contracts
- Absence of innovation in technology, research, and development
- Limited financial resources of U.S. engineering and construction firms

Similar to the manufacturing industry, the U.S. E&C industry experienced new competitors who slowly but steadily contributed to the decline of U.S. market share. Between 1966 and 1985, the market share of the U.S. E&C industry decreased to one-half its former strength. Prior to this decline, U.S. construction firms built over 50 percent of the projects worldwide. By 2003, the U.S. market share of global construction work was only 19 percent. In 2003, European construction firms built 60 percent of projects worldwide, Chinese firms built 6 percent, Japanese firms built 9 percent, and Korean firms built 2 percent (*Engineering News Record*, March 5, 2003). In Europe and Africa, French and U.S. firms generate the most revenue; in Asia, Japan and China generate the most revenue; in the Middle East, U.S. and Italian firms generate the most revenue, although Japanese firms are now securing more work; in Latin America, U.S. and Italian firms generate the most money; and U.S. firms generate the most revenue in the United States, followed by German and Swedish firms.

There are more Japanese firms in the top 10 firms in the manufacturing sector worldwide than firms from other countries; European firms dominate the building and transportation sectors; 5 of the top 10 firms in the petroleum and industrial sector are from the United States; and no one country dominates in the power, water, sewer, hazardous-waste, or telecommunication sectors (*Engineering News Record*, various issues). In the areas of construction management and program management, the top firms are from the United States, Europe, Egypt, China, Japan, Kuwait, and Canada (*Engineering News Record*, 2005).

During the later part of the twentieth century, most global E&C firms performed global work in Asia and the Middle East, but the demand for megaprojects in Asia and the Middle East declined because of a number of factors, including the fact that members of local E&C firms acquired the ability to design and build their own projects. However, Asia and the Middle East continue to provide projects that are of importance to the global E&C industry. There are increasing needs for construction projects in eastern European countries, India, and China, with a large portion of the work concentrated in transportation systems (27 percent of worldwide projects). The second-largest segment of global construction is the building sector (26 percent), and the third-largest segment is petroleum (19 percent) (*Engineering News Record*, March 5, 2005).

5.2.1 Global Contracting

Multinational contractors have some unique features that differentiate them from multinational corporations. Multinational contractors typically provide management and expertise for construction projects in return for a fee. Construction projects foster short-term involvement and presence in a country rather than require long-term commitments. An opportunity for corporate expansion and the possibility of capitalizing on a particular expertise are the main reasons constructors venture into foreign countries. Certain characteristics differentiate multinational contractors from multinational corporations, and they include the following (Ashley and Bonner, 1987):

- A lack of permanent involvement in direct capital investments, which means that the risks of expropriation and nationalization are lower during a construction project.
- Since multinational contracting involves the export of goods and services only, conflicts arising due to perceived depletion or misuse of national resources are excluded.
- Multinational contractors usually recover their returns in a much shorter period of time, typically ranging from 2 to 5 years.
- Most multinational contracts are project-specific in character, without any permanence in terms of paraphernalia such as offices, public relations offices, and permanent facilities.

In addition, members of global E&C firms are exposed to certain political risks that domestic contractors do not have to deal with on domestic projects. These include an extreme sensitivity to environmental factors, such as currency regulations, labor restrictions, and interim or ad-hoc host-government policies. The inability to insure against such risks makes the multinational contractor highly susceptible to these *microenvironmental risks.* "These types of risks can hamper project cash flows, labor costs, material costs, overhead costs and revenues" (Ashley and Bonner, 1987).

The changing political situation worldwide has fostered an emerging desire to create long-term financial commitments and presence in foreign countries (build-own-transfer and privatization projects). In addition to the risks specific to the multinational contractor, longer-term commitments expose multinational contractors to the same risks multinational corporations face. Therefore, multinational contractors with substantial business overseas should integrate foreign and domestic operations. In order to accomplish this, members of multinational construction firms should establish suitable control over foreign affiliates, understand and adapt to local conditions, and develop a global management strategy (Mallon, 1992).

There are several things that can be implemented to circumvent some of the risk associated with operating in foreign environments—to help reduce risks (Moore et al., 1992):

- Establish a suitable local partner
- Develop congenial relations with the host government, local business groups, labor unions, and local power groups
- Show appreciation of local business, culture, ethics, and operations

Certain factors influence relationships with foreign governments; therefore, a country-by-country *environmental analysis* is usually essential. Legal (patents, copyrights, labor, consumers, repatriation restrictions, foreign-investment restrictions, and commercial codes), social (population trends, life expectancy, education, income distribution, literacy, and predominant religion), political (type of government, age of system, number of political parties, international relations, nature of internal opposition, industrial policy, trade policy, monetary and fiscal policy, and the role of the government in the economy), and economic (rate of growth, manufacturing/GNP, inflation, unemployment, balance of payments, industry structure, and disposable income) conditions need to be analyzed (Davidson, 1982; Morgan, 1988).

5.3 The Global Dimension

Vanishing corporate geographic boundaries and the consequential global cross-migration of business are factors that all firms must face in the global marketplace. Domestic firms have to deal with an ever-increasing number of influences in order to operate effectively in the global environment, and these influences include things such as (More, 1988, p. 197):

- National control of international business
  - Foreign policy
  - Trade policies
  - Monetary policies
  - Foreign-exchange regulations
  - Foreign-investment controls—taxation, antitrust laws, immigration laws
  - Sectoral policies—laws and regulations
- Government and economic systems abroad
  - Formation and dissolution of alliances and blocks
  - International tensions and rivalries
  - Business trends
  - Cyclical changes
  - Economic growth
  - Inflation rates
  - Monetary stability
  - Consumption needs and preferences
  - Laws and regulations
- International business practices
  - Import and export channels
  - Foreign investments
  - Multinational production
  - Joint ventures
  - Turnkey projects

- Production-sharing agreements
- Management contracts
- Financial and insurance syndicates
- International business objectives
- Strategies and organizations

These influences paved the way for *global strategic management* to become increasingly important.

Global strategic management is the process of defining, developing, and administering a corporate strategy for worldwide business (Davidson, 1982). This definition stresses global rather than international management, suggesting a comprehensive approach to both domestic and global operations. In contrast to operations management, strategic management emphasizes an analysis of the environmental and internal factors that determine the position and profile of the organization in a dynamic environment.

The concept of global strategic management is especially important to multinational engineers and constructors because they are an omnipresent and continuously growing force (Ashley and Bonner, 1987). For contractors involved in global construction work, their strategic plan should include ways to deal with risks, because the inevitability of risks is an issue that will continually surface. Most global engineers and constructors are cognizant of the situation, and they usually adopt preemptive measures to overcome and/or cater to uncertainties.

However, with the emerging global dimension, and given the increasing influence of globalization in the E&C arena, firms presently classified as domestic firms are being forced to deal with some of the issues normally found only in the global E&C marketplace. As a result of declining overseas work, this predicament has been catalyzed by the entry or impending entry of foreign firms into the domestic U.S. E&C market, which is approximately 25 percent of the world construction market, which makes it an attractive marketplace for foreign firms. Many foreign countries have curtailed their construction megaprojects, which has left the United States a target for foreign infiltration.

5.3.1 Foundations for Global Strategic Management

Members of global E&C firms look to foreign markets when opportunities for growth in their domestic market are declining or to capitalize on the experience they have gained through specialization in a particular type of construction. Usually firms involved in global E&C are those that are well entrenched in their respective domestic markets. In the event that a firm decides to globalize its operations, it must be prepared to deal with foreign competition and chart out specific global strategies.

In order to determine the chances of success in a global environment, six criteria form the foundation for global strategies, including (1) policies, (2) market selection, (3) marketing-mix management, (4) sourcing strategies, (5) financial policies, and (6) organizational structures. Each of these criteria should be examined in relation to the types of policies that can be developed by a firm. Analyzing the foundations of global strategy helps to determine a preferred course of action. Three examples follow of possible courses of actions (Davidson, 1982):

- *Defensive strategy.* Monitoring for competitive threats in domestic markets, and preparing contingency plans for that prospect.
- *Reactive strategy.* Imitating strategies used by other firms before adopting a course of action.
- *Proactive strategy.* Initiating and following a contrived course of action based on an assessment of the factors mentioned earlier.

Each of these three strategies needs to be examined in relation to the criteria for global strategies. An example of this examination would be realizing that firms are able to increase their market share by cutting prices, offering favorable terms, and providing better services and financial support. Once a decision to enter foreign markets has been made, management should examine the key elements that were mentioned previously (legal, social, political, and economic) to determine how these different elements may affect their performance or increase their risk exposure.

5.4 Risks in Global Investment

Global engineers and constructors operate in an environment that traditionally contains a multitude of uncertainties and requires constant circumvention and adjustment. Any lapse in an understanding of or a sensitivity to the environment and its potential dynamics can seriously jeopardize profitability, market share, stability, and, consequently, preconceived returns on investment. Therefore, it is important to adopt a system for evaluating global risks. Many of the risks applicable to multinational engineers and contractors are different from those experienced by a typical multinational corporation (MNC). The following subsections address risks in the context of multinational contractors, but many of the concepts are applicable to multinational corporations.

5.4.1 Risk Definition and Classification

Risk refers to those potential outcomes of uncertainty that are unfavorable to a given condition or situation. The probability, frequency, and severity of the outcome(s) are prime motives for preemptive actions. According to the Construction Industry Institute (CII), uncertainties can be classified into *knowns, known unknowns,* and *unknowns* (Construction Industry Institute, 1989). Currency rate fluctuations represent the knowns, the possibility of war represents a known unknown, and the occurrence of natural calamities represents the unknowns. In addition to technical (the risk of using new technology), contractual (the possibility of breach and misrepresentation), and financial risks that typically plague the domestic E&C industry, the global engineer and constructor also have to deal with political risks. The CII postulates that certain factors can be used to determine the importance or threat of risk, and they include (Construction Industry Institute, 1989):

- The frequency of occurrence
- The amount of information available
- The ability to measure the consequences of loss
- The potential severity of the loss
- The manageability of the risk
- The variability of the consequences
- The potential effects of publicity in the event of a loss

A typical decision-making process may by based on these factors, but it is often also influenced by intuitive rationalizations, with greater aversion for the more catastrophic and less likely natural calamities than for the more recurrent but less serious losses.

5.4.2 Risk Management

According to the CII, a typical "total risk management program" applicable to construction firms includes the identification of risks, the development of measurement systems, and implementing ways of controlling risk. *Risk identification* refers to the function of identifying various sources of information and enumerating a "checklist" of factors that could be consequential to the risk-management program. Measurement systems may include traditional systems, simulations, analytic analyses, and discrete-event analyses. *Controlling risk* involves preplanning to prepare for the risk and containment, which involves contingency planning and training programs (Boehm, 1965; Construction Industry Institute, 1989).

In construction, a contingency allowance of approximately 5 percent of the total project cost is included in the estimated contract price to allow for the incidence of risk. Statistical simulation methods, such as the Monte Carlo technique (probability simulations), may be used to measure risk of the knowns and known unknowns, and analytic and discrete-event analyses offer mathematical models for risk measurement.

Risk can be controlled in a number of ways, for example:

- Avoiding risk or dropping out of the competition when the potential for profitability appears borderline given the expected rate of return.
- Sharing risk through joint ventures and partnerships or cost-per-work-hour contracts.
- Reducing risk by reducing the effects of risk through a study of the factors that contribute to its potency.
- Transferring risk by transferring certain portions of the contract to subcontractors.

- Insuring against risk is the traditional method for safeguarding against risk, but it carries a recurring premium with every project. In the global arena this is becoming prohibitive due to the increasing cost of insurance premiums, especially in high-risk countries.
- Accepting risk with or without contingency, which requires adding in a certain percentage of project costs as contingency funds; however, other measures need to be taken when there is strong competition.

Multinational contractors do not have to worry about the risks associated with operating in foreign environments unless they are awarded foreign contracts. In order to remain competitive in securing foreign contracts, multinational contractors are investigating alternative techniques for gaining entry into new markets, including project financing, privatization, build-own-transfer, joint ventures, and partnerships. Each of these alternatives is discussed in the following sections.

5.5 Project Financing

Financial aspects of the E&C industry are considered to have a strong bearing on future competitiveness. Venturing into new and possibly risky endeavors is an inherent characteristic of the E&C industry. The types of things that can be implemented to maintain future competitiveness include adoption of innovative methods of financing, such as privatization and build-own-transfer, to obtain projects. The provision or arrangement of financing for owners is considered to be an increasingly important aspect of securing construction contracts in the global marketplace. *Financial engineering,* or the management and provision of leverage to acquire projects, is a phenomenon rapidly gaining widespread application in the global marketplace.

Greater equity participation by contractors is another means of securing projects, and some contractors are becoming more comfortable with this concept. This is significant to the privatization of certain sectors of infrastructure reconstruction, including environmental cleanup, prisons, and roads.

Lump-sum negotiated contracts and an increasing incidence of long-term partnerships between owners and engineering and construction firms also appear to be a trend for future construction contracts. Although competitive pricing continues to be used when granting contracts, this may change in the future with the increasing popularity of the *strategic-partnering concept* being promoted by contractors. Strategic partnering involves a long-term relationship between owners and contractors that benefits both parties.

5.6 Privatization

The forces of global economic competition have forced a rethinking of national strategies to reduce the burden on federal governments and taxpayers without denying basic services. There is an increasing global trend toward reducing government involvement in the provision of basic services. One reason cited for this trend is the natural consequences of monopoly systems that create inefficiencies. Government involvement in national economics is steadily decreasing through deregulation, sale of government-owned assets, and privatization. This trend holds true for both developed and developing nations, and it is especially prevalent in eastern European nations.

Privatization as a strategy to improve economic performance is being practiced throughout the world. In spite of differing political beliefs among nations, the fact that the government is expected to provide certain basic services remains a universal phenomenon. A number of pressures are driving the move toward privatization: the pressure governments face as a result of poor performance and low productivity, the ideological pressures exerted by citizens to prevent governments from becoming too powerful (such pressures usually aim toward smaller, more efficient governments, requiring that governments spend more time governing and less time providing services), commercial pressures routinely exercised by private taxpaying businesses that support the opportunity for growth by providing services that a hard-pressed government may not be able to provide, and populist pressures fostered by the belief that people should have a greater choice in public services and that they should be empowered to decide their common needs without relying on bureaucracies. Proponents rationalize that privatization decentralizes power, strengthens traditional institutions, and reinforces a local sense of community (Fitzgerald, 1988; Savis, 1983).

The implementation of privatization concepts results in a number of different actions including (Hastings and Levie, 1983):

- The selling off of state holdings
- The sale of physical assets of the government
- Public issue of majority shares of a nationalized concern
- Joint public-private ventures
- Placement of shares with institutional investors
- Introducing competition in areas where public corporations had a monopoly market
- Permitting private contractors to provide services previously provided by the government
- Arranging for private financing of large government projects, with consequent loss of government equity

Although privatization is aimed at improving efficiency, expanding the equity base, and reducing the burden of the government, these objectives are not always achieved completely. Public enterprises, in addition to providing services, also have a social function to perform that can lead to a conflict of interest if privatization methods are implemented on projects. As a consequence of privatization, equity distribution is often misinterpreted because overall control remains with the majority shareholder no matter how small the percentage of holding (Hastings and Levie, 1983). Privatization is usually only used on successful public enterprises.

Thus, apart from losing revenues from successful enterprises, governments are stuck with unprofitable organizations. Finally, most privatized ventures tend to operate only in certain profitable sectors, completely rending unnecessary the social responsibility of the government (Donahue, 1989).

Four factors that influence privatization are the following (Linowes, 1988):

- *Full-cost pricing.* Those who want goods provided by the government should pay the full cost of having public goods and services provided to them. Usually, in government accounting systems, the cost of depreciation, interest on borrowed money, and overhead costs are not accounted for when charging consumers. Usually consumers are undercharged for the services provided by governments, which means that services financed through a different source. This violates the fundamental economic theory that prices should be set equal to costs, and consequently it puts a drain on governments.
- *Competition.* Production in the private sector that results from competition is likely to be more efficient and therefore less expensive. In the bureaucratic setup, prestige and power are relegated to size, not efficiency or competition. This violates the essence of the concept of efficiency.
- *Consumer satisfaction.* Consumers are likely to be more satisfied when they are provided with an opportunity to choose from a number of service providers. The market is structured to accommodate a dissatisfied consumer by having alternative sources available. The political process, in its inherent disposition of being a monopoly, is not focused on providing choices.
- *Entrepreneurship.* Under a government monopoly, the entrepreneur is denied the opportunity to innovate and better meet the demands of consumers for better service at a lower price. Privatization helps to unlock the innovative genius of the entrepreneur.

One possible drawback to some of these suggestions is the short-range planning philosophy that is fostered when a private developer is allowed to provide a service previously provided by the government. Initial costs may be reduced, but the long-term life-cycle costs may be increased by privatization. In addition, consumers may be faced with a situation in which the private supplier has a monopoly on the service.

Privatization is beneficial for the following reasons (Ramanadham, 1988):

- It reduces budgetary constraints on the government that occur due to losses incurred by public enterprises or their investment requirements, making funds available for other uses.
- It enhances the efficiency of the government apparatus through market discipline, reinforcement of competition by stimulating innovation, and elimination of governmental intervention.

- It improves the allocation efficiency of investments by increasing growth rates and developing money markets.
- It increases indigenous ownership by encouraging equity distribution, and it enhances productivity and commitment through incentive measures and employee stock ownership.

Two other objectives related to engineering and construction are achieving economies of scale and increased technical performance. In smaller communities, when services are needed for water supply, wastewater treatment, or solid-waste management, a private developer may be able to contract with different cities or towns within a particular region to offer the service through a joint regional project. Technical performance may be increased by the use of private developers, because they may possess technical expertise within their firms that local communities do not possess within their agencies.

Privatization allows the transfer of functions normally performed by the government to the private sector. It can either deal with the power to authorize public demand services or be related to the production of services. The level of private participation increases, from both functions being public to both functions being totally private. Such public-private partnerships are aimed at being mutually beneficial to the enterprise, the government, and the party undertaking to provide the product or service. The role of privatization may be interpreted in two other ways. The first way is *strategic privatization,* or the process of shrinking the *collective realm.* The second way is *tactical privatization,* aimed at cutting costs through competitive pricing using policies of the private sector.

In developed countries, privatization is viewed as either: (1) a philosophy of "selling off the state" (as is done in the United States) or (2) as a comprehensive proposal with four areas of government activity: (a) state industries, (b) state services, (c) state utilities, and (d) regulatory procedures (as is done in the United Kingdom). For many government projects, when governments perform the service it costs approximately 40 percent more than if it were done through private enterprises (Pririe, 1988).

In developing countries, privatization is a result of two factors: (1) poor performance by public-sector enterprises and (2) a severe shortage of capital. Decisions based on a categorization of public enterprises are followed in some African countries, including (1) enterprises to be retained, (2) enterprises whose objectives have been satisfied and therefore should be discontinued, and (3) enterprises that will function better as private firms (Swann, 1988). Sectoral demarcations involving decisions based on the importance of the industry to the economy and the possible effects if it is privatized are routine in small economies. Four global incentives for privatization in developing countries are as follow (Burton, 1989):

- Development of stabilization programs by the International Monetary Fund (IMF) for countries with balance-of-payment problems

- The influence of the World Bank for implementing privatization schemes in developing countries
- Changing emphasis on developmental policies with priority given to export expansion, diversification, and growth of private enterprises
- The role of multinational companies and their influence

Certain issues can be identified that affect decisions on whether to privatize an investment, and they include (Ramanadham, 1988; Hanke, 1985):

- Country issues
  - Socioeconomic conditions of the country and its effect on national policies
- Macro issues
  - Distribution impacts to identify potential losers and beneficiaries
  - Preference in capital ownership and share allotment techniques
  - Attitudes of the labor force, managers, and civil servants who would be affected by a decision to privatize
  - Policies on liberalization, foreign ownership, level of dependence on private capital, and national development and planning
- Financial matters
  - Impacts of debt servicing by the government, including sale of profitable enterprises, maximizing sale proceeds, restructuring costs, and tax waivers
  - Valuation and underwriting costs
  - Legal provisions, including liability, royalties, and conversion of equity holding
  - Procedures for utilization of income from the sale of the enterprise, with debt-servicing and tax-subsidy plans
- Strategies
  - Level and method of privatization
  - Defining roles and affixing levels of responsibility, including deregulation procedures, time-frame decisions, and monitoring efficiency procedures

The types of projects that are being privatized globally include:

- Highways and bridges
- City streets
- Municipal water systems
- Wastewater systems
- Solid-waste and hazardous waste disposal systems
- Air transportation
- Ports and waterways
- Prisons

Engineering firms will be affected if the concept of privatization is increasingly embraced, because government clients will be replaced by private-sector organizations. This type of a shift affects engineers who are currently employed by governments. As more work is moved to the private sector, either their jobs would be eliminated through restructuring or their expertise would be shifted into other areas.

5.7 Build-Own-Transfer (BOT)

Build-own-transfer (BOT) refers to a process similar to privatization whereby private organizations undertake to build and operate a facility that normally is constructed by government agencies. The ownership of facilities is returned to the government after a fixed period of time, and the revenues generated from operating the facility are used to repay lenders and to provide a profit to the firm who built the project.

This form of financing model was first pioneered in Turkey in 1984 and has been gaining acceptance throughout the world as one of the most popular global financing techniques. In a global economic scenario with limited budgetary resources, an urgent need for new infrastructure, a trend toward privatization, and a lack of external currencies, BOT projects are viable alternatives.

5.7.1 Procedural Aspects of BOT Projects

Contractors, in association with industrial partners, bankers, investors, or operators, create a multifunctional group with various duties. The BOT organization studies the market and social and economic effects and integrates the project into the environment. The construction group oversees production, installation, construction, training programs, operations, and maintenance. The lending company deals with all financial matters pertaining to the project. The operator prepares a joint-concession proposal that is signed by the agency authorized to award the contract and then puts the operation into effect.

If a BOT option is being considered for a project, an economic analysis should be performed to determine whether the rate of return would be over the minimum attractive rate of return for the company considering a project, including a risk investment of 1 to 2 percent of the total investment costs. A rate of return of 33 percent or greater is considered reasonable for investment in a BOT project. This includes investigating factors such as user price, competition from similar facilities, and the acceptability of contract conditions for all parties concerned with the project.

5.7.2 Feasibility Conditions for BOT

When BOT projects are implemented the involvement of each additional entity introduces potential problems that have to be resolved, and BOT projects become particularly complicated with the involvement of international agencies and, consequently, currency and trade-regulation issues. In general, for a BOT project to be feasible the following factors need to be considered (Ounjian and Carne, 1987):

- A strong government will to realize the importance and arrange to cater to public facility demand.
- A realistic and achievable economic equilibrium to reach the objectives is necessary and may entail provision of government subsidies.
- Technical realism is needed to incorporate state-of-the-art technology.
- Legal and administrative realism helps to arrive at negotiated contract conditions.
- Financial realism is needed, with consideration given to the duties of the government, the profit motive of the private enterprises, and the level of uncertainties involved in a project.
- Limiting competition to a reasonable level may be necessary for a fair chance of success.

5.7.3 Characteristics of BOT Projects

BOT projects have two major characteristics that differentiate them from limited-resource projects: government involvement and risk-return relationships. Government involvement in BOT projects usually requires enhanced roles for government participation. The creation of a concession company, ownership of the fixed assets, and control of revenues form the bulk of avenues for governmental involvement. In addition, BOT projects aim to diminish in size over time (from the perspective of the contractor), as opposed to a typical company that seeks to expand its operations.

Risk-return relationships characteristic of BOT projects differ considerably from typical construction projects. The main differences are significantly longer construction periods, longer life of the assets being constructed, low costs for operation and maintenance of the constructed facility, the stand-alone nature of the project (not with a project portfolio), and the nonavailability of funds during construction, which may increase project costs if there is accrued interest.

BOT projects create significant challenges to the private sector, as well as the government, including dealing with conflicting interests and motives, the possibility of hostility from a competing public utility, and exposure to greater risk for the contractor over a greater period of time.

5.7.4 Risks in BOT Projects

BOT projects are exposed to four major kinds of risks: project commercial, country commercial, political, and operational. Project commercial risks include risks that are inherent in the execution of the project: the risk of losing the project to competitors, failure to complete the project as scheduled, or risks associated with variations in the cost of operations. Country commercial risks involve the conversion of project revenues into hard currencies.

Political risks are related to the stability of the government, policies on repatriation of revenues, fluctuations in regulations, and the integrity of the government. Operational risks are ones that deal with default, insolvency, or situations requiring legal recourse. These types of risks are perpetuated by underinsured catastrophes, poor project economics, or incompetent management (Beharrel, 1989).

5.7.5 Trends in BOT Projects

Build-own-transfer projects represent an innovative and potentially limitless supply of global as well as domestic construction work; therefore, the number of BOT projects is expected to increase in the global arena. The World Bank has put together an innovative financing package whereby developing countries would have to cover only 30 percent of their loan requirements for BOT power projects. These packages are being used successfully in a number of developing countries and involving partnerships between various leading global construction firms, global financial organizations, and host-country governments.

Build-own-transfer strategies are used in developed countries to finance infrastructure reconstruction as an alternative or complement to privatization. The viability of this option was clearly demonstrated in the construction of the 15-mile toll road in the United States in Virginia that was completed at a cost that was less than half the estimated cost if it had been built by the state of Virginia (Mallon, 1992).

5.8 Joint Ventures and Partnerships

Joint ventures represent an innovative and increasingly important business strategy for remaining competitive. Most countries have been incorporating joint ventures into their legal and corporate framework in order to develop solutions with potential economic and anticompetitive effects. Many of the E&C firms involved in global engineering and construction already participate in joint ventures with foreign partners. Joint ventures with project-specific limitations or partnering-and-termination schedules are some of the most viable means of entering foreign markets. Joint ventures are considered particularly relevant to the opening of the eastern European and Chinese construction markets.

Joint ventures are defined as the integrating of operations between two or more firms in which the following conditions are present (Bradley, 1983):

- The enterprise is under the joint control of the parent firms.
- Each parent firm makes a substantial contribution to the joint enterprise.
- The enterprise exists as a business entity separate from the parent organizations.
- The joint venture creates significant new enterprise capability in terms of new productive capacity, new technology, new product or entry into a new market, and greater bonding capacity.

From a competitive standpoint, the complex nature of joint ventures places them somewhere between a contract and a merger, sharing some features of both. According to unofficial reports cited by the Organization for Economic Cooperation and Development, joint ventures have been implemented in most developed countries, primarily in the following areas (Committee for Economic Development, 1990):

- Research and development
- Natural resource exploration and exploitation
- Engineering, construction, and other services
- Product manufacturing (to enter new markets)

5.8.1 Participation Policies

Certain strategic choices determine the profile of the global strategies of a firm. Local partners often have a lot to offer global firms, including general local knowledge, managerial personnel, market potential, and access to distribution systems. At the same time, there are potential avenues for conflicts that include polices on pricing, dividend distribution percentages, export requirements, sourcing, and royalties. As a result, formal participation policies are necessary to iron out differences at the inception of the partnership. Motives for establishing joint ventures include complementary technology, to raise capital, risk sharing, economies of scale, and market power (Davidson, 1982).

Multifaceted partnering is another technique that would increase the competitiveness of firms. This requires that firms develop effective long-term plans to define their business strategies. Specialty firms, who are at the leading edge of technology, may form, as smaller companies reexamine their focus to develop the specialty services that are required by larger firms in partnering relationships. Firms partner and form alliances with other domestic firms and also with foreign firms. To increase their competitiveness, global firms maintain and enhance their foreign connections and hire multilingual foreign nationals as managers. To enhance global awareness managers are rotated through different areas within firms and possibly in other firms involved in the partnering arrangements.

5.9 Summary

Financing and finance-related issues form an important part of any global alliance program. To this end, most global firms incorporate financing of foreign affiliates within the capital-budgeting structure of the firm. Irregular funding requirements, cost optimization, and decisions on liquidity, political risk, tax repatriation, exchange risk, capital structure, and institutional relationships are considered factors in the capital-budgeting process. In many firms the choice between debt and equity, international and external sources, and local or foreign currency determines the key dimensions of a funding package. Given the complexities involved in rigorous analytical procedures, most firms rely on set guidelines and decision rules, with adequate provision for contingencies, to simplify funding and alliance decisions.

As the world increases in complexity, members of firms continue to investigate additional new and innovative forms of global alliances and financing techniques. In addition to investigating new methods, members of firms look for ways to perfect the current forms of risk management, project financing, privatization, BOT methods, joint ventures, partnerships, and global alliances.

DISCUSSION QUESTIONS

1. Explain the difference between multinational contracting and multinational corporations.
2. What are the five most important items that firms need to be cognizant of when operating in global environments, and why are these the five most important items?
3. Explain global strategic management.
4. What are the six criteria that form the foundation for global strategies?
5. How do construction companies account for global risks?
6. What are some of the aspects of financial engineering that can be used by contractors to secure projects?
7. What pressures are driving privatization throughout the world?
8. Explain how privatization projects are different from build-own-transfer (BOT) projects.
9. Explain what may happen when former Communist regimes privatize state-owned facilities.
10. Explain the four factors that influence whether facilities should be privatized or not.
11. Explain the benefits of privatization.
12. What issues influence decision makers when they are considering privatization?
13. What type of an engineering-economic analysis should be performed before undertaking BOT projects, and why should it be performed prior to undertaking BOT projects?
14. Explain the risks associated with undertaking BOT projects.
15. What is the motivation for forming joint ventures?

REFERENCES

Ashley, D. B., and J. J. Bonner. 1987. Political risks in international construction. *ASCE Journal of Construction Engineering and Management* 13 (3): 33–45.

Beharrel, S. 1989. The special legal considerations for BOT and other non-recourse finance. *Second International Construction Projects Conference.* Legal Studies and Services Ltd.

Boehm, J. E. 1965. *Obtaining International Business Data: International Impact on Engineering Managers.* New York: American Society of Mechanical Engineers.

Burton, D. F., Jr. 1989. Economic realities and strategic choices. *Vision for the 1990s: U.S. Strategy and the Global Economy.* Cambridge, Mass.: Ballinger Publishing Company.

Bradley, J. M. 1983. Joint ventures and antitrust policy. *Harvard Business Review* 95 (7): 27.

Committee for Economic Development. January 31–February 1, 1990. *Construction to and Beyond the Year 2000.* London: Construction Industry Sector Group of the National Economic Development Office.

Construction Industry Institute. 1989. *Management of Project Risk and Uncertainties.* Austin, Tex. Construction Industry Institute.

Davidson, W. H. 1982. *Global Strategic Management.* New York: John Wiley and Sons.

Donahue, J. D. 1989. *The Privatization Decision,* Basic Books, Inc., Publishers, New York, New York.

Fitzgerald, R. 1988. *When Government Goes Private,* Universe Books, New York, New York.

Hanke, S. H. 1985. The literature for privatization. *The Privatization Option: A Strategy to Shrink the Size of Government.* Washington, D.C.: The Heritage Foundation.

Hastings, S., and H. Levie. 1983. *Privatization?* London: Spokesman Bertrand Russell House.

Linowes, D. 1988. Privatization: Towards a more effective government. *Report of the President's Commission on Privatization.* Washington, D.C.: Library of Congress.

Mallon, J. C. 1992. Verifying cost and schedule during design. *Journal of the Project Management Institute* 23 (1): 38–41.

Moore, C. C., D. C. Mosley, and M. L. Slagle. 1992. Partnering: Guidelines for win-win project management. *Project Management Journal* 23 (1): 18–21.

More, R. A. K. 1988. Supplier/user interfacing in the development and adoption of new hardware/software systems: A framework for research. *IEEE Transactions on Engineering Management* 35 (3): 190–196.

Morgan, J. 1988. Partnering for world class suppliers. *Purchasing* 105 (1): 49–80.

Mosly, D. C., M. L. Moore, and D. R. Burns. 1990. The role of the OD consultant in Partnering. *Organization Development Journal* 7 (3): 43–49.

Ounjian, M. L., and B. Carne. 1987. A study of the factors which affect technology transfer in a multilocation, multibusiness unit corporation. *IEEE Transactions on Engineering Management* 34 (3): 194–201.

Pirie, M. 1988. *Privatization.* Wildwood House Ltd, Hants, Great Britain.

Ramanadham, V. V. 1989. Privatization: The U.K. experiences and developing countries. *Privatization in Developing Countries.* Worcester, U.K.: Billings and Sons, Ltd.

Savas, E. S. 1982. *The Efficiency of the Private Sector,* Chatham Publishing, Inc., Chatham, New Jersey.

Swann, D. 1988. *The Retreat of the States: Deregulation and Privatization in the U.K. and the U.S.* Worcester, U.K.: Billings and Sons Ltd.

Thamhain, H. J. 1989. Validating technical project plans. *Journal of the Project Management Institute* 20 (4): 43–50.

Chapter 6

Global Construction Financial Techniques

6.1 Introduction

Survival in the global engineering and construction (E&C) environment requires members of firms to incorporate innovative financing techniques into their repertoire of acceptable financing methods. Various innovative financing strategies have been implemented during the past few decades that allow E&C firms to undertake projects that would have been difficult or impossible to finance and build with traditional methods. To assist individuals in their selection of appropriate financing strategies, this chapter discusses global financing techniques being used in the global E&C marketplace.

Operations-level personnel sometimes question the importance of knowing about financing techniques, especially if they are not directly involved with the financial aspect of projects. However, knowing how a project is being financed is crucial to understanding whether a project will be successful or not, and it helps in identifing the constraints that E&C personnel must operate under while working on projects.

This chapter does not cover financial analysis or engineering-economic analysis, because feasibility studies are done prior to engineering and construction personnel being assigned to projects. However, it does provide insight into the complex world of global financing. The last section of the chapter explains different options for global payment methods, because E&C personnel are responsible for approving and processing payments for subcontractors and suppliers for materials and services.

Prior to the 1980s, financing methods for global E&C projects were classified into several diverse but stable areas. During the 1990s and early 2000s, political and social unrest altered the global financial situation, and financing techniques were modified to suit newly emerging financial markets. The phrase *innovative financing* was coined to represent new, unconventional financing methods (Shapiro, 2002). Every year, innovative techniques are introduced into the financial marketplace, but only a small number of global E&C firms are taking advantage of these new financing programs (Beenhakker, 2000). In order for E&C firms to remain competitive in the global marketplace, members of these firms should reexamine their traditional financing policies to determine whether alternative financing techniques can be incorporated into their corporate strategies.

The development and implementation of innovative financing techniques are a result of supply factors, particularly those associated with external constraints (regulatory or other) that have been imposed on financial intermediaries. Other influences that foster financial innovation are a lack of institutional support and the fact that the financial environment has been altered, and investors are creating new ways to maintain or increase their existing incomes (Ben-Horin and Sibler, 1977).

Several conditions have a direct bearing on financial innovations, including (1) the global regulatory environment, (2) competition, and (3) vacillating monetary-exchange rates (Duffy and Giddy, 1981). The financial marketplace is also affected when restrictions are placed on the international transfer of funds because such restrictions cause investors to attempt to circumvent controls (Kuemmerle, 2004).

As a result of the changing financial environment of the last two decades, innovative financing techniques have been evolving throughout the world. Many of the methods referred to as *innovative* are unusual only when viewed from the perspective of Western standards, because several of them are merely adaptations of financing techniques that are prevalent in other cultures. New techniques are only limited by the imaginations of financers and the willingness of members of firms to gamble on new methods. This examination of global financing techniques is limited to a discussion of the financing methods that are applicable to global E&C projects.

6.2 Countertrade

Countertrade financing is a program that has gained in popularity as an acceptable form of E&C financing. Countertrade is categorized as an innovative financing technique only in Western cultures or in countries where its use is relatively new. In the Middle East, South America, and Africa, countertrade has always been an acceptable form of commerce, because it has its roots in the earliest financial transactions. The basic premise of countertrade is the exchange of goods or services for the goods or services of others. The innovative aspect of countertrade that is becoming more prolific is the concept of tangible goods being traded for services from companies based in other countries (Tuller, 1992; World Bank, 2002).

Numerous transactions involving countertrade take place within the global E&C marketplace. When a country does not possess the cash required to pay for new projects, it may negotiate to pay for the project with goods or services, such as natural resources or items manufactured within the country. In addition, countertrade is used as a means whereby companies who have had their funds blocked by government regulations may continue operating without using cash to pay for goods or services.

In order for countertrade contracts to be successful, the company that receives goods as payment must be capable of finding a party who will, in turn, buy the goods for cash. In effect, countertrade contracts force companies in developed countries to act as agents for goods from less developed countries. Several less developed countries are benefiting from the present world economy, because E&C companies are willing to enter into countertrade agreements for their services rather than render their company idle because of a lack of overseas contracts.

Some firms have in-house trading companies that market the goods they have acquired through countertrade to companies from other nations. Purchasing a trading company also allows E&C firms to diversity into new areas and to continue operating by using countertrade as a viable means of financing projects. In one example of a countertrade agreement, shiploads of bananas were traded for the construction of a dam in a developing country. The bananas, in turn, were traded to another country for steel that was then sold for cash to pay for the initial project. If the World Bank is involved in any way with a project, countertrade is not permitted as a financing method because the policies of the World Bank prohibit financing methods that minimize the outflow of hard currency (World Bank, 2005).

Another obstacle to countertrade financing is the scarcity of companies or countries having experience in negotiating countertrade contracts, and this inexperience makes some companies reluctant to enter into countertrade agreements. Before the recent resurgence of countertrade agreements, widespread use of the concept was limited to eastern Europe, Indonesia, Japan, and China. If E&C companies are interested in pursuing contracts in these markets, they need to investigate countertrade-financing methods.

6.3 Cofinancing with the World Bank

In the global marketplace, the World Bank has been associated with private creditors in an assortment of E&C projects. Traditionally, the World Bank and private creditors enter into separate loan agreements with companies who want to borrow funds or with countries that intend to use the funds to pay for E&C services. Other cofinancing methods are evolving that increase the involvement of private creditors in the global marketplace.

Cofinancing methods have the potential to (1) enlarge the flow of capital to developing countries, (2) increase lending volumes, with additional security created by interaction with the World Bank, and (3) increase the involvement of multinational companies in new areas of development (Pelosky, 1999). The World Bank participates in financing methods such as (1) direct financing participation in the loans of private creditors referred to as *A/B modalities*, (2) contingent participation in the latter stages of private-creditor loans (depending on the volatility of interest rates) called *vertical financing*, and (3) guaranting the late maturities of private loans (Pelosky, 1999).

6.3.1 A/B Modalities

One method of cofinancing sponsored by the World Bank is *A/B modalities*, which are a method where the World Bank continues financing "A" loans (World Bank loans) and then directly participates in "B" loans, which are the province of commercial creditors. A/B modalities allow commercial creditors to rely on the World Bank to furnish accurate information to lending institutions pertaining to prospective clients. In addition, the World Bank investigates and determines the feasibility of proposed construction projects.

Since commercial banks are primarily concerned with the ability of a government of a country to pay off its loan according to the terms of their contracts, funding A/B modality agreements allow commercial banks to rely on the expertise of the World Bank. When the World Bank becomes involved in a lending agreement, commercial lenders are more willing to lend additional capital because their risks are reduced by the knowledge of potential clients that members of the World Bank possess because of their expertise in this area.

6.3.2 Vertical Cofinancing

A second financing method associated with the World Bank is *vertical cofinancing*, which is a method whereby commercial banks handle the initial loan period at market interest rates, and then the Multinational Development Bank takes over loans if the maturity period increases beyond a set time period. The benefit of vertical cofinancing to commercial lenders is twofold: Cofinancing overcomes the weaknesses of cross-default clauses, and it forces the Multinational Development Bank to disclose all relevant information pertaining to the project, the borrower, or both (Pelosky, 1999).

The principal attraction of vertical cofinancing lies in the lending enterprise of the World Bank. Additional advantages include "(1) diversification of funds, (2) expansion of regional banks, (3) exposure to new markets, (4) the introduction of countries to commercial bank lending, (5) a chance to engage in projects in certain sectors that would interest multinational companies, and (6) the methods which prepare countries to support large-scale business operations" (World Bank, 1982, p. 73).

6.3.3 Swap Financing

Swap financing is a financing procedure that takes advantage of the ability of someone else to acquire financing terms that are unavailable to a particular company or individual. The swap takes place if two people or two firms have opposite existing liability positions, financing needs in one country or in two different countries, or both. In order to initiate a financial swap, two parties engage in a transaction in which they do not wish to be financially obligated and then swap their obligation for the financial contract they would like to attain (U.S. Department of Commerce, 2002).

Several objectives are achieved by swap-financing methods, including (1) a reduction in financing costs, (2) elimination or reduction of funding uncertainty, (3) the creation of otherwise unavailable long-term forward-exchange cover (which protects the investment against fluctuations in foreign exchange rates), and (4) attainment of limited or inaccessible long-term funding in a particular currency (Militello, 1984; Das, 1993).

Financial swaps are performed in many different ways. As long as the swap serves both of the parties involved, the permutations are unlimited. Two forms of financial swaps dominate the market: coupon swaps and long-term forward-cover swaps.

6.3.4 Coupon Swaps

Coupon swaps, or *interest-rate swaps,* develop when parties interested in obtaining funding cannot obtain a particular interest rate. A coupon swap entails the exchange of a *coupon,* or interest payment stream, of one configuration for a coupon obligation with a different configuration on essentially the same principal amount. If an investor wants funding that is financed at an attractive fixed rate but can only locate variable interest rates in the desired range, a swap can be arranged with another investor who is looking for variable interest rates. In this type of situation, the first investor would enter into a contract for the variable interest rate, and then he or she swaps rates with a second investor who has a fixed-rate coupon. The first investor makes interest payments on the coupon of the second investor.

Although coupon swaps have the advantage of being off-the-balance-sheet transactions, they do have one major drawback—each party is obligated to the original creditor. This means that if one party defaults on its original loan, the party currently making the payments on the loan may not have any avenues of recourse.

6.3.5 Forward-Cover Swaps

A long-term *forward-cover swap* takes place if an investor is unable to locate a particular maturity or is unable to buy forward cover to compensate for long-term fluctuations in monetary exchange rates. Transactions require an investor to swap with someone who has a long-term currency need, thereby offsetting the currency flow of the currencies in question. In order to perform this type of swap, an intermediary is required (such as a bank), and the intermediary becomes involved in the deal to facilitate the transaction, which increases transaction costs.

Swap financing is not common in the financial marketplace, because it is contrary to the traditional method of companies borrowing only in currencies for which they have a clear need. Swap-financing techniques imply that investors should borrow in the market where they have the best comparative advantage, regardless of their cash flows, international presence, or currency needs (Militello, 1984).

6.4 Project Financing

Project financing is defined as a "financial method whereby the project being financed is treated as an entity separate and distinct from the company sponsoring the project" (Mao, 1982, p. 24). If a company decides to utilize project-financing techniques instead of traditional financing methods, the E&C project under consideration is set up as a separate entity legally distinct from the parent company. The parent company can invest equity in the project, but project financing is most advantageous if the project is largely financed by debt.

If a project is being analyzed to evaluate its creditworthiness, the parent company is not considered in the evaluation if it is using project-financing techniques, because the financial evaluation of the project is based solely on the merit of the project. The principal advantage of project financing is that project debt does not appear on the balance sheet of the parent company; therefore, project-financing techniques come under the category of off-the-balance-sheet financing methods.

Although project financing has grown out of both *spin-off* and *segmental financing*, it has carved its own distinctive niche in the financial marketplace. One major justification for project financing is that the net present value of a large project is increased, and a greater borrowing capacity exists for the project (Mao, 1982).

Various arguments have been made both for and against project-financing methods. On the positive side, it appears that lenders will lend more money per dollar of project asset and at lower interest rates if project-financing techniques are employed on a project. This advantage exists because creditors know that they will have prior claim on the cash flow of the project for repayment of the principal and interest on the loans. The primary benefits of project financing are lower costs (interest rates) and greater credit responsibility.

If creditors assume that if a project stands on its own financially, it undermines their position, this could result in higher interest rates or a restriction on the amount of money creditors are willing to lend for projects. In general, most creditors do not have access to a detailed analysis of their prospective clients; therefore, they cannot determine the effects of commingling funds within firms.

Two other financing methods analogous to project financing are spin-off financing and segmental financing.

6.4.1 Spin-Off Financing

Spin-off financing methods require a distribution of equity to a new venture, which in a manner spins off the venture and creates a separate market for the common stock of the subsidiary or division. Spin-off methods are used to create a new stock holding that eventually appreciates in value.

6.4.2 Segmental Financing

Segmental financing takes place when a company transfers part of its assets to one of its wholly owned subsidiaries, and the subsidiary borrows against the newly acquired assets. Segmental financing creates a separate market for the debt, which is not recorded on the balance sheet of the parent company.

So far, three forms of project financing have dominated the financial marketplace: (1) nonrecourse, (2) project financing with completion guaranteed, and (3) project financing with guarantees beyond completion.

6.4.3 Nonrecourse Financing

Nonrecourse financing methods are suited to a variety of projects, but their use is more prevalent in the following situations: (1) Either a project has high equity or the customer is willing to make advance payments on output to be delivered in the future; (2) there is a guaranteed sale, take, or pay arrangement where the customers unconditionally agree to buy the product or output; or (3) the project is utilizing tested technology, thereby reducing the risk involved in the project.

6.4.4 Completion Guarantees

Completion guarantees require the involvement of the parent company, and the parent company either guarantees a certain completion date for the project or covers the cost of project overruns with equity investment.

6.4.5 Guarantees beyond Completion

Guarantees beyond completion are used when the lender will be exposed to market risks associated with the sale of the output from the project after its completion. In this type of a situation, lenders will seek a guarantee from the project sponsors indicating that the sponsors will provide funds to meet the obligations of the project if, at some point after the completion of the project, the project is short of cash to meet operating expenses and debt service.

Several companies have been experimenting with *nonrecourse-financing* techniques. Two successful examples include a project on the island of Sumatra, Indonesia, and a project on the island of East Kalimantan, Indonesia, that were developed by a consortium of companies from the United States, Japan, and Indonesia. Indonesia has been involved in nonrecourse financing because Pertamina (the government energy company) was not allowed to make speculative or business-development-type loans because of the high amount of debt owed by the Indonesian government.

Financing for the expansion of a liquefied natural gas plant on the island of Sumatra was secured using nonrecourse-financing techniques. A firm that is an oil producer provided funding for the project and also signed a guarantee to buy the liquefied natural gas product. Even if the oil producer did not take the liquefied natural gas product, it was responsible for making payments based on a minimum yearly quota. The oil company arranged for the income from the plant to go directly to a brokerage house, which then made payments to the oil company before the government of Indonesia realized its revenue funds. This arrangement reduced the risks involved with repayment of the loan.

On the island of East Kalimantan, Indonesia, a consortium of companies from the United States, Japan, and Indonesia developed a financing package based on project-financing methods. Rather than providing financing for the project, the consortium put the project out to bid on the world market. A number of lenders submitted bids on the liquefied natural gas project, and eventually the project was financed by an uninvolved third party.

Since Taiwan was going to be purchasing the liquefied natural gas product from the plant in Indonesia, the government of Taiwan signed a contract guaranteeing repayment of the loan whether it purchased the liquefied natural gas product or not, thus creating a new form of nonrecourse contract. The consortium that created the project viewed this loan as a risk-free loan. The consortium acted as a go-between for Taiwan and the lenders who secured the loan. A third party handled the loan payments, which in this case was a major bank. The loan did not appear on the books of the Indonesian government, nor did it show up on the books of the consortium, instead it was carried on the books of the bank. This type of financing allows countries such as Indonesia to work within their own loan restrictions.

In addition to the global financing techniques mentioned previously, investors often create new methods that cannot be categorized into an existing financing classification. Although some innovative financing methods are not considered to be major breakthroughs, several deserve attention when innovative financing techniques are considered for the global construction marketplace.

6.4.6 Note Swaps

Several companies have been swapping their debt holdings for equity by trading their common stocks for bonds. This method produces an immediate tax-free gain and also eliminates the long-term interest bearing debt of a company (Clemente, 1982).

6.4.7 Floating-Rate Notes

Floating-rate notes are used by issuers to obtain long-term funds during periods of high interest rates for the life of the issue (Clemente, 1982). With floating-rate notes, the cost of debt automatically falls when interest rates fall. The convertibility features of these notes are a hedge against increasing interest rates. Floating-rate notes (certificates of deposit) allow corporate borrowers to obtain funds directly from nonbank sources at interest rates that vary with short-term bank rates. Floating-rate notes assure banks of longer-term availability of funds.

6.4.8 Zero-Coupon Bonds

Zero-coupon Eurobonds have found a market in countries such as Japan and those under Islamic law. The Japanese prefer zero-coupon bonds because capital gains are not taxed in Japan. Zero-coupon bonds have gained popularity in Islamic countries because Islamic laws do not permit the receipt of interest payments. Members of U.S. firms are interested in zero-coupon bonds because zero-coupon bonds are not registered with the Securities and Exchange Commission; thus companies have a measure of freedom in their reporting techniques.

6.4.9 Rollover Credits

Rollover credits, or *revolvers,* were created to allow Eurobanks to make long-term loans without facing interest-rate risk. Funds are committed to borrowers for a medium-term period (usually 3 years), with the interest rate changing every 3 to 6 months in relation to the prevailing short-term rates in the Eurodollar-interbank market. With this method, customers avoid the risk of a nonrenewable loan (Duffy and Giddy, 1981).

6.4.10 Exchange-Participation Notes

Exchange-participation notes are a method whereby principal repayment claims of debtors are related to a real, measurable ability of debtors to pay (Bailey, 1983). Exchange-participation notes require central banks in debtor nations to issue notes to private and official lenders on a pro-rata basis to replace existing amortization schedules. With exchange-participation notes, central banks accept responsibility for repaying loans, and the banks act as payment and collection agents. One major advantage to the central bank concept is that exchange-participation notes may be sold in secondary financial markets.

6.5 Global Payment Methods

In the global arena, there are many different currencies and regulatory processes that should be followed in order to import construction materials. Since currencies fluctuate simultaneously, it is difficult for firms to keep track of their payment streams to other countries in native-country currencies (a problem the European Union has tried to solve with the introduction of Eurocurrency).

One method for avoiding the problem of fluctuating currencies is to make financial payments for materials used on construction projects in local currencies. Several different methods can be used to make payments for materials, including cash in advance, letters of credit, credit insurance, bills of exchange, and open accounts, and these methods are discussed in the following subsections.

6.5.1 Cash in Advance

If there is political instability in a particular country, or if little is known about the credit history of a buyer, suppliers could require that cash be paid in advance when an order is placed for construction materials. For construction firms, this creates an added financial burden if the construction firm will be paid through progress payments as work is completed on the project. Having to pay cash up front could force construction firms to secure short- or medium-term loans in the local currency for convenience or currency-exposure-management purposes.

One common method of securing short-term funds is through bank overdrafts that are a line of credit from a financial institution, where checks are drawn against a line of credit up to a specified amount. Usually a line of credit will be granted only if there has been a previous relationship between the banking institution and the company seeking the line of credit or if the company can meet the requirements for credit approval set by local banking institutions.

Another short-term financing technique is called *discounting,* and it is performed in the following manner: A manufacturer sells goods to a company on credit and draws up a bill for the buyer that is payable in a set amount of time. The buyer endorses the bill, and then the bill is taken to their financial institution, and after it is accepted by the financial institution, it is called the *buyer's acceptance.* The manufacturer then takes this bill to its local bank, which accepts the bill for a fee. The bill is then sold at a discount to the bank of the manufacturer or to a money-market dealer. The interest rate varies with the terms of the bill and the general level of local money-market interest rates.

For medium-term funds in local currencies, some companies will use *renewable overdrafts,* which are a method that can be used if funds have been deposited in a local bank. The bank will allow firms to write checks in excess of their deposits up to an agreed-on amount; once the overdrafts are repaid, the firms can again write more checks up to the agreed-on amount. The fees charged for renewable overdrafts vary, depending on the terms of the renewable draft.

Another method for acquiring medium-term funds is *bridge loans* (which are also renewable loans that include promissory notes) that are used to provide interim financing while a borrower obtains medium- or long-term financing. The terms for bridge loans are usually set up with a local lending institution at the start of a project so that funds will be available immediately when materials need to be ordered for a job site.

Other medium- and long-term loans can be negotiated on the basis of the expected cash flow to be generated by a project. Rather than the loan being evaluated on the liquidity position of a firm, the loan is based on the ability of a firm to generate cash flow to service its liabilities.

Other traditional methods of securing long-term loans include: (1) bonds, which have different restrictions depending on where they will be purchased (in certain countries, such as Japan, they are sold mainly to financial institutions); (2) pension and insurance plans, which cannot be used in France or Italy because their governments supply compensation for their citizens; and (3) lease financing, where the value and use depends on the tax regulations relating to depreciation write-offs in a particular country (the governments of some countries do not allow write-offs for depreciation of capital assets).

6.5.2 Letters of Credit

If cash in advance is not required, then *letters of credit* (LOCs) can be used to order construction materials and supplies. Letters of credit are addressed to sellers, and they are written and signed by the bank acting on behalf of a buyer. The bank promises to honor drafts drawn on the bank if the seller conforms to the conditions as set forth in the letter of credit. Letters of credit eliminate credit risk because any seller may investigate the financial credibility of a bank, even in a situation where the seller would not be able to verify the credit of a potential buyer. Letters of credit also reduce the chance that payments will be delayed or withheld because of exchange controls or other political acts.

Letters of credit also guard against preshipment risk since sellers have letters of credit during the manufacturing phase. If prepayment is required, a firm is better off depositing money into a bank and having the bank issue a letter of credit as it is easier for the bank to recover its funds if a seller does not make a shipment.

Although letters of credit are in widespread use, some firms might use credit insurance or advanced financing for global trade. Banks and underwriters prefer these methods because they increase their lending capacity. Credit insurance is a viable alternative to letters of credit, but it introduces another party into transactions between buyers and lending institutions—the insurer—and the rates, fees, and terms may be more variable than those charged by banks, because they can be set based on different criteria related to the history of the buyer rather than predetermined bank rates (Morris, 2004).

6.5.3 Bills of Exchange

Bills of exchange (drafts) are another payment method, and they are an unconditional order in writing that is signed by a supplier and addressed to the purchaser, ordering him or her to pay on demand or at a fixed time. The bill of exchange provides evidence of a financing obligation, it reduces the cost of financing, and it can be used as a *sight draft* that is payable at some specified future date.

6.5.4 Consignment

Consignment is used to secure construction materials for projects that are outside of a country where the materials are manufactured and sold. Consignment requires that when the goods are shipped, they are not sold to a firm directly by the supplier; rather, the supplier retains title to the goods until they have been installed and paid for by a third party (the owner).

6.5.5 Open Account

If there are no political or financial concerns related to the country in which a project is being built or the country from which the materials are being supplied, then an *open account* can be used to pay for materials. Open accounts refer to situations where materials are shipped, and then the firm receiving the materials is billed later, and the recipient pays for the materials based on terms previously negotiated with the supplier.

6.6 Bills of Lading, Commercial Invoices, and Consular Invoices

All materials and supplies that are shipped between countries should include bills of lading, commercial invoices, and consular invoices (if required). *Bills of lading* are the contracts between carriers and shippers in which carriers agree to ship goods from a port of shipment to a port of destination, and they should include, at a minimum, the associated costs, the terms for shipping the materials, and the shipping and delivery dates.

Commercial invoices contain an authoritative description of the merchandise shipped, which helps to prevent materials from becoming lost during shipment or switched for inferior materials after fabrication. Commericial invoices should include a list of the following:

- Quality
- Grades
- Price per unit
- Total value
- Names and addresses of the exporter and the importer
- The total number of packages or bundles
- Any distinguishing marks
- The payment terms
- Other expenses, such as transportation and insurance charges
- The name of the carrier shipping the materials
- The ports of departure and destination
- Any export permit numbers required

Consular invoices are similar to commercial invoices, but they are presented to the local consul in exchange for a visa.

6.7 The U.S. Export-Import Bank and Private Export Funding Agencies

The U.S. Export-Import Bank is the only U.S. government agency dedicated to financing and facilitating exports of materials and supplies manufactured in the United States. It provides loans for the export of U.S. goods if private capital is not available in the amount required by the manufacturer, and it supplements rather than competes with private capital. In order to secure financing from the U.S. Export-Import Bank, there must be a reasonable assurance of repayment, and the fees and premiums charged are based on the risks covered by the loan.

The Private Export Funding Corporation was created in 1970 by the Banker's Association for Foreign Trade, and it mobilizes capital from private institutions for financing the export of large capital assets by U.S. firms.

6.8 Summary

The global financing techniques cited in this chapter are only a few of the multitude of financing methods available for use on global E&C projects. Each year new techniques will be tried and either adopted or rejected, paving the way for the development of even more creative financing techniques. It is important for individuals and corporations to realize that funding sources for E&C projects are virtually unlimited if members of companies are willing to take a chance on creativity in their financing methods and if they are able to convince others of the viability of new methods.

Not all of the global financial methods presented in this chapter are suitable for every type of E&C project. Members of each corporation have to examine the alternatives available to determine which global financing technique fits the financial style of their firm and whether the alternative selected is appropriate for use in a particular country.

A major obstacle to proposing innovative financing techniques is convincing E&C personnel to no longer use traditional methods that have been considered industry standards for decades. Unfortunately, the global situation dictates the need for additional funding sources; therefore, if innovative financing techniques are ignored, the possibility arises that someday in the near future there may be a shortage of funds for financing global projects.

If financial managers hope to maintain their competitive advantage in the global marketplace, they need to be constantly adapting their financial styles to suit the changing world environment. In order for E&C firms to compete on a global basis, they should include innovative financing techniques in their repertoire of viable financing methods.

The objective of this chapter was to provide insight to E&C personnel who are working on global projects on the restrictions that financing techniques create and why there could be limited resources for the construction of a project. This chapter provided information on countertrading and cofinancing with the World Bank, including vertical cofinancing, swap financing, coupon swaps, and forward-cover swaps. Project financing was discussed in terms of how it can be used to acquire funds for construction projects. Variations of project financing were also presented, including spin-off, segmental, and nonrecourse financing. Other financing methods discussed in this chapter include completion guaranteed, guarantees beyond completion, note swaps, floating-rate notes, zero-coupon bonds, rollover credits, and exchange-participation notes.

This chapter also discussed alternatives for paying subcontractors and suppliers in the global environment, including cash in advance, letters of credit, bills of exchange, consignment, and open accounts. The last two sections of this chapter explained bills of lading, commercial invoices, consular invoices and how the U.S. Export-Import Bank participates in funding U.S. exports.

DISCUSSION QUESTIONS

1. Explain the difference between traditional financing methods and *innovative financing techniques.*
2. Explain why innovative financing techniques have been developed and implemented on global construction projects.
3. Develop a scenario for using countertrade to finance a construction project in China.
4. Explain how countertrade agreements benefit the host country where a construction project is being built.
5. What is the benefit of involving the World Bank in the cofinancing of construction projects?
6. Explain why any firm would use a coupon swap rather than obtaining financing directly from one financial institution.
7. Explain how project financing can be used to finance a construction project even if the private firm that is going to construct the project cannot obtain financing based on the firm's financial status.
8. Explain nonrecourse financing and how it was used on the two projects in Indonesia.
9. Why are zero-coupon bonds used in some countries and not in other countries?
10. Explain why it is important for E&C personnel to know how a project is being financed when they are working on it.
11. Explain the difference between renewable overdrafts and bridge loans.
12. Explain why letters of credit are a no-risk way for suppliers to be paid by constructors.
13. How does consignment benefit constructors if it is used on construction projects to pay suppliers?
14. Why should commercial invoices be created for every construction material that is shipped to construction job sites?
15. Which of the financing techniques in this chapter helps to protect constructors against currency exchange-rate fluctuations, and how does it do so?

REFERENCES

Bailey, N. A. 1983. Safety net for foreign lending. *Business Week,* January 10.

Beenhakker, Henri. 2000. *The Global Economy and International Financing.* New York: Quorum Books.

Ben-Horin, M., and W. Sibler. 1977. Financial innovation: A linear programming approach. *Journal of Banking and Finance* 1 (2): 277–296.

Clemente, H. A. 1982. Innovative financing. *Financial Executive* L (4): 14–19.

Das, Satyjit. 1993. *Swap Derivative Financing: The Global Reference to Products, Pricing, Applications and Markets.* New York: McGraw-Hill.

Duffy, G., and I. Giddy. 1981. Innovation in international markets. *Journal of Banking and International Business Studies* 12 (Fall): 33–50.

Kuemmerle, Walter. 2004. *Case Studies in International Entrepreneurship: Managing and Financing Ventures in the Global Economy.* New York: McGraw-Hill.

Mao, J. C. T. 1982. Project financing: Funding the future. *Financial Executive* 1 (8): 23–28.

Militello, F. C., Jr. 1984. Swap financing: A new approach to international transactions. *Financial Executive* 52 (10): 34–39.

Morris, Gregory. 2004. *Relegating Letters of Credit to the Dustbin: Credit Insurance and Advanced Financing.* Kennesaw, Ga.: The Axion Group.

Ohmae, Kenichi. 2005. *The Next Global State: The Challenges and Opportunities in Our Borderless World.* Philadelphia, Pa.: Wharton School Publishing, University of Pennsylvania.

Pelosky, R. J., Jr. 1999. *Managing and Organizing Multinational Corporations.* New York: Pergamon Press, pp. 205–517.

Shapiro. A. 2002. *Multinational Financial Management,* 499–511. New York: John Wiley and Sons.

Tuller, Laurence. 1992. *The McGraw-Hill Handbook of Global Trade and Investment Financing.* New York: McGraw-Hill.

U.S. Department of Commerce. 2002. *A Guide to Financing Exports,* 5–18. Washington, D.C.: U.S. and Foreign Commercial Service.

World Bank. 1982. Cofinancing with the private sector. *Finance and Development* 19 (2): 72–73.

World Bank. 2002. *World Bank Global Development Finance 2002: Financing the Poorest Countries.* Geneva, Switzerland: World Bank Publications.

Chapter 7

Global Legal Issues for Engineers and Constructors

7.1 Introduction

The colonization of nations throughout the world created a common foundation of legal theories. The affects of colonization still influence citizens of conquered nations even after they have gained independence from their colonial overseers, sudden or violent regime changes, terrorist attacks, and kidnappings increase the complexity, cause instability, and create uncertainty for global engineering and construction (E&C) professionals.

If construction workers are from countries other than where a project is being constructed, they might be subject to the laws of their native country in addition to host-country laws. In the global E&C arena, it is not always clear which legal system has the right to settle disputes; therefore, international contracts usually specify that international arbitration or exclusive jurisdictions will be used to settle construction claims and disputes.

This chapter provides information on legal issues that affect contractual relationships on global E&C projects, including regional legal issues, international contract clauses, claims and change orders, international arbitration, terrorism and kidnapping insurance, regime changes, and liability issues. In addition, international contract clauses are discussed, and information is provided on how specific clauses affect global E&C projects and the ramifications of contract clauses when judges interpret them in foreign legal systems.

Although E&C professionals work in the global arena, the contracts they use are referred to as *international contracts* rather than global contracts because legal contracts are between members of two nations not the entire world.

7.2 International Conventions

Engineering and construction personnel and their clients should be familiar with the international conventions that are enforced in the jurisdictions where they are constructing projects. One resource that contains information that is pertinent to international contracting is the *Legal Guide for Drawing Up International Contracts for the Construction of Industrial Work,* which is published by the United Nations Commission on International Trade and Law (UNCITRAL), and another useful resource is the *Martindale-Hubbell International Law Digest* (UNCITRAL, 1988; Martindale-Hubbell, 2000). Examples of legal conventions include: the *Convention on the Recognition and Enforcement of Foreign Arbitral Awards,* commonly known as the *New York Convention,* which requires legal systems throughout the world to honor international arbitration awards; the *United Nations Convention on Contracts for the International Sale of Goods,* which includes information on how to draft international contracts; and the *International Commercial Terms* (INCOTERMS), which is published by the International Chamber of Commerce and includes information on terms that are applicable to international contracts (O'Hare, 1980; UNCITRAL, 1988; Murphy, 2005). The definitions used in the INCOTERMS were written for sales contracts, but they are also applicable to E&C contracts.

7.3 Regional Legal Issues

Whenever a nation formed a colony, the legal system that was established was similar to the legal system of the colonizing nation. However, having similar legal systems does not guarantee that disputes would be settled in the same manner in both legal systems. Even within nations, there is a high degree of variability in the outcomes of legal cases (Stokes, 1978; Knutson, 2005). But being familiar with the origins of legal systems is advantageous when attempting to settle disputes that arise on E&C projects. Country-specific laws and a directory of international law firms are included in the *Martindale-Hubbell International Law Digest* (Martindale-Hubbell, 2000).

7.3.1 Comparative Legal Systems

Civil law and *common law* are the two most prevalent legal systems. Civil law is based on the ancient Roman legal system, and the common-law legal system was implemented during the Norman Conquest to unify England (Katz, 1986).

7.3.1.1 Civil-Law Jurisdictions

Examples of civil-law jurisdictions include France and Spain and their former colonies (Wolf, 2000). In civil-law jurisdictions, judges are more involved in litigation processes than in common-law jurisdictions. Civil law does not rely on the use of previous cases (*precedent law*) to plead current cases; rather, cases are evaluated based on existing laws. Therefore, it is easier to prove that an arbitration clause is valid in civil-law legal systems than in common-law legal systems (Stokes, 1978; Knutson, 2005). In civil-law legal systems, awards for *liquidated damages* are usually based on actual damages, not *punitive* (to punish) *damages.*

7.3.1.2 Common-Law Jurisdictions

Examples of common-law jurisdictions include Great Britain and the former colonies of Great Britain, such as the United States, India, and Malaysia (Knutson, 2005). In common-law systems, an adversarial process occurs between opposing lawyers, and judges maintain a passive role. Civil-law jurisdictions typically emphasize the intent of contracts; common-law legal systems focus on literal interpretations of contract clauses (Knutson, 2005).

Examples that illustrate common-law systems include the British and the U.S. legal systems. Even though both systems are common-law systems, they use different terminology. In the United States, attorneys who represent clients are referred to as *lawyers for the plaintiff,* and the lawyer defending someone is the *lawyer for the defense.* In the British legal system, attorneys who gather facts for a case are referred to as *solicitors,* and the attorney who delivers the argument is called a *barrister.* In the United States, the burden of proof is on the person who files a lawsuit, and in the British legal system the burden of proof is on the defendant.

In the United States, hundreds of people may be called to courthouses as potential jurors, and they are questioned by lawyers; and a certain percentage of them are rejected by the lawyers. In the British legal system, the names of potential jurors are written on pieces of paper, put in a box, and 12 names are drawn out of the box. The United States is predominately a common-law country, but the state of Louisiana has been influenced by civil law, because it was originally a colony of France (Katz, 1986).

In the United Kingdom an enemy alien cannot sue in an English court; however, a plaintiff may sue an enemy alien. An *enemy alien* is defined as someone or an organization from a country that is involved in a war between the United Kingdom and the foreign nation, and the person (or organization) is physically located in the nation that is at war with the United Kingdom. A citizen of a foreign nation who is legally living in the United Kingdom is not considered to be an enemy alien (Hill, 1998). The legal system in Canada is unique in that it has a civil-law system in Quebec and common-law systems in the nine other provinces.

Asian countries are examples of jurisdictions that are neither common law nor civil law. The legal system in Japan is derived from ancient Japanese law, but it was influenced by French civil law. The Japanese are extremely averse to litigation, as is demonstrated by the fact that the per capita litigation rate in Japan is one-twentieth the rate of litigation in the United States (Katz, 1986). In Japan, there are eight engineers for every lawyer, and in the United States, there are eight lawyers for every engineer, which demonstrates major differences in the two societies. In China, conciliation is a preferred method of dispute settlement instead of litigation, because "the basic goal of Chinese social philosophy is to attain harmony and mediation (and conciliation) is compatible with this Confucian ideal" (Redfern and Hunter, 1986; Katz, 1986, p. 246).

7.3.1.3 Islamic-Law Jurisdictions

In the Middle East, contracts usually require owners to withhold a certain amount of the contract price to ensure that a project is completed satisfactorily, and this is called retention. Performance bonds are rarely used in the Middle East, because owners do not want to have to sue a bonding company if it is in a foreign nation. Middle Eastern E&C firms frequently mandate that the records kept by contractors have to be in Arabic, and they must be physically located in the host country.

Corporate taxes in Middle Eastern countries can be as high as 60 percent, but they usually are between 30 and 50 percent. If a company is in a joint venture with a Saudi Arabian firm, their corporate taxes could be exempt for up to 5 years, and the foreign firm is eligible for loans from the Saudi Industrial Development Loan Fund (Stokes, 1978). Government projects in Saudi Arabia have to follow the *Government Tender Law* of Saudi Arabia, and the government of Saudi Arabia does not allow labor unions or strikes (Al-Jarallah, 1983; Stokes, 1978).

7.4 International Contracts

Domestic E&C contracts are similar to international contracts, but international contracts also contain additional clauses that address situations that arise that are related to the global nature of international projects. This section includes information on the types of clauses that should be included in international contracts and provides explanations on how and why some of these clauses affect international projects. Table 7.1 provides a summary of the clauses that should be included in international contracts, along with a brief explanation of each clause (Stokes, 1978; UNCITRAL, 1988).

In order to keep money in the local economy, international construction contracts may be divided into two contracts: one for work done within the host country and the other for work done outside of the host country (Lantis, 2005). One commonly used standard format international contract is published by the International Federation of Consulting Engineers. In the United States, the American Institute of Architects (AIA), the Associated General Contractors (AGC), and the Associated Building Contractors have standard contracts.

7.4.1 Technical Standards and Inspections

Local legal jurisdictions often regulate the technical aspects of E&C projects, as well as safety and pollution emissions by enforcing local standards. Contracts should specify *technical standards* that are internationally accepted, as well as familiar to local contractors. Local subcontractors and suppliers might not be able to participate in a project or the cost of a project might increase if the standards being used are not well-known standards. Local inspection and testing institutions should be used for projects that are in unfamiliar locations. If governments restrict access to facilities where equipment and machinery are being manufactured, then owners should specify a testing agency that has access to these facilities.

Government authorities usually inspect construction and test facilities to ensure that they comply with safety, health, and environmental regulations, so it is im-

Table 7.1 Synopsis of Common International Contract Clauses

Clause Title	Highlights
Technical standards and inspections	Safety: Use local regulations. Environment: Local regulations. Technical: International, local, or other country. Testing: Can use national or international testing institutions. Inspections: Municipal governments for safety, health, and environmental compliance.
Confidentiality	Local government might require all documents for a project, to be submitted to them which compromises confidentiality. It is hard to enforce these clauses because parties to the contract are subject to the laws of different nations.
Patents and trademarks	Only protected in the country where they are issued.
Currency clauses	Contractors are paid in local currencies, with a multiplier for currency-exchange rates. Can specify that payments be made in a different currency. Subcontracts may have to be paid in the local currency.
Language	If there are versions of a contract in different languages, the original version controls.
Local subcontractors and suppliers	Contracts may require the use of local subcontractors and suppliers. Use local subcontractors and suppliers to maintain goodwill.
Transporting equipment	There could be size and weight restrictions, and permits may be required.
Laws of host nation	Owners should provide local building permits; owners should help contractors get visas and local work permits for their personnel. Customs duties on imported equipment. Could be duties on some exported items. Import duties on construction equipment may be required, even if it will be exported at the end of the job. Contracts should state who will pay import and export fees. There could be transit taxes on items shipped. Some items are illegal to import or export. Check with respective governments. If duties increase after a contract is signed, contractors should file change orders.
Liens, subcontracting	Owners may require proof of payment to subcontractors and suppliers. Find out if the host government allows liens to be placed on property if subcontractors or suppliers are not paid by the prime contractor. There could be local limits on how much work may be subcontracted. Governments should require a certain portion of contracts to be performed by national personnel.
Bankruptcy of contractors	Owners could be required to make back payments to subcontractors if they take over subcontracts.

Table 7.1 (*Continued*)

Clause Title	Highlights of Clauses
Liquidated damages	Not enforced if they are punitive damages.
	Amounts could be reduced by the legal system if there was partial performance.
Hardship	Allows renegotiation of contracts based on hardship.
Index and currency	If prices for products rise faster than inflation rates, this clause allows for renegotiation of contracts.
Exemption	One party cannot seek damages if the other party does not perform when items in the exemption clause occur.
	Examples include severe weather, natural disasters, civil strife, riots, war, destruction of the subject matter, fire, flooding, and the failure of a government to approve a project.
Termination clause	Owners can terminate a contract for any reason if this clause is included in the contract.
	Some governments do not allow termination unless it is stated in the contract.
	Contracts should state specific conditions for termination.
	Contracts should allow contractors to terminate contracts if owners do not pay, they order a suspension of work, or they file for bankruptcy.
	Some countries do not allow termination due to bankruptcy.
	Some countries do not allow bankruptcies.
	Proceedings similar to bankruptcy may be called: receivership, liquidation, insolvency, assignment of assets, or reorganization.
	If a contract is terminated, it may void contractual obligations such as dispute-settlement rights and confidentiality.
Spare parts and maintenance after construction	Contractors may have to provide spare parts and repair facilities for a set period of time.
	Check whether owners or contractors pay if equipment is shipped out of a country for repairs.
Choice of law and choice of forum	Allows parties to select which legal system is used for settling disputes related to the contract. If there is no choice of law clause, but if there is an arbitration clause, arbitrators decide which laws apply to contracts.
	Could be limited to a country with a connection to the contract.
	Could be different legal systems used to settle different legal issues.
	May stipulate an exclusive jurisdiction and a specific court.
International contracts	See subheading 7.4 on international contracts.

portant to be familiar with host country standards that could be enforced on local construction projects. If governments change inspection and testing requirements after a contract has been signed, then contractors are entitled to a change order that reflects the new cost of complying with government regulations.

7.4.2 Confidentiality

If the governments of host countries require that all project documentation be submitted to them, this limits the level of confidentiality that can be maintained on projects. *Confidentiality agreements* should be entered into before contract negotiations begin to maintain confidentiality during the negotiations. Confidentiality agreements are difficult to enforce if parties to the agreement are from different nations, because in the global arena, attorney-client confidentiality is not guaranteed, which creates problems for foreign nationals, as well as domestic firms, that are competing for projects in foreign countries.

7.4.3 Patents and Trademarks

Patents and trademarks are only protected in the country in which the patent or trademark is registered. There are no *global patents,* and the laws of one nation do not extend beyond its borders into other countries. Engineering and construction personnel are reluctant to use proprietary processes or products in foreign nations unless they have had a previous (positive) relationship with a client. This cautiousness may reduce the ability of an E&C firm to compete for projects based on advanced technical capacity.

7.4.4 Use of Local Suppliers

Contractors could be required to use local subcontractors and suppliers to generate employment in the host nation, or owners might specify the use of local subcontractors and suppliers to earn and maintain goodwill in the local community.

7.4.5 Packaging and Transportation of Equipment and Materials

Shipping agencies have regulations restricting the size and weight of items to be shipped, which precludes the use of large construction equipment on some projects and on the shipping of hazardous materials. *Transportation permits* are required for shipping large construction equipment, materials, and machinery, and contracts should specify who is responsible for obtaining these permits.

7.4.6 Laws Governing Engineering and Construction Personnel

When working in any country, E&C personnel are subject to the laws of the host nation, as well as federal laws of their native country, such as laws governing the bribery of foreign government officials. International treaties, such as the Kyoto Protocol, have to be followed by members of the E&C industry when they are designing and constructing projects if host nation governments have ratified the treaties (see Chapter 9).

7.4.7 Local Construction Permits, Visas, Customs, and Duties

International contracts should indicate who is responsible for securing permits and licenses and for paying customs duties and transit taxes required for projects. Customs duties are levied on materials and equipment imported into a country, and duties are imposed on items exported to other countries. Whoever receives the items being shipped usually pays for custom and transit fees.

7.4.8 Subcontracting

In legal systems where contractual relationships are only between contractors and subcontractors (not subcontractors and the owner), owners cannot recover losses from subcontractors, only from contractors. Owners should protect themselves from *liens* by requiring contractors to submit written proof when they pay subcontractors or by requiring contractors to provide proof of a payment bond. Liens are a legal claim against the property that can be filed by subcontractors so that they are paid for their services when the property is sold. If contractors order materials from foreign suppliers, they might have to pay taxes in the host country and in the country of origin of the materials.

Host-country governments could require a certain level of participation by national contractors on domestic projects, and owners set requirements on how much of a contract may be performed by subcontractors. Owners may specify that specific subcontractors have to be used on projects, and prime contractors should request that their liability be limited if they use the specified subcontractors (UN-CITRAL, 1988).

7.4.9 Liquidated Damages

Assessment and enforceability of *liquidated damages,* which are actual damages paid to owners on a daily basis when projects are not completed on time, varies throughout the world. Liquidated damage awards could be reduced or eliminated if they are merely being used to punish a contractor for not completing a project on time. Owners must prove that they sustained an actual loss in order to collect liquidated damages. Reviewing prior court decisions from a local court provides insight into how the local legal system has settled cases involving liquidated damages. Liquidated damages attempt to fix the limit of damages that would be incurred in the event that completion of a construction project is delayed. This predetermination of damages, which is based upon a precise dollar amount for each day of delay, eliminates the need to litigate actual damages.

With respect to assessing intent of the parties for the damages not to be penal, courts will generally allow parties the freedom to contract and to reach agreements that will not be set aside as penal in nature. The legal theory for this is that when a contract is entered, which provides for liquidated damages, the parties have a right to fix the amount by contract, and, if reasonable, the courts will not abrogate such provisions by declaring them to be a penalty.

One problem with this view is that public construction contracts are frequently from documents that the contractor must take or leave. There is no freedom to negotiate the amount of liquidated damages. Nevertheless, this issue of unequal bargaining power among the parties is generally not considered by the courts in a typical evaluation of a liquidated damages clause.

Liquidated damages should be set at a reasonable and specific amounts. If they are too high they will likely be declared penal in nature and hence unenforceable. The setting of damages at unrealistic levels can cause contractors to generate delay claims. To reduce litigation regarding liquidated damage clauses, it would be advisable to provide in the contract a general description relative to how the amounts were calculated. During litigation, the courts will be guided by seeking fair compensation for injuries sustained and will attempt to place the injured party in the position it would have been in had the delays not occurred. When the contracting parties keep this principle in mind in the setting of liquidated damages, the courts will generally allow the provisions to stand.

7.4.10 Hardship Clauses

The United Nations Commission on International Trade Law (UNCITRAL) defines *hardship* as "a change in economic, financial, legal or technological factors that causes a serious adverse economic consequence to a contracting party, thereby rendering more difficult the performance of his contractual obligations" (UNCITRAL, 1988, p. 242). In other words, some type of hardship makes performing the contract more difficult but not impossible to perform. A *hardship clause* allows owners and contractors to renegotiate a contract based on the conditions that create a hardship. Hardship clauses are not allowed in all legal systems and some legal systems may not uphold such a clause when it is challenged in a court of law. Contractual parties can agree to renegotiate the contract even if a hardship clause is not in the contract, but the presence of a hardship clause makes it easier to request a renegotiation of the contract.

7.4.11 Index and Currency Clauses

When prices for certain products increase faster than the average inflation rate, having an *index clause* allows for automatic adjustments to payments when there are fluctuations in exchange rates. *Currency clauses* are also used to account for fluctuations in exchange rates.

7.4.12 Exemption Clauses

Exemption clauses may be used in a manner similar to impossibility clauses. Items that might be included in an exemption clause include abnormally severe weather or natural disasters; civil strife, riots, or war; destruction of subject matter; fire; and flooding. In international contracts, this clause may include interference by the host government, acts of terrorism, coups, or host-government agents not approving projects.

7.4.13 Termination Clauses

Owners are allowed to terminate contracts at their convenience in some countries, even if there is no clause in the contract that permits them to do so. Other governments do not allow such *unilateral termination* unless there is a *termination clause* in the contract. Some legal systems require judicial consent for owners to terminate contracts without termination clauses. If termination clauses are used, contracts should state the conditions under which the owner may terminate a contract.

If a termination clause in included in contracts, then contractors can terminate contracts if owners do not provide prompt payment, if they order a long suspension of work, or if they file for bankruptcy. Parties to contracts may not be able to terminate contracts solely on the basis of the bankruptcy of other parties because bankruptcy laws of a particular nation control in these situations. Bankruptcy proceedings exist throughout the world, except in India, and they could be called *receivership, liquidation, insolvency,* the *assignment of assets,* or *reorganization* (UN-CITRAL, 1988).

In foreign legal systems, if a contractor is terminated, he or she could loose the legal right to use dispute-settlement techniques and confidentiality agreements that are included in the contract unless the contract states that these agreements are valid even if either party terminates the contract.

7.4.14 Choice of Law, Choice of Forum, and Exclusive Jurisdiction Clauses

Choice of law clauses, choice of forum clauses, and *exclusive jurisdiction clauses* indicate which legal system has jurisdiction over contracts. If contracts do not have any of these clauses but they contain an arbitration clause, then arbitrators have the right to choose which laws will apply to a contract, and these laws might be a combination of laws from different legal systems (Stokes, 1978). The legal system selected does not have to be from the host nation or the native country of any of the parties to the contract.

Legal restrictions could be placed on contracts due to the choice of laws that governs such items as job-site safety standards, environmental protection, the importing and exporting of materials, taxes and duties, and the transfer of ownership of property. Local legal systems could restrict the choice of law to a place that has a connection to a contract, such as the host nation or the legal systems of the native countries of owners or contractors (UNCITRAL, 1988). Choice of law clauses cause confusion when contracts specify different choices of law for each legal issue that could arise during performance of a contract (Rubino-Sammartano, 2001).

The parties writing the contract should make sure that the legal system used in the country where the project will be built will uphold an *exclusive jurisdiction* clause.

7.4.14.1 Calvo Clauses

Calvo clauses, which are named after Carlos Calvo who was a nineteenth-century Argentinean diplomat, are exclusive jurisdiction clauses that specify that the courts of a host nation will be used to settle disputes (Bondzi-Simpson, 1990). Calvo clauses came into practice in 1927, and they were developed by the United States–Mexican General Claims Commission to force plaintiffs to use local jurisdictions before they file a claim in foreign courts (Stokes, 1978).

7.4.15 Legal Issues in International Joint Ventures

If an *international joint venture* (IJV) is going to be formed, both parties should understand the financial and legal status of the other firms that are a party to a joint venture, and minutes from meetings of the board of directors of IJV partners provide insight into how companies are managed and whether there is any pending litigation against them.

International joint venture agreements should include a warranty clause that waives the liability of partners if there is any pending litigation or omission of facts that might cause litigation. International joint venture bylaws may be used in common-law countries but not in civil law countries. How the IJV will be managed should be put in writing along with who has decision making power and who has veto power (Wolf, 2000)

When forming an IJV, potential partners should verify that the contracts their potential partners have with suppliers and customers will be valid and remain in force after formation of an IJV. Licenses held by IJVs could be terminated when a company is sold to prevent the transfer of proprietary data or information to competitors (Wolf, 2000).

7.5 Claims and Change Orders

Legal issues throughout the world are complicated because courts interpret contract clauses differently, and the governments of the home nations of employees and owners might become involved in contractual disputes along with the governments of nations where firms are incorporated.

7.5.1 Adjustment and Revision Index Clauses

An *adjustment of price* occurs when a contractor is paid a different amount from what is stated in the contract when the scope of work is changed and a *revision of price* is used to compensate contractors for changes that arise due to fluctuating exchange rates and inflation. A *revision index clause* changes the amount of payments to the contractor based on a specific index. If an index clause is included in a contract, an algebraic formula may be included for revising payments based on a price index (UNCITRAL, 1988).

7.5.2 The Defense of Contractors by Home Nations in Contractual Disputes

The legal theories of *denial of justice* and *expropriation* are involved when the government of one party to a contract acts on behalf of that party because the nation is damaged if the company is damaged in a lawsuit. Under denial of justice, governments try to make sure that their citizens receive due process of law, which requires access to the legal system and a fair trial, when they are involved in a lawsuit in another country (Stokes, 1978). For a denial of justice claim to be valid, someone has to demonstrate that the judicial process is being delayed or that there is prejudice or hostility involved in the legal proceedings that makes it impossible for a foreign national to receive a fair hearing. Foreign nationals have to seek help from their native country prior to using a host-country legal system (Stokes, 1978). Usually foreign ministers or state-department officials negotiate a settlement instead of taking cases to an international court. In the United States, the Hickenlooper amendment allows the U.S. government to suspend aid to a foreign country that seizes property that is owned by U.S. citizens, but this law is used primarily as a negotiating threat (Stokes, 1978).

Expropriation occurs when a host nation seizes the property of a foreigner, such as construction equipment, without proper compensation. Foreign governments do not even have to seize or transfer the title of property for a case of expropriation to arise; they only have to not allow an owner to use his or her property (Stokes, 1978).

7.5.3 Typical Causes of Claims and Change Orders

In construction processes, time is a valuable commodity and owners may require projects to be completed within a stipulated time frame, or by a specified date. This is generally provided for by a contract clause indicating "time is of the essence" where the contractor is obligated to complete the work by the specified date. Where there is no time-of-the-essence clause, courts require completion within a reasonable period of time. By virtue of the need for timely completion, should one party be delayed or believe that they have been delayed by the other party, they may proceed against that other party in order to recover damages. Such damages, which do reflect financial losses, serve as a catalyst for the high incidence of construction-delay claims (Yates and Epstein, 2006).

Construction claims in the global E&C industry can be submitted for the same reasons that they are submitted in domestic-court systems. Examples of problems that arise on international contracts that lead to claims being filed for extra payments include:

- Local regulations for testing structures could be changed after the contract has been executed, and if this occurs a contractor may submit a claim for any additional costs associated with the new testing requirements for the project.
- Contractors cannot finish projects in developing countries on schedule because they do not have enough laborers, equipment, technically competent employees, or supervisors or because of delays by subcontractors (Scott, 1993).
- Owners could file a claim if projects are delayed because contractors have not imported enough employees or did not arrange subcontracts in a timely manner.
- Guarantees could be voided if contracts do not specify how change orders affect guarantees (UNCITRAL, 1988).
- Owners do not provide prompt payment, which allows contractors to submit a claim for interest on late payments.
- Interest rates for late payments are specified in contracts and applicable local laws also stipulate interest rates, and the two interest rates are in conflict (some legal systems do not allow interest to be paid on late payments, which is the case in some Islamic countries).

7.5.4 CPM Scheduling

The critical path method (CPM) scheduling specifications for a project provide an opportunity to significantly reduce the incidence of delay claims. The process from submission/approval to regular updating through preparation of as-builts should be meticulously detailed and should require the involvement of not just the contractor and the contractor's consultant, but of the subcontractors, suppliers, owner, architect, and engineer. The key to the process should be total commitment and cooperation of all parties even though it may be time consuming and somewhat costly.

The CPM submission should require that the contractor, key subcontractors, and key suppliers have input in preparing the detailed CPM arrow diagram and related narrative reports. Each activity on the diagram should include the following at a minimum (Yates and Epstein, 2006):

- Description
- Trade
- Location
- Duration
- Budget
- Manpower (crew size)

- Equipment
- Special resources

These categories provide proper information and data for the accurate monitoring of each and every activity on the project. The specifications should also provide for a minimum and maximum number of activities on the overall project as well as minimum and maximum durations for each activity. Once this information is submitted to the owner, it should be reviewed in careful detail. The approved schedule will be used for the implementation of the work as well as to establish a baseline for the contractor's as-planned schedule.

The CPM specification should expressly require sign-offs by the owner, architect, engineer, suppliers of owner-furnished equipment, etc. Meetings should be held, as necessary, to finalize acceptance of the schedule. All parties should be required to sign off and approve the schedule. The owner and owner's representatives should review all aspects of the schedule, including the activity durations, budgets, etc., since these may become component factors in loss of productivity claims.

The CPM schedule should provide for updating during the construction phase. It should be understood that the original schedule is a plan or tool to be used by the participants, and not a rigid document; it should be changed and updated as job conditions warrant. The specifications should provide for, or require, that various actions be taken as part of the updating process.

7.5.5 Use of Productivity Factors

Loss of productivity on construction projects is frequently difficult to define, therefore, a specification should be provided to address this issue. The contractor should be required to provide data regarding their actual productivity on the project as it proceeds and these records should be submitted to the owner for verification. Owners should specify, if applicable, manuals of productivity that will be utilized to determine objective standards for productivity. The figures in these manuals can be adjusted to allow for the particular contractor's productivity on this, or on other projects, where variances are documented. The key is to provide a standard of productivity by which to compare actual productivity. The failure of the contractor to comply with these requirements could be tied into a witholding of progress payments (Yates and Epstein, 2006).

7.5.6 Documentation of Delays

Another important opportunity to reduce or mitigate the possibility of construction-delay claims is to properly address problems and to thoroughly document job progress as it occurs. Specifications should provide for certain required actions by the owner and the contractor in the area of project documentation including the following (Yates and Epstein, 2006, p. 22):

- Project Correspondence—Owners should be required by the specifications to respond to all questions raised by a contractor promptly or within a specific time frame. Failure to do so should weigh heavily in computing compensable excusable or time extensions.
- Diaries—Should be kept by all parties as a requirement of the specifications.
- Daily Logs—Specifications should require an official daily log prepared by the contractor and submitted to the owner for approval. The owner should then review it and approve or require modification.
- Progress Photographs and Videotapes—Specifications should require dual input and approval by owner and contractor of photographs, videotapes, descriptions, and narrative reports.
- Payroll, Cost, Manpower, and Equipment Records—These should be mandated by specifications. Failure to submit such documentation should be tied into a withholding of a portion of progress payments.
- Minutes of Meetings—Specifications should require periodic meetings and stipulate who shall attend (contractor, subcontractor, A/E, owner, etc.).
- Progress Reports—Should be required as part of the CPM updating procedure. These reports should be linked to CPM schedule changes.

7.5.7 Notice Provisions

Contract clauses should require proper notice relative to delays and constructive changes; the time frames and content of all notice requirements should be highlighted by the contractor during bid preparation to assure compliance. Clauses should require acknowledgement by the owner in a timely fashion of receipt of such notification. Specifications should also include time requirements for an owner to respond to change orders, time extensions notices, and requests so that a project will not be unnecessarily delayed while such matters are reviewed by the owner.

7.5.8 Exculpatory Clause—"No Damages for Delay"

The "no damages for delay" clause has been utilized for a considerable amount of time, yet the state of the law remains somewhat unsettled at this time and tends to vary from jurisdiction to jurisdiction. In an effort to relieve owners from "liability for delay" damages on construction projects, many contracts now contain a "no damages for delay" clause. This is an exculpatory clause that purports to excuse an owner from contractural liability for damages due to delays caused by the owner.

7.6 Dispute-Resolution Techniques

In the global arena, various methods are used to resolve disputes, including negotiation, mediation, conciliation, arbitration, litigation, and dispute-review boards.

7.6.1 Negotiation

Negotiation is a common dispute-settlement technique in the global E&C industry because negotiation is fast, inexpensive, and helps to maintain business relationships. In countries where nonadversarial relationships are the accepted way of doing business, negotiation is the preferred method for settling disputes (some Asian countries).

7.6.2 Mediation and Conciliation

Mediation and *conciliation* are similar in that they do not result in a binding decision. In mediation, someone acts as an intermediary to facilitate communication between parties in order to bring about a successful negotiated settlement. In conciliation, someone provides a recommended settlement after hearing arguments from each side. Then the parties have the option of either accepting or not accepting the proposed settlement (Redfern and Hunter, 1986).

7.6.3 Arbitration

Arbitration provides an international forum in which to settle disputes and the International Arbitration section of this chapter (see 7.7) provides more information on international arbitration.

7.6.4 Litigation

Litigation takes place when one party sues another party in a court of law within a specific jurisdiction. Since there is no international court of law, the court of a specific jurisdiction is chosen by the plaintiff, and that is where the case will go to court (Redfern and Hunter, 1986). In order to avoid a situation where more than one court hears a specific case, contracts should specify an exclusive jurisdiction.

7.6.5 Dispute-Review Boards

Another, less frequently used alternate dispute-resolution (ADR) technique is *dispute review boards* (DRBs). Dispute-review boards have been used in Europe for the past 25–30 years, and they are gaining acceptance in the United States in the 2000s. In order to prevent disputes from escalating to the point where they may be settled by arbitration or litigation, DRBs could be used on construction projects. Dispute-review boards are a panel of independent arbitrators who are chosen by contractors and owners, and the board members are hired at the beginning of a project to review the contract and make recommendations for settling disputes and claims on a regular basis. Decisions of the DRB are not binding, and contractors and owners still may use other ADR methods to settle disputes and claims (Vorster, 1993).

In addition to analyzing contract documents before they render a decision, DRB members should conduct site visits. On some construction projects, DRB members have offices at job sites, and they conduct regularly scheduled weekly meetings to settle disputes before work progresses any further. Presentations are given by contractors, the owner's representatives, or the DRB members before the board makes its decisions.

In a case study that was conducted while a project was using DRB to settle disputes, it was noted that whoever provided the most polished, well-documented presentation usually had DRB decisions made in their favor. Recommendations that were derived from the case-study project for improving DRB techniques include the following (Duran and Yates, 2000, p. 36):

- Special meetings between the contracting parties and the construction manager are necessary whenever there is sufficient evidence about potential disputes.
- If a project-management oversight contractor is used, he or she should audit the effectiveness of the DRB and tabulate records of the final costs of settled disputes compared with original requested costs. Final costs should be compared with costs arrived at through negotiations by both sides rather than the cost of the potential litigation or of unsettled disputes.
- The DRB must be careful about allowing subcontractors to participate in hearings. The three-party agreement is with the owner, constructor, and board members, not with subcontractors. The master construction agreement is executed between owners and contractors, not subcontractors.

7.6.6 Cardinal Changes to Contract

If contracts are changed more than a certain percentage (usually 10 percent), it is considered to be a *cardinal change,* and it may void a contract if a cardinal change clause is included in a contract. Before doing work in a country, it is important to know the extent to which an owner is allowed to change the scope of work without voiding the contract because contractors could end up performing an unreasonable amount of extra work if there is no limit to the number of changes allowed during the execution of a contract.

7.7 International Arbitration

During international arbitration, some legal systems allow parties to obtain interim relief, or arbitrators may change or add requirements to contracts. Arbitration awards need to be enforceable, but the enforceability of awards is affected by government restrictions.

There are several reasons why arbitration is the preferred dispute-resolution technique rather than litigation including the following:

- It is less formal than court hearings, and it does not require attorneys.
- Parties are allowed to select arbitrators who are experts in the fields of architecture, engineering, or construction.
- Arbitrators normally respect the contract choice of law whereas court systems do not always honor the contract choice of law.
- Contracts can specify the location of arbitration proceedings and the language of the proceedings.
- Arbitration proceedings are less disruptive than litigation, and the results of the proceedings are confidential. If legal proceedings are used to settle a dispute, there is no confidentiality.
- Arbitration proceedings may be conducted at job sites.
- Settlements can be reached in a few weeks, not the months or years required in legal systems.

Judicial challenges to arbitration clauses and to arbitration awards are allowed in some legal systems. Arbitration awards could be overturned if (UNCITRAL, 1988):

- The arbitrators were incompetent;
- A party was unable to thoroughly explain his or her case;
- The arbitration tribunal was not conducted in accordance with the contractual requirements;
- The arbitration award did not follow public policy.

7.7.1 Agreement to Arbitrate and Arbitration Clauses

In order for arbitration to take place, all parties to the agreement must agree to arbitrate either by including an agreement to arbitrate in the original contract or by an agreement that is created when a claim cannot be settled through negotiations. Some legal systems make it a requirement that all parties must agree to arbitration when a dispute cannot be settled through negotiation (UNCITRAL, 1988). In these cases, parties should provide a written agreement to arbitrate at the beginning of the arbitration hearing, even if they have already agreed to arbitrate by having signed a contract with an arbitration clause.

Contracts usually specify an odd number of arbitrators, such as three because an odd number of arbitrators guarantees that there will not be split decisions and having three or more arbitrators also provides arbitrators with diverse backgrounds. If arbitration awards are not paid it could result in breach of contract claims (UNCITRAL, 1988).

The contract or an agreement to arbitrate should specify the location of the arbitration proceedings, because local jurisdictions regulate arbitration tribunals within their jurisdiction. Arbitration proceedings should be located in a jurisdiction where awards will be enforced, such as in a jurisdiction where one of the parties or their assets are located, or in a jurisdiction within a nation that has an international convention with one or more of the nations involved, or in a jurisdiction with arbitration regulations that are well suited for international cases. Other issues that should be considered in selecting the location for arbitration proceedings are whether a location is convenient for the parties involved in the dispute, the availability of facilities, and the availability of support from members of international arbitration institutions.

If arbitration proceedings are in the home jurisdiction of a defendant, then awards are easier to enforce than if the proceedings are in the home jurisdiction of a plaintiff. Specifying that more than one language be used in arbitration hearings is useful, but paying for translators increases the cost of the proceedings.

7.7.2 Authority of Arbitral Tribunal to Order Interim Measures

Some jurisdictions allow arbitrators to order interim measures before issuing an award, and in jurisdictions where interim measures cannot be set by arbitrators, the court system may order interim measures even if the case is going be settled by arbitration (UNCITRAL, 1988). However, if a party sues in a court of law to secure an interim measure, it could void that party's right to use arbitration proceedings (Knutson, 2005).

7.7.3 New York Convention

The *Convention on the Recognition and Enforcement of Foreign Arbitral Awards* is commonly known as the *New York Convention*. The New York Convention was written in 1958 but not used widely until the 1970s. The New York Convention requires defendants to prove that an award is invalid, as opposed to the 1927 Geneva Convention, which requires plaintiffs to prove that an award is valid. The New York Convention only applies to the enforcement of an award and not to the hearing stage of a dispute. Arbitrators have to follow the procedures outlined in arbitration agreements, and if they are not followed, then awards could be invalidated based on the New York Convention (Chukwumerije, 1994).

7.7.4 Ad Hoc versus Institutional Arbitration

Under *ad hoc* arbitration, contractual parties write their own rules for the arbitration process. *Institutional arbitration* occurs when parties agree to follow the rules of a particular arbitration institution, and neither party can change the rules. An advantage of institutional arbitration is that the institution used will provide staff to conduct arbitration hearings (Redfern and Hunter, 1986). Some examples of arbitration institutions include the American Arbitration Association (AAA), the Inter-American Commission of Commercial Arbitration (IACAC), the International Centre for the Settlement of Investment Disputes (ICSID), the International Chamber of Commerce (ICC), the International Arbitration Association (IAA), the

London Court of International Arbitration (LCIA), and the Stockholm Chamber of Commerce (SCC).

7.8 Anticorruption Legislation

In the United States, the federal Foreign Corrupt Practices Act of 1977 prohibits *questionable and illegal payments* to foreign officials by U.S. citizens, but the act does not apply to foreign nationals who receive bribes (*National Law Journal,* July 12, 2004). The U.S. Sarbanes-Oxley Act of 2002 complements the U.S. Foreign Corrupt Practices Act, as it pertains to bribes and accounting fraud used to hide bribes.

The Convention on Combating Bribery of Foreign Public Officials of the Organization of Economic Development (OECD) is a 35-nation treaty that is designed to reduce international bribery. Under this convention, Australia, Japan, Mexico, New Zealand, South Korea, and many of the European countries have stopped allowing firms to write off the bribery of foreign officials on their taxes. The United Nations Convention against Corruption is a 35-nation treaty that is also being used to reduce international bribery.

The American Society of Civil Engineers International Activities Committee, Task Committee on Global Principles for Professional Conduct (with 137,000 members, 15,000 of which are outside of the United States), in conjunction with the International Federation of Consulting Engineers, the Pan American Federation of Consultants' Institution of Engineers, the Japan Society of Civil Engineers, the National Institute for Engineering Ethics (United States), the United Kingdom Institution of Civil Engineers, and the World Federation of Engineering Organizations, along with 66 other engineering societies, have all been addressing issues of bribery, fraud, and corruption with worldwide agreements of cooperation related to "mutually acceptable principles and guidelines for the procurement of services and execution of work worldwide—zero tolerance for bribery, fraud, and corruption" (*American Society of Civil Engineers Magazine,* December 2004, p. 1).

7.9 Kidnapping and Ransom Insurance

Kidnapping and ransom (K&R) *insurance* provides benefits to companies if any of their employees are kidnapped and held for ransom. There are 15,000 reported kidnappings per year, 70 percent of which are resolved through the payment of ransom, and only 10 percent result in a rescue without ransom payments (*Financial Times,* July 29, 2003).

Executives of E&C firms do not mention that their firms have K&R insurance because that might result in their employees being targeted for kidnapping. Public disclosure of a K&R insurance policy typically voids the insurance policy (*Insurance Day,* October 22, 2003). Governments of host nations are reluctant to pay ransom because it encourages more kidnappings. Kidnappers can be prosecuted on charges of money laundering in addition to kidnapping in some nations such as the United States and Japan (*Japan Economic Newswire,* September 30, 2004).

7.10 Changing Governments

Changing government regimes in a violent or dramatic manner presents particular risks when millions of dollars worth of heavy construction equipment is being used on projects. For example, the violent Iranian regime change in 1979 resulted in 6000 claims, worth an aggregate total of more than $75 million (U.S. dollars) (Amin, 1983). The international claims tribunal had difficulty processing arbitration claims from contractors because their contracts included exclusive jurisdiction clauses stating that disputes would be settled in Iranian courts (Amin, 1983). The *United Nations Compensation Commission* was established after the Iraq invasion of Kuwait to examine claims and verify their validity (Rubino-Sammartano, 2001).

Force majeure (acts of god or beyond anyone's control) *clauses* can be used by contractors during periods of political instability. Another possible defense for contractors is the legal *theory of impossibility* which, results in clauses that state that no one can perform the work due to its destruction or that no one has the ability to perform the work in its present state (Murphy, 2005).

7.11 Liability Issues

In the global arena, there are different types of risks, and this section discusses them along with providing suggestions for how such risks might be mitigated on global E&C projects.

7.11.1 Sources of Risk

Risks are created by technical, financial, social, and political uncertainties (Williams, 1992). Decision risk is highest during contract negotiations, and as construction projects progress, the risks associated with decisions declines (Moran, 2001; Wearne, 1992).

The laws applicable to the *transfer of ownership of property,* including real estate, may be specified by local jurisdictions where the property is located at the time of the transfer. In most legal systems, building materials become a part of the project at the time the materials are installed, as it would be difficult to move them and reinstall them in the same manner and with the same quality. However, legal systems vary as to the specific time that equipment becomes part of a structure.

Some legal systems require that equipment be permanently installed, yet other legal systems only require temporary installation for the transfer of ownership to occur. If an applicable law is not clear as to when transfer of ownership takes place, then the contract should state when it occurs. For example, if international commercial terms (INCOTERMS) are specified in the contract, then they will govern the transfer of risks associated with ownership (UNCITRAL, 1988).

7.11.2 Risk of Loss

Legal systems either assign risk of loss to the party in possession of the property or with ownership. In legal systems that assign risk with ownership, contractual parties are allowed to specify their own terms regarding transfer of risk with ownership. The risks associated with loss of or damage to the tools and equipment of a contractor normally are borne by contractors, and this risk normally does not extend to owners (UNCITRAL, 1988).

7.11.3 Taxation within a Host Country

To avoid having contractors inflate their overhead to avoid *taxes in a host country,* some countries require a gross-receipts tax on international contracts because it is difficult to audit international construction overhead. Requiring joint ventures with local companies helps to track income taxes of expatriate firms. Host-country taxes may be less if money is not transferred out of a host nation. Contractors can reduce their taxes by becoming incorporated in a nation that has a tax-reducing treaty with the host nation (Stokes, 1978).

7.11.4 Insurance

Lending institutions might require that contractors obtain advice about reducing risks from their insurance companies before they start overseas projects. If multiple insurance policies are used on a project, all of the policies should be evaluated to determine whether there are any gaps in coverage for some risks and double coverage for other risks. Host-nation government officials could mandate that foreign contractors be insured by an insurance firm that is located within a host country and that all insurance claims be paid in local currencies (UNCITRAL, 1988).

7.11.4.1 Professional Liability and Structural-Defect Insurance
Designers could be required to show evidence of having insurance against errors and omissions when they perform work in foreign nations. Contractors could be held responsible for defects for 10 years after project completion, and *professional liability and structural-defect insurance* covers this risk (UNCITRAL, 1988).

7.11.4.2 Political-Risk Insurance
Political-risk insurance is available through private companies, such as Lloyd's of London, and through some governments such as the U.S. government (Kangari

and Lucas, 1997). The *U.S. Overseas Private Investment Corporation* (OPIC) assists with feasibility studies and loans, and it also provides political risk insurance that protects firms if local currencies are inconvertible, if projects are expropriated, or if they sustain any losses when there are wars or revolutions (Stokes, 1978).

7.11.4.3 Workers' Compensation Insurance

Companies could be required by governments to provide insurance to compensate employees who are injured at work. Governments that provide workers' compensation insurance pay either a lump sum or a monthly stipend to injured workers or the families of deceased workers. Owners could require workers' compensation insurance to protect them from being sued if an employee is injured on a job site. Expatriates should check to see if they are covered by their workers' compensation insurance if they are working overseas.

7.11.5 Contractor's Guarantee and Performance Bonds

Government officials could mandate that contractors have to provide *standby letters of credit* instead of performance bonds. If an owner provides documents proving that a contractor failed to perform, such as an arbitration award to the bank holding the standby letter of credit, the bank will issue funds up to a predetermined limit to owners so that they can perform corrective actions (UNCITRAL, 1988).

7.11.6 Risk Associated with Currency Variations

When governments devalue their currencies, it create problems for owners and contractors; therefore, it is important to monitor currency fluctuations and the events that cause them. Daily television news broadcasts, such as the British Broadcasting Company (BBC) News, the Deutsche Weller News (German news broadcast in English), and weekly financial newspapers are sources that provide information on world events and economic news.

If contractors are paid in their native currency, then owners are assuming the risk of exchange rate fluctuations, and if contractors are paid in the currency of a host nation, then contractors are assuming exchange rate risk. If contractors pay their subcontractors in native currencies, then contractors are assuming exchange rate risk. Both owners and contractors assume exchange rate risk if contractors are paid in the currency of a different nation. One option for avoiding these types of risks is for owners to pay contractors with the currency of the financing institution supporting a project.

In lump sum and unit price payment contracts, contractors can reduce their risks if they are paid in the same currency that they use to pay their subcontractors and suppliers; therefore, contractors might request that owners pay them in multiple currencies. When contractors are paid in a cost-reimbursable system, it is beneficial to them if they are paid in the same currency that they use to pay their suppliers and other costs. Contracts should not specify two currencies and then allow one party to choose which one will be used without a currency adjustment, because the party that is allowed to choose might be *unjustly enriched* if they obtain more money than they are entitled to from the other party.

It might be difficult for contractors to export their profits back to their home nation because of host-country laws limiting the international transfer of funds. One alternative is to exports goods out of the host nation instead of money, but this is risky if there is no market for the goods being exported or prices decline for the products being exported back to a home country (Stokes, 1978).

7.11.7 Local Labor Laws and Risks Associated with Using the Local Work Force

National and local labor laws control the hiring and firing of members of a local labor force, and these laws could restrict the firing of employees. In Germany, employers are required to pay 2 years worth of wages to each worker who is laid off. One method for reducing this risk is to use local companies that provide temporary workers who remain employees of the local firm and not the contractor. Another local labor law might be a mandate to use a certain amount of local labor, even if the local labor is unskilled labor (Stokes, 1978).

Government officials or owners might require foreign companies to train local labor, but if the owner is the government, contractors should try to negotiate for compensation to cover some of their training costs.

7.12 Summary

Legal issues and claims in the global arena may be a decisive factor in whether projects realize a profit; therefore, having a basic understanding of different legal systems is essential when operating in a global environment. Engineering and construction personnel should be familiar with whether the legal system they will be operating under in a foreign country will be similar to or different from the legal system in their native country (either common law, civil law, or another type of legal system). International contracts are similar to domestic contracts, but they include additional clauses that address international issues.

This chapter presented information on the types of legal and contractual issues that arise during the execution of international contracts. It explained differences between civil law and common law and how countries evolved into using these two forms of law. If a person knows which legal system is being used in a country, it helps him or her try to settle contractual disputes.

In this chapter, international contract clauses were discussed in terms of the clauses that are different in international contracts versus domestic contracts, and a table is provided that includes standard international contract clauses along with a brief explanation of what they mean to someone involved in an international contract. A section on claims and changes to international contracts was included that also provided examples of situations that could cause claims to arise on international contracts such as labor disputes or payment delays. Information was also provided on the involvement of home nations in contractual disputes in cases of expropriation or denial of justice.

International arbitration, agreements to arbitrate, and arbitration clauses were reviewed, along with the authority of arbitration tribunals to order interim measures in arbitration cases. Anticorruption laws were only discussed briefly because they vary from country to country. A discussion on kidnapping and ransom insurance was included in this chapter, but a more detailed presentation on kidnapping is provided in Chapter 13. Global liability issues, sources of risk, and insurance were discussed in this chapter, along with a discussion on changing government regimes and how changing governments affect construction projects.

DISCUSSION QUESTIONS

1. Explain the difference between civil and common law legal systems.
2. Explain why it is easier to prove that an arbitration clause is valid in civil law legal systems rather than common law legal systems.
3. Explain why citizens of the United Kingdom can sue enemy aliens, but enemy aliens cannot sue a citizen of the United Kingdom in a court of the United Kingdom.
4. Explain why Chinese citizens prefer to use conciliation to settle contract disputes rather than litigation.
5. Explain the legal benefits of forming a joint venture with a Saudi Arabian engineering or construction company before doing business in Saudi Arabia.
6. Which five clauses from Table 7.1 always should be included in international contracts, and why are the five clauses selected more important to include than the other clauses in Table 7.1?
7. How could confidentiality agreements be enforced on international contracts?
8. When E&C personnel are working in foreign countries, which legal system governs their actions and why?
9. Explain how a hardship clause benefits owners and contractors if it is included in international contracts.
10. Explain how exemption clauses are similar to impossibility clauses.
11. Explain the conditions under which owners and contractors can terminate contracts if a termination clause is included in contracts.
12. Explain choice of law clauses.
13. Explain the difference between an adjustment of price and a revision of price.
14. Explain the difference between denial of justice and expropriation.
15. What would be a reasonable percentage that a contract could be changed by an owner without it being a cardinal change and why?
16. Why is international arbitration a better way to settle international contract disputes than litigation?
17. How might the organizations listed in the section on anticorruption legislation be able to reduce corruption on E&C projects?
18. Why was it so difficult for contractors to settle their contractual disputes in Iran after the Iranian revolution in 1979?
19. Explain political risk and how it affects E&C contracts.

20. Discuss which currency should be used to pay a foreign contractor if he or she is performing work in a host country using subcontractors that are from a country that is not the host nation and why.

REFERENCES

Al-Jarallah, M. I. 1983. Construction industry in Saudi Arabia. *ASCE Journal of Construction Engineering and Management* 109 (4): 355–368.

American Society of Civil Engineers. 2004. ASCE members work to reduce corruption worldwide. *American Society of Civil Engineers Magazine.* Reston, Virginia. December 2004.

Amin, S. W. 1983. Iran-United States claims settlement. *The International and Comparative Law Quarterly* 32 (3): 750–756.

Bondzi-Simpson, P. E. 1990. *Legal Relationships between Transnational Corporations and Host States.* Westport, Conn.: Greenwood Publishing Group Inc.: Quorum Books.

Chukwumerije, O. 1994. *Choice of Law in International Commercial Arbitration.* Westport, Conn.: Quorum Books.

Duran, J., and J. K. Yates. 2000. Dispute Review Boards—One View. *Cost Engineering Journal* 92 (1): 31–37.

Hill, J. 1998. *International commericial disputes.* 2nd edition. London: LLP Limited.

Insurance Day. 2003. Kidnap risk on agenda for multinationals. Section: New Analysis. London, United Kingdom: Informa Publishing Group, Ltd. October 22, 2003.

Japan Economic Newswire. 2004. "Suspected Kidnapper of Japanese Charged with Money Laundering." Manila, Philippines: Kyodo News Service, International News Section.

Kangari, R., and C. Lucas. 1997. *Managing International Operations.* New York: American Society of Civil Engineering Press.

Katz, A. N. 1986. *Legal Traditions and Systems: An International Handbook.* Westport, Conn.: Greenwood Press.

Kelleher, E. and Yee, A. 2003. "Economic Growth Boosts the Demand for Kidnap Insurance: Services Are for More than Colombians and Oil Companies," *The Financial,* p. 28.

Knutson, R. 2005. *FIDIC: An Analysis of International Construction Contracts.* Amsterdam, Netherlands: Hague-Kluwer Law International.

Lantis, E. D. 2005. Telephone interview. May 9.

Martindale-Hubbell. 2000. *Martindale-Hubbell International Law Digest.* Chicago: R. R. Donnelley and Sons.

Moran. J. 2001. *International Political Risk Management: Exploring New Frontiers.* Washington, D.C.: World Bank Group Multilateral Investment Guarantee Agency (MIGA).

Murphy, O. J. 2005. *International Project Management.* Mason, Ohio: Thomson Higher Education.

National Law Journal. 2004. Oil company bribery suit settles: Government lays out blueprint for successor liability in resolving case. *National Law Journal,* July 12, p. 5.

O'Hare, C. W. 1980. Cargo dispute resolution and the Hamburg rules. *International and Comparative Law Quarterly* 29 (parts 2 and 3): 219–237.

Redfern, A., and M. Hunter. 1986. *Law and Practice of International Commercial Arbitration.* London: Sweet and Maxwell.

Rubino-Sammartano, M. 2001. *International Arbitration: Law and Practice.* Amsterdam, Netherlands: Hague-Kluwer Law International.

Scott, S. 1993. Dealing with delay claims: A survey. *International Journal of Project Management* 11 (3): 143–153.

Stokes, M. 1978. *International Construction Contracts.* New York: McGraw-Hill.

UNCITRAL. 1988. *Legal Guide on Drawing Up International Contracts for the Construction of Industrial Works.* Vienna, Austria: United Nations Commission on International Trade Law.

Vorster, M. C. 1993. *Dispute Prevention and Resolution in Construction with Emphasis on Dispute Review Boards.* Austin, Tex.: Construction Industry Institute.

Wearne, S. H. 1992. Contract administration and project risks. *International Journal of Project Management* 10 (1): 39–41.

Williams, T. M. November 1992. Risk management infrastructures. *International Journal of Project Management* 10 (4): 5–10.

Wolf, R. C. 2000. *Effective International Joint Venture Management.* New York: M. E. Sharpe.

Yates, J. K., and A. Epstein. 2006. Avoiding and minimizing construction delay claims in relational contracting. *Journal of Professional Issues in Engineering Education and Practice* 132 (2): 1–12.

Chapter 8

International Engineering and Construction Standards

8.1 Introduction

Standards are an important aid in the process of homogenizing the world, while incorporating its variations and international standards are a major component of global engineering and construction. This chapter discusses standards, international standards, and the International Organization for Standardization (ISO). Definitions are provided for standards, including technical standards, nongovernment standards, government standards, international standards, consensus standards, quality standards, and environmental standards. The development process for standards is explained, along with a discussion on who participates in the development of international standards and how they are used in the engineering and construction (E&C) industry.

8.2 Definition of Standards

Standards, as cited in codes, handbooks, or manuals, are methods and procedures that are part of the technical specifications used in the E&C arena, and they set minimum requirements, exact measurements, processes, and other requirements that need to be satisfied to be in conformance with a contract. A *standard* is something that is accepted as a basis for comparison, and it is a procedure or product used as a reference to determine the quality of similar procedures and products. It is established by determining the technical and nontechnical specifications of the procedure or product (McKechnie, 2005). The main basis for the technical requirements in building codes is the standards that are referenced and incorporated into contracts. Standards are used to evaluate qualities, characteristics, or properties of materials, and they are defined as "having the quality or qualities of a model, gauge, pattern, or type; generally recognized as excellent and authoritative; generally used and regarded as proper for use" (McKechnie, 1972). In some countries standards may be required by law because they are the common language used for communication between designers, constructors, vendors, fabricators, and regulatory agencies. Standard components, materials, and products are less costly; they provide savings in the cost of engineering materials and in time; and they foster improvements in quality.

8.3 Technical Standards

The basic intention of *technical standards* is to achieve a recognizable and acceptable level of quality in an industry. Standards help to set common goals pertaining to quality that members of an industry strive to achieve. Enhancing communication to achieve quality in design and construction is one of the main purposes of standards. Factors of cost, materials of construction, and material specifications are the main elements of the requirements for standards, which, in turn, establish an acceptable level of quality.

According to the U.S. Department of Energy (DOE) Technical Standards Program, a *technical standard* is a "document that establishes uniform engineering and technical requirements for processes, procedures, practices, and methods. Standards may also establish requirements for selection, application, and design criteria of facilities systems, components, and other items" (U.S. DOE, 1993, p. 1). Standards document proven methods and techniques to allow a better understanding between those who develop the standards and those who use them (DOE, 1993). Standards differ from specifications in that they are a long-term application, whereas specifications are used for unique or one-time applications.

8.4 Consensus Standards

A consensus approach is used when standards are developed by an organization. *Consensus* means "substantial agreement by concerned interests according to the judgment of a duly appointed authority, after a concerted attempt at resolving objections. Consensus implies much more than necessarily unanimity" (Gross, 1989, p. 33).

Consensus standards in the United States, as well as industry standards, are developed by a voluntary system that includes representatives from the industry, government, producers, consumers, institutions, and individuals. The system is referred to as *voluntary* because participants are not paid, and the standards they produce are used voluntarily in building regulations and contracts (Gross, 1989). The procedures used to develop standards in other countries are quite different from the methods used in the United States (Gross, 1989). In countries other than the United States, there is greater governmental involvement, and the processes are less rigorous than those used in the United States for the establishment of consensus. In many standards organizations, there is limited opportunity for participation, particularly by outsiders, no matter how affected an outsider might be by the development of standards.

8.5 Government Standards

In addition to standards that are developed by private organizations, governments also develop and adopt standards. Government standards account for 55 percent of U.S. standards. The largest developer of standards in the U.S. government is the Department of Defense, with 50,000 standards. Federal specifications and standards published by the U.S. General Services Administration (GSA) are used by federal and state agencies to procure common products, and federal test methods are used in many industries. Over 10,000 standards have been developed by U.S. federal agencies, such as the Occupational Safety and Health Administration (OSHA), the Environmental Protection Agency (EPA), the Postal Service, and others.

For many government standards, it is difficult to distinguish whether they are truly standards or simply regulations (Breitenberg, 1989). Although some federal standards are voluntary, many become mandatory when they are referenced in legislation or regulations or when they are invoked in contracts as a condition of sale to government agencies. Nongovernment standards may be mandated in this manner.

8.6 Nongovernment Standards

According to the U.S. Department of Energy (DOE), a *nongovernment standard* (NGS) is "a standardization document (also known as voluntary, consensus, and industry standards) developed by a private sector association, organization, industry association, or technical society which plans, develops, establishes, or coordinates standards and related documents" (U.S. DOE, 1993, p. 2). Standards are put into practice by being referenced in DOE orders, contractual documents, specifications, drawings, and work-authorization statements. Nongovernment standards are maintained by the technical society or trade organization that was involved in their development.

8.7 International Standards

In the twentieth century, the United States developed many of the standards that were used throughout the world. In the United States, currently over 100,000 standards are actively maintained by different government agencies and private organizations. Countries in favor of international standards are now discarding some U.S. standards, which until a few years ago were used internationally. Currently, the United States has limited representation in the international community for standardization. Other countries are now adopting the International Organization for Standardization (ISO) standards, which puts pressure on the United

States to become more involved with the ISO standards. Clients throughout the world are stipulating international standards in their contractual arrangements with contractors, and some of the countries in the European Union and the Pacific Rim region are contributing millions of dollars to the development of international standards.

International standards are used to influence or regulate the types of goods and services that flow between nations. One of the main purposes of international standards is to work toward harmonizing the technical regulations of individual nations. Organizations such as the American Society of Mechanical Engineers (ASME), the American Society of Testing and Materials (ASTM), the American National Standards Institute (ANSI), the British Standards Institution (BSI), the Canadian Standards Association (CSA), and the Deutsche Institute produce standards that are used in many parts of the world, but they are all national organizations. Many of the standards from these organizations are incorporated into international standards through the ISO. The technical work produced by the ISO is published as international standards, and the ISO standards cover all areas except those that are related to electrical and electronics engineering, which are covered by the International Electrotechnical Commission (ISO, 2005).

The ISO has committees that develop guidelines, definitions, and documentation related to its standards, and there are three main ISO publications (ISO, 2005):

1. *ISO Guide 30: Terms and Definitions Used in Connection with Reference Materials*
2. *ISO Guide 31: The Contents or Certificates of Reference Materials*
3. *The ISO Directory of Certified Reference Materials* (which is an English or French guide to sources of certified reference materials throughout the world, with materials classified according to domain)

International standards related to construction materials and building account for 7 percent, or about 1000, of the more than 15,000 ISO standards; but many of the other ISO standards also have a direct or indirect influence on the E&C industry, including testing, mechanical systems, fluid systems, electrical engineering, railway engineering, paint, materials handling equipment, chemical technology, petroleum, metallurgy, wood technology, environment, and civil engineering. Twelve percent of the standards used internationally pertain to engineering and construction.

Despite the existence of the ISO and other such organizations, there is no international regulatory agency that oversees the implementation of and compliance with international standards. An engineering and construction client bidding internationally may or may not demand compliance with an international standard. Furthermore, registration for compliance to the ISO quality management and environmental management standards is performed by independent registrars and not internationally designated organizations or countries.

In the United States, in addition to the federal government, there are hundreds of organizations that develop and prepare standards. This is in contrast to countries that have one centralized agency responsible for preparing national standards. In the United States, of the over 100,000 standards that are maintained by the standardization community, 45 percent were developed by the private sector. Private organizations that develop standards are categorized into three areas:

1. Scientific and professional
2. Trade associations
3. Standards developing organizations

8.8 International Organization for Standardization (ISO)

The ISO was formed in 1947 to promote the development of standardization, to facilitate the international exchange of goods and services, and to foster cooperation between intellectual, scientific, technological, and economic activities (ISO, 2005). The deficiencies caused by each country in the world using different standards are being recognized internationally, and an immense effort to harmonize existing standards and to develop new ones has been undertaken by the ISO. As long as ISO standards exist, the European Union is committed to using them. Therefore, some of the standards prepared by the European Committee for Standardization (CEN) are being put forth to the ISO for adoption as world standards. The U.S. government (or U.S. private firms) has no input into the European Committee for Standardization.

There are currently 91 countries with committees involved in the ISO, most of which are funded by their governments, and Table 8.1 provides a list of these countries. In the ISO, a one country, one vote rule prevails, but block voting is not uncommon.

There are many stages or processes that occur before a standard becomes an international standard. The processes for developing standards vary from country to country and from one standard developing organization to another. Once an organization has developed a standard, it can submit it to the ISO for review and possible inclusion as an international standard. A professional society or association cannot submit its standards directly to the ISO for consideration. An official representative to the ISO must submit standards from organizations within his or her country. The United States has only one official representative to the ISO, which is the American National Standards Institute (ANSI).

The ISO does have numerous committees and technical subcommittees that allow participation by individuals from different business sectors. Members of organizations can either hold an observer (O) status or be a participating (P) member on a task committee (TC). In order for members of organizations to register as participating members on an ISO technical committee, the official representative of their country has to first designate an *accredited standards developer* as the tech-

Table 8.1 ISO Member Bodies

Algeria	Hungary	Philippines
Argentina	Iceland	Poland
Armenia	India	Portugal
Australia	Indonesia	Qatar
Austria	Iran, Islamic	Romania
Azerbaijan	Republic of	Russian Federation
Bahrain	Iraq	Saint Lucia
Bangladesh	Ireland	Saudi Arabia
Barbados	Israel	Serbia
Belarus	Italy	Singapore
Belgium	Jamaica	Slovakia
Bosnia and Herzegovina	Japan	Slovenia
Botswana	Jordan	South Africa
Brazil	Kazakhstan	Spain
Bulgaria	Kenya	Sri Lanka
Canada	Korea, Democratic	Sudan
Chile	People's Republic	Sweden
China	Korea, Republic of	Switzerland
Colombia	Kuwait	Syrian Arab Republic
Congo, The Democratic	Lebanon	Tanzania, United
Republic of	Libyan Arab	Republic of
Costa Rica	Jamahiriya	Thailand
Croatia	Luxembourg	The former Yugoslav
Cuba	Malaysia	Republic of Macedonia
Cyprus	Malta	Trinidad & Tobago
Czech Republic	Mauritius	Tunisia
Côte-d'Ivoire	Mexico	Turkey
Denmark	Mongolia	USA
Ecuador	Morocco	Ukraine
Egypt	Netherlands	United Arab Emirates
Ethiopia	New Zealand	United Kingdom
Fiji	Nigeria	Uruguay
Finland	Norway	Uzbekistan
France	Oman	Venezuela
Germany	Pakistan	Viet Nam
Ghana	Panama	Zimbabwe
Greece		

nical advisory group (TAG) member. The United States has over 200 technical advisory groups.

When an organization is registered as a participating member of a TAG, it is required to vote on draft international standards (DIS) and to participate in meetings. Societies and associations can comment on standards through their official representatives if they are registered as participating members on ISO technical committees.

The ISO does not fund participation on its committees or subcommittees; therefore, committee participants must be self-funded or sponsored by firms, a consortium of firms, a professional society, or a trade organization. The extensive cost of funding a representative excludes many countries from participation in the ISO.

8.8.1 ISO Liaisons

Intergovernmental organizations help to implement ISO standards when they use them in intergovernmental agreements. These organizations are closely associated with the work of the ISO by the following four mechanisms (ISO, 1991, p. iii):

- International organizations are allowed to submit proposals for the preparation of ISO standards in a new field in the same way as ISO member bodies.
- International organizations may be granted *liaison status* with technical committees and subcommittees. Liaison status has two categories: A (effective contribution to the work) and B (provided with information only). Liaison A status confers the right to submit papers, attend meetings, and participate in discussions.
- International organizations are invited by the ISO to comment on relevant drafts of standards.
- Technical committees are instructed to seek full and, if possible, formal backing of the main international organization liaison for each ISO standard in which these organizations are interested.

There are over 500 international organizations that are liaisons to the ISO technical committees and subcommittees, which results in thousands of individual liaisons.

8.9 The Development Process for Standards

There are 10 main stages that standards developed by an organization have to go through before being adopted as an ISO standard. Figure 8.1 shows the ten stages for the development of an ISO standard. Before a standard is submitted to the ISO for consideration, it has to first be created by an organization, society, or association. The following paragraphs outline the steps required to have a standard reviewed and accepted by the ISO.

Standards are created through participation by individuals on standard committees within organizations and professional societies. International standards usually begin when there is a *new work item proposal* (NP) for technical work in an area where there is no existing technical committee. If the proposed work is closely related to the work of an existing technical committee, it is assigned to the existing committee by the Committee of Action of the International Engineering Council (IEC) or the Technical Board of the ISO. The IEC central office or the

10) Audit passed

9) Audit performed

8) Submit manual for approval

7) Select and meet with assessor

6) Compile quality manual

5) Define and implement new procedures

4) Establish a program

3) Identify what needs to be done

2) Review procedures for the standard selected

1) Evaluate your organization's needs for a better quality management system

Figure 8.1 Ten steps to develop ISO standards.

ISO central secretariat conducts a survey of the national member bodies to determine interest in the proposed standard.

If a two-thirds majority approves of forming a technical committee, and there are at least five member bodies willing to actively participate, the ISO creates a new committee. The committee consists of interested individuals or individuals sponsored by their firms. Committees meet only when it is necessary to discuss committee drafts (CDs) or if there are other issues that cannot be handled through correspondence. The committees operate under the consensus procedures mentioned earlier.

The standards developed by committees should be evaluated for use based on their *technical* and *quality performance* (Termaat, 1994, p. 5). To achieve technical performance, a standard should be written to the latest level of technology, and a standard should not be a device to protect less efficient organizations or construction methods or to ensure business for a particular firm, standards developer, or professional assessor.

Quality performance standards set forth minimum levels of quality that must be achieved when products are produced or processes are performed to create a structure. Quality standards can include issues such as environmental, health, and safety responsibilities.

8.9.1 Issue Raised to ISO Council

As shown in Figure 8.1, once the idea for a standard has been fully developed, it is forwarded through an appropriate representative to the ISO. For a standard to warrant consideration, there must not be an existing ISO standard that addressses the same item, and the standard has to have technical merit.

8.9.2 Formation of Technical Committees, Subcommittees, and Working Groups

Members of ISO technical committees, subcommittees, and working groups are nominated by the ISO from the ranks of industry, government, labor, and individual consumers. These members work together through the different stages of the development of a standard until the standard becomes a *draft international standard* (DIS).

8.9.3 Technical Committee Drafts

Committee members meet several times over the course of a year to review an item until it is incorporated into a *committee draft* (CD). Through consensus procedures, the technical committee members agree on the committee draft before it becomes a DIS.

8.9.4 Approval of International Standards

Once a committee draft is approved by the appropriate technical committee, it is submitted to the ISO as a DIS. In order to become a DIS, two-thirds of a committee or subcommittee have to approve it. In addition to the voting procedures, ISO has a process for technical challenges. Even with a 75 percent favorable vote, a technical challenge will stop approval of a standard until the deficiencies or inaccuracies have been cleared up.

Not all of the countries who are ISO members can vote on DIS. Only 41 countries are voting members of ISO, with the remaining countries having only observer status. Countries with observer status can provide input into standards development through technical committees, subcommittees, and working groups.

8.10 International Technical Standards

The ISO has 40 categories of *international technical standards,* and these categories are listed in Table 8.2. The 40 categories contain numerous subcategories, and Table 8.3 provides a sample of the subcategories under the major category of civil engineering standards to demonstrate the types of items that are included within categories.

Under each of the subheadings, there are specific standards, and the specific standards may be purchased directly from the ISO at the ISO Web site at www.iso.org. The standards may be located by investigating the appropriate categories and finding a desired standard or by searching through the ISO catalogue of standards at www.iso.org/iso/en/prods%2Bservices/catalogue/CatalogueListPage.CatalogueList. The telephone number for the ISO is +41 22 749 01 11 (Geneva, Switzerland), and the fax number is +41 22 733 34 30.

Table 8.2 ISO Categories of Technical Standards (ICS Field Numbers)

ICS Number and Category	ICS Number and Category
01 Generalities, Terminology, Standardization, Documentation	45 Railway Engineering
03 Services, Company Organization, Management, Quality, Administration	47 Shipbuilding and Marine Structures
07 Mathematics, Natural Sciences	49 Aircraft and Space Vehicle Engineering
11 Health Care Technology	53 Materials Handling Equipment
13 Environment, Health Protection, Safety	55 Packaging and Distribution of Goods
17 Metrology and Measurement, Physical Phenomena	59 Textile and Leather Technology
19 Testing	61 Clothing Industry
21 Mechanical Systems	65 Agriculture
23 Fluid Systems and Components	67 Food Technology
25 Manufacturing Engineering	71 Chemical Technology
27 Energy and Heat Transfer Engineering	73 Mining and Minerals
29 Electrical Engineering	75 Petroleum and Related Technologies
31 Electronics	77 Metallurgy
33 Telecommunications, Audio and Video Equipment	79 Wood Technology
35 Information Technology, Office Machines	81 Glass and Ceramics Industry
37 Image Technology	83 Rubber and Plastics Industry
39 Precision Mechanics	95 Paper Technology
43 Road Vehicles Engineering	91 Construction Materials and Building
	93 Civil Engineering
	95 Military Engineering
	97 Domestic and Commercial Equipment, Entertainment, and Sports

In the United States, the ISO standards may be purchased through national standards organizations such as: American National Standards Institute (ANSI) (phone: 212-642-4980; e-mail: ansioline@ansi.org); the American Society for Testing and Materials (ASTM) (phone: 610-832-9585; e-mail: service@local.astm.org); and Global Engineering Documents (phone: 800-624-3974; e-mail: globalcustomerservice@ihs.com).

Table 8.3 ISO Technical Standards for Civil Engineering—93 (ICS Numbers)

ICS Number and Category	ICS Numbers and Category
93.010 Civil Engineering in General	93.080 Road Engineering
93.020 Earthwork, Excavation, Foundation, Construction,	93.100 Construction of Railways
93.025 External Water Conveyance Systems	93.110 Construction of Ropeways
93.030 External Sewage Systems	93.120 Construction of Airports
93.040 Bridge Construction	93.140 Construction of Waterways, Ports, and Dykes
93.060 Tunnel Construction	93.160 Hydraulic Equipment

8.11 ISO 9000 Quality Management System Standards

The *ISO 9000 series of standards* are specific quality management standards, in contrast to most other ISO standards, which are either performance or product standards. "The ISO 9000 standards do not apply to specific products. They are generic system standards that enable a company, through a mix of internal and external audits, to provide assurance that it has a quality system in place that will enable it to meet its published quality standard" (Breitenberg, 1989, p. 24).

The ISO system does not apply directly to purchasing, assessment of subcontractors, inspection, or testing of purchased components. Companies do have to have procedures for verification and maintenance of components or procedures for tracking damaged or lost components (Saunders, 1992).

The ISO 9000 series of standards were recognized from the beginning as being significantly different from what might be called *normal* engineering technical standards, such as those for units of measurements, terminology, test methods, and product specifications. The ISO 9000 concept is that certain generic characteristics of management practice could be usefully standardized, providing a mutual benefit to both producers and users.

The ISO Technical Committee (TC) 176, which deals with quality systems, produced five international standards in 1987. These five international standards are known as ISO 9000, 9001, 9002, 9003, and 9004. In addition to these standards, ISO has a standard numbered 8402 that contains terminology and definitions that help with the selection of an appropriate quality management program. The ISO standards are reviewed, reaffirmed, or revised at least every 5 years.

The following provides a brief description of the ISO 9000 standard series (ISO, 2005):

ISO 9000 (ANSI/ASQC Q 90): Quality Management and Quality Assurance Guidelines for Selection and Use. ISO 9000 provides an overview of the 9000 series of standards, and it explains fundamental quality concepts, defines key terms, and provides guidance on selecting, using, and (if necessary) tailoring ISO 9001, 9002, and 9003 for use. It also discusses which standard is appropriate for a company.

ISO 9001 (ANSI/ASQC Q 91): Quality Systems: Model for Quality Assurance in Design/Development, Production, Installation and Servicing. The ISO 9001 covers all elements listed in ISO 9002 and 9003. In addition, it addresses design, development, and servicing capabilities. It is typically used by companies involved in the entire product cycle.

ISO 9002 (ANSI/ASQC A 92): Quality Systems: Model for Quality Assurance in Production and Installation. ISO 9002 addresses the prevention, detection, and correction of problems during production and installation. It generally applies to industries where the work is based on designs and specifications supplied by the customer.

ISO 9003 (ANSI/ASQC Q 93): Quality Systems: Model for Quality Assurance in Final Inspection and Test. ISO 9003 addresses requirements for the detection and control of problems during final inspection and testing. It generally applies to companies with relatively simple production services that may be adequately assessed by inspection.

ISO 9004 (ANSI/ASQC Q 94): Quality Management and Quality System Elements: Guidelines. ISO 9004 provides guidance for a supplier to use in developing and implementing a quality system and in determining the extent to which each quality system element is applicable. The ISO 9004 examines each of the quality system elements (cross-referenced in the other ISO 9000 standards) in greater detail and can be used for internal and external auditing purposes.

The five ISO quality management standards were developed from the goals of the ISO 9000 series, which are to foster (Gross, 1989, p. 66):

- *Universal acceptance.* Standards should be widely accepted and adopted worldwide.
- *Compatibility.* All standards should work well with the existing documents and supplements, both now and in the future.
- *Flexibility.* Any supplements can be combined as required to meet virtually any industry and product need.

8.11.1 ISO 9000 Compliance Requirements

Registration and compliance with the ISO 9000 series of standards is voluntary unless clients stipulate compliance by including the ISO standards in contracts or by citing them as a part of the approval process for products.

8.12 ISO 14000 Environmental Management Series of Standards

The ISO has environmental management system standards that are called the *ISO 14000 series of standards,* and they include 26 different standards related to environmental management systems. The ISO 14000 series of standards are standards related to implementing an environmental management system, they are not environmental standards. The ISO 14000 series addresses issues related to environmental performance evaluations, environmental labeling, and life cycle assessments (LCAs).

The ISO 14000 series of standards are used to address the environmental aspects of the processes of an organization along with its products and services and to provide methods for controlling and improving environmental performance. A certification process for ISO 14000 is available that is similar to the ISO 9000 certification process. Additional information and a complete listing of the 26 ISO 14000 series of standards are available at the ISO Web site at www.iso.org or www.iso14000.org.

The ISO also has a standard that provides guidance on social responsibility, which was developed by the ISO Technical Committee 207, the committee responsible for environmental management. The ISO also has over 350 international standards that relate to the monitoring of the quality of air, water, soil, noise, and radiation, and the ISO environmental standards have been used in some countries as the technical basis for environmental regulations (ISO, 2005).

8.13 ISO 4217 Global Currency Codes and Names

The ISO Standard 4217 provides a list of every currency in the world, along with an alphabetical and a numerical code for each currency. Samples of currency codes are listed in Table 8.4 (ISO, 4217, 2005).

8.14 The ISO 9000 and 14000 Registration Process

It is important to note that the ISO 9000 series of standards does not register individual products, as it is only registering the *quality system.* Having the quality system registered does not mean that the product conforms to a set of requirements. If suppliers are interested in obtaining *product certification,* it may or may not be available from the registrar they have selected to certify their quality system (Breitenberg, 1989).

Quality-system registration, which is sometimes inaccurately referred to as *certification,* requires that a firm periodically submit to an assessment and audit related to the adequacy of its quality system. The audits are performed by a third party called a *quality system registrar.* If the system being audited conforms to the ISO 9000 standards, the registrar issues a certificate of registration. Unfortunately, the ISO 9000 standards may be interpreted differently from one registrar to another.

The registration process for ISO 9000 may be performed by *quality-standards-accredited registrars* (QSARs). If a company is registered by a qualified registration organization in the ISO/EU system, the registration it receives should be considered valid by customers regardless of where the registration organization is located or where the supplier or the customer is located (Eicher, 1993).

Another ISO standard, the ISO 10011, provides guidelines on how quality-systems auditing is to be performed, and it defines the parties concerned with an

Table 8.4 Samples of ISO Currency Names and Codes

Entity	Currency	Alphabetic and Numeric
Afghanistan	Afghani	AFN-971
Argentina	Argentine peso	ARC-032
Austria	Euro	EUR-978
Bangladesh	Taka	BDT-050
Bosnia and Herzegovina	Convertible marks	BAM-977
Brazil	Brazilian real	BRL-986
Canada	Canadian dollar	CAD-124
Chile	Chilean peso	CLP-152
China	Yuan renminbi	CNY-157
Columbia	Columbian peso	COP-170
Congo	CFA franc BEAC	XAF-950
Croatia	Croatian kuna	HRK-191
Egypt	Egyptian pound	EGP-818
Ethiopia	Ethiopian birr	ETB-230
Ghana	Cedi	GHC-288
Hong Kong	Hong Kong dollar	HKD-344
India	Indian rupee	INR-356
Iraq	Iraqi dinar	IQD-368
Japan	Yen	JPY-392
South Korea	Won	KRW-410
Malaysia	Malaysian ringgit	MYR-458
New Zealand	New Zealand dollar	NZD-554
Pakistan	Pakistan rupee	PKR-586
Peru	Nuevo sol	PEN-604
Russian Federation	Russian rubble	RUB-643
Serbia and Montenegro	Serbian dinar	CSD-891
Turkey	New Turkish lira	TRY-949
United Kingdom	Pound sterling	GBP-826

audit. It describes the roles and responsibilities of everyone involved, it shows the logical phases of the audit, and it deals with reporting the audit results and follow-up corrective action. Parts 2 and 3 of this three part standard describe the qualifications quality systems auditors should have and the management of audit programs.

Countries in the EU are required to accept the conformity assessment results if they are performed by designated *notified bodies* (NBs), which are organizations that have received official sanction. The EU has *mutual recognition agreements* (MRAs) that recognize auditors from countries outside the EU. For nonregulated products in the EU, buyers or procurement authorities may require registration to show compliance with an ISO standard. This requirement is driven by the marketplace rather than a regulatory requirement.

A firm may evaluate its own quality system through self-audits. Other options are to have second-party evaluations, which could be conducted by someone such as the buyer or an evaluation done by third-party evaluators and registrars.

8.15 Advantages and Disadvantages of International Standards

There are many valid reasons for implementing international standards for both engineering design and quality processes. According to the document, *ISO Benefits of Standardization,* the main criteria for developing and implementing international standards include (ISO, 1982, p. 21):

- Improvements in universal technical communication and mutual understanding
- Facilitation of the international exchange of goods
- Removal of technical barriers to trade
- The transfer of technology.

Standards for the design of products reduce variety, which limits the range of choices, but the tradeoff in benefits is in increased interchangeability and worldwide availability.

Quality standards, or *basic standards,* are hard to assess because the benefits are widespread in society, and their relationship to particular products is indirect. The types of gains that are inherent with the use of quality standards are *time gains* and *transfer gains,* and they increase in proportion to the increased implementation of standards (ISO, 2002). The tradeoff to achieve higher quality is usually balanced by increases in production costs; therefore, these two variables have to be analyzed to determine the optimal quality level that can be achieved by using standards.

In order for each firm to determine whether the benefits of international standardization are of a magnitude that would justify their implementation, the following typical items should be assessed to determine if savings will be realized by standardization (ISO, 1982, p. 38):

- Larger manufacturing and purchasing quantities
- Reduced ordering costs, storage space, and handling requirements
- Reduced safety storage (backup stock) in view of shorter delivery times
- Reduced tied-up capital in stores and in production
- Use of standard stock items in lieu of new items
- Retrieval times in design and engineering
- Reference to standards in lieu of detailed specifications of drawings

There are several main advantages to using ISO standards, including:

- Protection of worldwide market share.
- Competitive advantage over other companies that are not fluent with international standards.
- Access to facilities in which to do business worldwide.
- Participation in the international standards development process allows companies to help shape these standards

However, even if a company is not seeking ISO certification, there are still many advantages that can be achieved by implementing the ISO 9000 standards. Overall, the ISO 9000 guidelines are part of total quality management (TQM). Another advantage is that the independent audit by the registrar can be trusted by customers. Other advantages can result from "meeting, or exceeding, the expectations of customers with a high-quality product, including reducing scrap, rework, and overhead costs; examining and improving materials flow and production planning; reducing inspection of manufactured goods; utilizing materials well; and improving turnaround time" (Eckstein and Balakrishnan, 1993, p. 68).

The benefits of being registered to the ISO 9000 series of standards are the following (Lofgren, 1991, p. 37):

- Reduction in the number and scope of rapidly increasing second-party audits by customers
- Use of the assessment process as a tool for improving suppliers' operations
- Use of registration as a marketing tool to demonstrate supplier commitment to quality
- Access to markets that require quality system registration

8.15.1 Disadvantages of ISO 9000 Standards

Major disadvantages to using ISO 9000 standards include the following:

- There will be additional internal costs to revise standards and to modify work processes and procedures.
- If significant work process changes are necessary, this may temporarily disrupt business and affect employee morale.
- The addition of external costs for audits and registration.

Many other disadvantages result from misunderstandings about the ISO 9000 series of standards. It needs to be understood that ISO 9000 standards do not guarantee quality, nor do they ensure that a given plan will satisfy customer requirements. Once ISO 9000 registration is achieved, it means that a company is following its predetermined quality standards. But what needs to be realized is that the standards set by the company may not be superior standards. For example, "ISO says that you must set your own requirements, and we will see if you're following them. If management says that 60 percent on-time shipments is fine, . . . that's okay with ISO people" (Grund and Mullin, 1993, p. 51). There is also the problem of variability within the required standards. "While ISO specifies companies meet or exceed stated product specifications, it does not prevent variability within the specifications of products: a given product specification could change

and still meet ISO requirements as long as it meets minimum stated standards"
(Grund and Mullin, 1993, p. 32).

8.16 Summary

Engineers and constructors need to understand the importance of the standards-development process because it influences how and what they may design and build in the global marketplace, and they should become involved in the development of standards. The implementation of international standards helps to increase the quality of products, and costs may be reduced if designs are easier to interpret. There are better processes, and there are broader markets for products. Technical barriers to trade are reduced when international standards are specified, and universal technical communication and mutual understanding are also improved through the use of international standards.

The ISO 9000 series of standards are the most prevalent international registration processes used by the construction industry, and having ISO certification increases the competitiveness of E&C firms in the global arena. Unfortunately, there is a great deal of misunderstanding about what ISO 9000 certification means in the E&C industry. Certification only indicates that an E&C firm follows the quality program it has set forth in its company manuals, not that a firm has a superior quality program.

This chapter discussed international standards, the ISO, international government standards, nongovernment standards, ISO standards, international technical standards, consensus standards, quality management standards, environmental-management standards, currency codes and name standards, and the development process for standards. The ISO 9000 registration process was explained, and the advantages and disadvantages to ISO registration were discussed in relation to E&C firms.

DISCUSSION QUESTIONS

1. Explain the difference between a technical standard and a quality standard.
2. Explain how international technical standards are implemented in the E&C industry.
3. Explain the difference between government and nongovernment standards.
4. Discuss who develops international standards and how they are developed by this organization.
5. Explain how international standards are enforced globally.
6. Explain how individuals can provide input into the development of international standards.
7. What are ISO liaisons, and how do they participate in the development of international standards?
8. Using Tables 8.2 and 8.3, discuss which categories of international standards would apply to the building of a 12-story building.

9. Explain what the ISO 14000 series of standards are and how they would be used in the global E&C industry.

10. Discuss where engineers and constructors would use the ISO Standard 4217 in contracts, plans, and specifications.

REFERENCES

Breitenberg, M. 1989. *Questions and Answers on Quality, the ISO 9000 Standard Series, Quality System Registration, and Related Issues.* Washington, D.C.: U.S. Department of Commerce.

Eckstein, A. L., and J. Balakrishnan. 1993. The ISO 9000 series: quality management for the global economy. *Production and Inventory Management Journal* 33 (7): 66–71.

Eicher, A. L. 1993. ISO 9000: Origins, evolution, and market relevance. *ISO Forum Application Symposium Proceedings* 3 (10): 32–37.

Gross, J. G. 1989. International harmonization of standards: done with or without us. *The Building Office and Code Administrator* September–October: 46–47.

Grund, H., and R. Mullin. 1993. Can registration stand in for supplier auditing? *Chemical Week* November 10: 51–52.

International Organization for Standardization (ISO). 1982. *ISO Benefits of Standardization.* Geneva, Switzerland: International Standards Organization Council Committee on Standardization Principles.

International Organization for Standardization (ISO). 1991. Introduction. *ISO Liaisons.* Geneva, Switzerland: International Organization for Standardization.

International Organization for Standardization (ISO). 2005. *The International Organization for Standardization Catalog.* Geneva, Switzerland: International Organization for Standardization.

Lofgren, G. 1991. Quality systems registrars. *Quality Progress* May: 35–37.

McKechnie, J. L., ed. 2005. *Webster's New Universal Unabridged Dictionary.* New York: Simon and Shuster.

Saunders, M. 1992. ISO 9000 and marketing in Europe: should U.S. manufacturers be concerned? *Business America,* April 20: 24–25.

Termaat, K. 1994. Introducing SSM into both corporate and industry culture. In *Proceedings of the American National Standards Institute,* 139–145. Washington, D.C.: American National Standards Institute.

U.S. Department of Energy (DOE). 1993. Public inquiries PA-5. Washington, D.C.: DOE.

U.S. Department of Defense. 2006. Source. Washington, D.C.: DOD. http://www.dod.org.

Chapter 9

Global Environmental Issues of Concern to Engineers and Constructors

9.1 Introduction

In achieving success, both industrially and economically, members of different nations need to be cognizant of the effects of their detrimental actions and strive to preserve the environment for future generations through sustainable development and the elimination of nonecofriendly facilities. Figure 9.1 is an example of how the environment is negatively impacted by construction operations. This figure is a picture of a strip-mining operation that has removed the top of a mountain to extract copper. A similar type of operation is being used in Canada to extract oil from tar sands. While governments of nations are improving the standard of living for their citizens, they are also compromising air quality and releasing toxins into the environment from industrial facilities, construction, and manufacturing plants. Figure 9.2 shows a liquified natural gas plant that is flaring off excess gas into the environment. The World Health Organization (WHO) has estimated that air pollution endangers over 1 billion people in urban areas, that over 2 million people die each year from waterborne diseases, and that, by the year 2015, over half of the people in the world will not have safe drinking water (World Health Organization, 2005).

Global environmental restrictions affect the engineering and construction (E&C) industry by regulating the amount of pollution the industry can generate and by restricting the use of hazardous materials. Construction operations should be carried out with materials that are environmentally friendly and, at the same time, structurally safe. Construction materials are normally selected based on their structural integrity, including their strength, stiffness, and durability, but there is growing concern over the amount of energy it takes to produce construction materials and to transport them to construction job sites (Garner, 2000). The materials that require the most energy to produce are paint, glass, metals, and plastic, and even a small reduction in their use would result in measurable improvements to the environment.

This chapter explains the United Nations Framework Convention on Climate Change (UNFCC), known as the *Kyoto Protocol Treaty*, as well as other treaties

Figure 9.1 Copper strip mining operation, Salt Lake City, Utah.

Figure 9.2 Environmental degradation—flaring of liquefied natural gas.

and protocols that the UNFCC has implemented to try and reduce greenhouse gas emissions and hazardous wastes. A discussion is provided on how the Kyoto Protocol affects E&C projects throughout the world and how it is being implemented and monitored globally. The construction industry creates a large percentage of hazardous wastes; therefore, different alternatives that are being used to reduce the generation of hazardous wastes are also discussed, along with specific examples of environmental requirements and compliance issues in countries such as Great Britain, Germany, China, India, South Korea, and Australia.

9.2 The United Nations Framework Convention on Climate Change (UNFCCC)

Environmental concerns related to global climate change led to the formation of the United Nations Framework Convention on Climate Change (UNFCCC) and development of the Kyoto Protocol, which is a treaty that introduced measures for controlling global climate changes caused by *greenhouse gases* (GHGs) in industrialized and developing countries. The Kyoto Protocol established baseline principles and commitments that countries that are a party to the convention have to follow in order to help reduce GHGs.

Table 9.1 provides a list of the processes that emit GHGs. Scientists have indicated that GHGs may be responsible for depleting the *ozone layer* that surrounds the earth. The ozone layer protects the surface of the earth from the damaging ultraviolet light rays that are emitted by the sun and if the ozone layer is compromised, it may cause climate changes throughout the world, such as increasing temperatures and melting of the polar ice caps. Greenhouse gases include the (UNFCCC, 2005a) following:

- Carbon monoxide (CO_2)
- Nitrous oxide (N_2O)
- Hydrofluorocarbons (HFCs)
- Perfluorocarbons (PFCs)
- Sulfur hexafluoride (SF_6)

Table 9.1 Sources of Greenhouse Gases

Fuel combustion	Mineral products	Agricultural soils
Energy industries	Chemical industry	Prescribed burning of
Manufacturing industries	Metal production	savannas
Construction	Other production	Field burning of
Transportation	Solvents	agricultural residues
Fugitive emissions from fuels	Agriculture	Solid waste disposal on land
Solid fuels	Enteric fermentation	Wastewater handling
Oil and natural gas	Manure management	Waste incineration
Industrial processes	Rice cultivation	

9.2.1 The Kyoto Protocol

The Kyoto Protocol is an amendment to the UNFCCC, and it formalizes the intentions of the UNFCCC, which is an international agreement that sets binding targets for the reduction of GHG emissions by industrialized countries by the year 2012. It is available in Arabic, Chinese, English, French, Russian, and Spanish. The goals of the Kyoto Protocol include (UNFCCC, 2005a):

- Changing consumer patterns
- Protecting and promoting human health conditions
- Promoting sustainable humane settlement development
- Protecting the environment, air, water, and ecosystems and combating deforestation and managing wastes

The Kyoto Protocol sets specific targets for reducing GHG emissions that each country that has ratified the protocol has to meet by 2012. The emission targets for each country represent a percentage of GHGs that ranges from 5 to 8 percent below 1990 emission levels (Jeong, 2001). Emission targets vary by country, with targets of 8 percent reduction in emissions set for the European Union, Switzerland, and most central and eastern European states; 6 percent for Canada; 7 percent for the United States; and 6 percent for Hungary, Japan, Poland, New Zealand, Russia, and the Ukraine. Since Norway, Australia, and Iceland produce low levels of GHGs, they may increase their emissions by up to 1 percent in Norway, 8 percent in Australia, and 10 percent in Iceland.

The European Union may balance its emissions targets between countries by allowing countries with low emissions to increase their emissions and reducing emissions in countries that have high levels of GHG emissions. GHG emissions increased in developed countries between 1990 to 2000 by 8.2 percent, except for eastern Europe and the former Soviet Union, which saw an overall drop in GHG emissions during that time period due to the decline of their economies.

The premise behind the target emissions in the Kyoto Protocol is not to restrict growth in economies that are in transition (EITs), which include the former Soviet Union and central and eastern European nations, and developing countries but to limit emissions in developed countries. Economies in transition are also allowed to choose a different base year from 1990 because they may not have measurements for that year. Any country may choose a base year of either 1990 or 1995 for the emissions of hydroflourocarbons (HFCs), perflourocarbons (PFCs), and sulfur hexafluoride (SF_6). *Sustainable development* is part of the goal of the Kyoto Protocol, and it is defined as development that meets the needs of the present without compromising the ability of future generations to meet their own needs.

The Kyoto Protocol became effective in February 2005 when 55 countries throughout the world ratified (approved and formally accepted). By the end of 2005, 84 countries had ratified the Kyoto Protocol, and they are listed in Table 9.2 (UNFCCC, 2005c). The United States and Australia have not ratified the Kyoto Protocol, although the United States produces the highest level of GHGs of any

Table 9.2 Countries That Have Ratified the Kyoto Protocol

Antigua and Barbuda	El Salvador	Maldives	Romania
Argentina	Estonia	Mali	Russian Federation
Austria	Fiji	Malta	St. Vincent/Grd.
Belgium	Finland	Marshall Islands	Seychelles
Bolivia	Germany	Mexico	Slovakia
Brazil	Greece	Micronesia	Slovenia
Bulgaria	Guatemala	Monaco	Solomon Islands
Burundi	Honduras	Netherlands	Spain
Canada	Indonesia	New Zealand	Sweden
Chile	Ireland	Nicaragua	Switzerland
China	Israel	Norway	Thailand
Cook Islands	Italy	Panama	Trinidad and Tobago
Costa Rica	Kazakhstan	New Guinea	Turkmenistan
Croatia	Latvia	Paraguay	Tuvalu
Cuba	Liechtenstein	Peru	Ukraine
Denmark	Lithuania	Philippines	United Kingdom
Ecuador	Luxembourg	Poland	Uruguay
Egypt	Malaysia	South Korea	Uzbekistan

nation in the world, followed by China, Russia, India, Japan, Germany, Brazil, Canada, the United Kingdom, Italy, Korea, Ukraine, France, Mexico, and Australia. Fifteen countries produce 70 percent of the world's GHGs.

Table 9.3 contains a list of the target emission reductions or increases that each country has to meet by 2012. The target emissions are a percentage of 1990 emissions (UNFCCC Web site at www.UNFCCC.org, 2005).

There is a *clean-development mechanism* in the Kyoto Protocol that permits industrialized countries to meet part of their emissions targets through the use of *emissions credits* earned by sponsoring GHG-reducing projects in developing countries (Elliott, 1998). *Joint implementation* allows developed countries to invest in clean technology that reduces GHGs in other developed countries, and both countries are awarded emission credits. Some countries can meet their emission targets by *emissions trading,* whereby countries may sell their emission credits or debt to other countries.

Countries are allowed to counterbalance GHG emissions by removing GHGs from the atmosphere using *carbon sinks,* such as reforestation, which is the process of planting trees that absorb carbon monoxide and other pollutants from the air. Countries may bank their emissions credits for use in other years or sell them to other countries. Emissions credits in Europe may be sold on the European Climate Exchange (ECX) and in the U.S. on the Chicago Climate Exchange. If a city or country is able to reduce their carbon dioxide emissions by more than the targeted percentage, they may sell their emissions credits on the exchanges (Lavelle, June 5, 2006).

The techniques and methods for estimating GHG emissions by sources and the removal of emissions by sinks have to be approved by the Intergovernmental Panel

Table 9.3 Percentage of 1990 Greenhouse Gas Emission Reductions or Increases Required by 2012

Country	Percentage of 1990 GHG Emissions	Country	Percentage of 1990 GHG Emissions
Australia	108	Liechtenstein	92
Austria	92	Lithuania[a]	92
Belgium	92	Luxembourg	92
Bulgaria[a]	92	Monaco	92
Canada	94	Netherlands	92
Croatia[a]	95	New Zealand	100
Czech Republic[a]	92	Norway	101
Denmark	92	Poland[a]	94
Estonia[a]	92	Portugal	92
European Community	92	Romania[a]	92
Finland	92	Russian Federation*	100
France	92	Slovakia[a]	92
Germany	92	Slovenia[a]	92
Greece	92	Spain	92
Hungary[a]	94	Sweden	92
Iceland	110	Switzerland	92
Ireland	92	Ukraine[a]	100
Italy	92	United Kingdom and	92
Japan	94	Northern Ireland	
Latvia[a]	92	United States of America	93

[a] Economies in transition.

on Climate Change, and the Conference of the Parties (countries that have ratified the Kyoto Protocol) also have to approve any proposed methods. The Conference of the Parties determines the consequences of a country not meeting its emissions targets by the year 2012, and if the Kyoto Protocol is not extended beyond 2012 or there is no other treaty in place by 2012, then countries will be able to continue to produce GHGs at toxic levels.

9.2.2 The Basel Convention

The UNFCCC also developed and implemented the *Basel Convention*, which controls the transboundary movement of hazardous wastes and their disposal. A total of 164 countries have agreed to minimize the generation of hazardous wastes, and 95 countries have banned the export of hazardous waste materials from developed to developing countries, including toxic, poisonous, explosive, flammable, ecotoxic (toxic to the environment), and infectious wastes. Some countries require prior notification before any other country is allowed to export a hazardous waste to their country. Prior approval also has to be obtained from nations where hazardous wastes will be in transit (UNFCCC, 2005b).

9.2.3 The Rio Declaration

The UNFCCC facilitated the development and implementation of the *Rio Decla-ration*. The Rio Declaration requires countries to enact environmental legislation that facilitates the exchange of environmental information, including *environmental impact statements* (EISs) or environmental impact assessments (EIAs), the results of decisions related to the environment, and the results of judicial and adminis-trative proceedings between countries that share natural resources (UNFCCC, 2005a). International EIAs are used during decision-making processes to evaluate the potential physical, biological, cultural, and socioeconomic effects of a project and its alternatives.

9.2.4 The Stockholm Convention

The *Stockholm Convention* is a treaty that was developed by the UNFCCC to reduce global production, use, and release of 12 of the most harmful chemicals, called *persistent organic pollutants* (POPs).

9.2.5 Environmental Compliance

The UNFCCC is developing international compliance methods and penalties for noncompliance for the Kyoto Protocol. In the international arena, the enforcement of international treaties is difficult unless countries submit to voluntary compli-ance. If a treaty is an *international customary law,* this means that it is a law that is defined by the International Court of Justice (a general practice that is accepted as law), and government officials enforce it because they see themselves as legally obligated to do so. Treaties and customs (such as the Kyoto Protocol Treaty) are called *hard laws,* and they may be enforced through economic sanctions set by international legal systems. *Soft laws* are nonbinding laws that are based on inter-national diplomacy and customs, and countries enforce them because they fear retribution by other countries if they do not enforce them.

The *International Network for Environmental Compliance and Enforcement* is a network of government and nongovernmental practitioners from over 100 coun-tries who are trying to raise awareness on complying with environmental standards and regulations. This network of individuals is trying to develop methods for en-forcing standards and trying to increase cooperation between nations to strengthen the capacity, implementation, and enforcement of environmental regulations (Of-fice of Environmental Affairs—Environmental Protection Agency, 2005).

9.3 Effects of the Kyoto Protocol on the Engineering and Construction Industry

The Kyoto Protocol will affect the engineering and construction (E&C) industry because the construction materials that are currently being used in the industry might be eliminated, such as fly ash and pressure-treated lumber, and the treaty requires that heavy construction equipment generate fewer toxic emissions. The Caterpillar Company in the United States has developed heavy construction equipment that use diesel engines, and this improves fuel efficiency by 5 to 10 percent, which reduces toxic emissions. The purchase price for fuel-efficient construction equipment is 2 to 3 percent higher than the cost of conventional heavy construction equipment, but operating costs are less due to the increased efficiency.

When using construction materials, members of E&C companies should consider *life cycle waste-treatment processes,* which are referred to as *environmentally sound management,* which is a process where materials are considered for inclusion in a construction project based on how they are manufactured, stored, transported, treated, used, reused, recycled, recovered, and their final disposal (UNFCCC, 2005b).

Other mechanisms for reducing GHG emissions that relate to engineering designs and construction include sustainable architecture and urbanism that minimizes heating areas, site areas, and surface areas through the use of densified housing that uses a smaller proportion of a building site for housing by having multiple stories and as this creates more open space and it reduces heating requirements. Structures are located close to property lines to create larger green spaces on one side of the structure. Other sustainable urbanism techniques include improving traffic circulation patterns and the restriction of road traffic by increasing the availability of public transit systems to protect the natural environment and rural landscapes (Gauzin-Muller, 2002)

Specific mechanisms to help conserve energy that may be incorporated into engineering designs include the following (*Newsweek,* 2005):

- Efficient water heaters, insulation, and roof systems
- Solar-controlled glazed windows, or smart windows, that become less transparent on sunny days, blocking heat rays (reducing cooling costs by 30%)
- New generation of fluorescent lighting (which use one-third of the power of incandescent bulbs) or semiconductor-based Liquid Energy Displays (LEDs) that are five times more efficient than incandescent lighting
- Programmable thermostats
- Green roofs (lawns, gardens, or trees planted on roofs)
- Microgenerators that recycle waste heat from water heaters to create electricity
- Minipower stations (fueled by weeds and agricultural wastes)
- Tankless water heaters
- White tile roofs that reflect the sun
- External insulation on masonry structures, which keeps structures cooler than internal insulation in hot climates
- Snap together heating, ventilation, and air-conditioning (HVAC) ducts that eliminate 90 percent of leaks
- Industrial air-conditioning units that use desiccant to dehumidify the air, which allows the compressor to operate more efficiently

- Light towers for solar electrical generation
- Meters that automatically turn down appliances during peak use hours
- Fuel cells that produce waste heat that can be captured and used in water heaters
- Electrovoltaic grid–connected solar panels (made by Sharp, Sanyo, and American Evergreen Solar and Sun Power) that are integrated into electrical rooftop systems (flat rooftops)
- Composting toilets (such as those used in India)
- Biomass heating systems (stoves that burn corn husks)
- Insulated panels in the construction of floors or photovoltaic panels (solar panels)
- Wind turbines to supplement solar panels
- Thicker walls to keep temperatures comfortable year round
- Triple-glazed windows that add extra insulation
- Wind-driven insulation systems that feed fresh air into a structure and that extract heat from outgoing air
- Conservators, such as thermal storage floors that face south and that trap light and the warmth of the sun and inject it into heating systems
- Solar panels on roofs that are used to run electric cars

The last five conservation techniques were installed during the construction of an 84-acre development called the *Zero Energy Drive* (BEDZED) in South London, England, and they have substantially reduced energy consumption. Buildings in the United States and Europe consume approximately 40 percent of all energy and produce the same proportion of carbon dioxide emissions as the transportation sector, but buildings consume 10 percent more energy than the transportation sector (*Newsweek,* September 20, 2004, p. E16). In Europe, governments issue grants and tax breaks to firms that incorporate energy-efficient techniques into structures.

A European Union EU directive requires house builders, landlords, and home sellers to have energy-efficiency certificates for their structures that list the energy-efficiency rating of each structure. The government of Austria provides grants for passive houses that expend only minimal energy, and in Sweden fluid-filled pipes use the heat in the earth to heat homes. The EU also uses energy labeling for household appliances.

9.4 Global Environmental Management

The International Organization for Standardization (ISO) publishes 350 international environmental standards. The ISO also has a series of environmental management standards called the *ISO 14000* series that addresses environmental management systems. Information about the ISO 14000 series of standards is provided in Chapter 8.

9.5 Country-Specific Environmental Issues

The minimum standards for energy efficiency were updated in the early 2000s in Austria, France, Japan, New Zealand, and the United Kingdom for roofs and walls to limit heat loss and to set minimum levels of thermal efficiency for furnaces and water heaters. If work is to be performed in these countries, the new guidelines should be researched by contacting the appropriate government agency in each country.

To locate the appropriate governmental agency that deals with environmental requirements and regulations in each country, see the New York University Web site www.law.nyu/edu/library/foreign_intl/intenvironment/html. This web site contains a list of government sponsored environmental agencies throughout the world.

In the United States, the Environmental Protection Agency has an International Affairs Program, which is called the Office of International Affairs (OIA; www.epa.org), that provides international environmental information. In the EU, there is the European Commission Environment Directorate (www.eurunion.org), and in other countries, such as France, Italy, South Korea, Portugal, Chile, Guinea, and eastern European countries, environmental issues are regulated by agencies that are called the ministry of the environment. In other regions of the world, there are a number of different agencies that regulate the environment including (New York University Web site, 2005):

- The Ministry of Environmental Protection (Russia)
- The Department of the Environment and Heritage (Australia)
- The Ministry of the Environment and Forests (India)
- The Environmental Agency (Japan and the United Kingdom)
- The National Commission on Medio Ambiente (Chile)
- The Ministry of Tourism and Environment (the Congo)
- The Ministry of Natural Resources (Kenya)
- The Ministry of State for Environmental Affairs (Egypt)
- The Ministry of Municipal and Rural Affairs and the Environment (Jordan)
- The Meteorology and Environmental Protection Administration (Saudi Arabia)

Construction and demolition wastes are a large percentage of the total hazardous wastes generated in most countries (50 percent in the United States, or 325 million tons in 2003, and 38 percent in Australia). Construction wastes include concrete, tiles, brick, soil, mortar, plaster, insulation, carpets, and paper. Demolition wastes include wood, plastic, metal, wire, concrete, cardboard, brick, insulation, asphalt, tar, paving stones, gravel, ballast, soil, gravel, and rock.

The following subsections discuss techniques that have been implemented in different countries throughout the world that have helped reduce the amount of hazardous waste generated by the construction industry. The following subsections also provide samples of some of the environmental issues that affect different countries.

Since environmental issues are unique to each country, and they vary depending on the level of industrialization in each country, this section focuses on environmental factors within specific countries that affect engineering and construction.

9.5.1 China

This sections describes the types of environmental issues that China is facing due to rapid industrialization.

9.5.1.1 Air Quality

Air quality in the People's Republic of China is poor, because high levels of trisodium phosphate (TSP) are released into the atmosphere from paint and washing detergents. Large quantities of phosphates damage lakes, rivers, and streams (Dwivedi and Jabbra, 1998). According to the Chinese State Environmental Control Network, 85 percent of the major cities in northern China have exceeded acceptable levels of sulfur dioxide (SO_2) by 30 percent. Some of the effects of high concentrations of SO_2 in the air include acid rain and an increased risk of people developing lung cancer.

9.5.1.2 Water Quality

A major problem in northern China is a shortage of safe water. Seventy-eight percent of urban river water and 50 percent of the subsurface water in major cities is polluted by improper disposal of wastes that leach into water systems.

9.5.1.3 Environmental Policies

Laws on environmental protection for the People's Republic of China were enacted in 1989, and they established a legal foundation for environmental management (Solange et al., 2003). Environmental policies in China include requiring environmental impact assessments (EIAs), pollution-discharge fees, a discharge-permit system, an environmental responsibility system, required assessments of urban environmental quality, and central pollution control.

Members of the construction industry are required to follow the environmental policies set by the government, but the enforcement of policies is sporadic in China. For example, in Shanghai, the Division of Development and Construction Administration only has eight officials to supervise 500 to 600 construction projects, whereas one official normally supervises 40 to 70 projects. Engineers and constructors are required to follow certain procedures that are set forth by the government of People's Republic of China to conform to environmental policies.

9.5.2 India

Major environmental problems in India are air pollution, water pollution from industrial and domestic effluent, soil erosion, deforestation, degradation of land due to increases in salinity and alkalinity, soil and water pollution due to the excessive use of pesticides and fertilizers, and improper agricultural practices. Natural resource–extraction activities such as mining and metallurgy, aggregate production, and other manufacturing industries that generate products for the construction industry create hazardous wastes.

9.5.2.1 Government Reforms and Policies

India established a National Committee on Environmental Planning and Coordination (NCEPC) in 1972 and a Department of Environment (DOE) in 1980. India has national standards for the abatement of pollution, and they are implemented by the central and state pollution-control boards. Environmental audits are required to monitor and evaluate effluents and emission control (Dwivedi and Jabbra, 1998).

Penalties for violating the requirements as set forth in the Prevention and Control of Pollution Act for water include jail terms of up to 3 months or fines of up to 5000 rupees or both. The largest fine is about US$111 if the exchange rate is 45 rupees per dollar, as it was in 2005. Under the Prevention and Control of Pollution Act for air, jail terms may be for 3 months, and fines could be up to 10,000 rupees. Violations of the Environmental Protection Act of 1986 include jail terms for up to 5 months or fines up to 100,000 rupees (Dwivedi and Jabbra, 1998).

9.5.3 Germany

In Germany, three environmental laws affect construction: (1) the Waste Disposal Act of 1972, (2) the Waste Avoidance and Waste Management Act of 1986, and (3) the Closed Substance Recycle and Waste Management Act of 1986. These laws have created a situation in which the disposal of waste is not allowed as all waste must be redirected to other locations and used as secondary raw materials. Germany has a 3R principle—reduce, reuse, and recycle.

Some European countries are using labeling systems for construction materials that indicate the amount of energy required to produce the materials, and in Germany they call this process the *blue angels* (Euring and Ashworth, 1993, p. 71). Conventional construction materials, such as concrete, wood, and brick, require low energy levels to produce compared with other construction materials. A steel I-beam that has the same strength as a wood beam requires six times more energy to produce than wood beams.

The construction industry in Germany was producing 63 percent of the total waste in the country during the 1990s, so members of the industry volunteered to reduce waste by half by 2005, and their efforts resulted in the construction industry recovering 70 percent of its waste per year by 2005. Germany also requires "green" roofs, and companies are taxed if they do not install impermeable drainage systems in new structures.

9.5.4 South Korea

In South Korea, construction waste is 49 percent of the total waste generated in the country. Ninety percent of the construction waste generated in South Korea is from concrete, asphalt, and soil, because concrete and asphalt are the primary components of structures. The waste problem in South Korea is being mitigated by a process that involves a sliding scale of rates charged for the disposal of materials. Mixed waste cost US$160 per ton (160,000 Korean won per ton) for disposal, but if the mixed waste is separated by type of material, then the disposal of concrete only costs US$16 per ton.

9.5.5 Great Britain

In Great Britain, many of the structures built prior to 1985 were painted with paint that contains lead, and the water pipes installed before 1985 also contain lead. Lead is used for roof and pipe flashing, leaded lights, paints, and lining pipelines, and it becomes toxic in soft water regions. Almost half the pipes that carry water in Great Britain contain some lead, and the water that moves through pipes there contains one-fifth of the allowable amount of lead. In Scotland, over 50 percent of the households have water that exceeds allowable lead concentrations per liter based on the maximum permitted upper limit set by the European Council Directives (Euring and Ashworth, 1993). Lead solders used for pipe joints and lead-based paints, varnishes, and wood stains are also hazardous to the environment, and lead is harmful to children if it is ingested in paint chips or if it is in paint used on drinking containers.

Four million of Britain's 4.5 million council homes (low income homes) built before 1985 have asbestos in their roofs and walls, as do 80 percent of metropolitan schools and colleges and 77 percent of school service buildings (Forester and Skinner, 1987; Euring and Ashworth, 1993). Vermiculite was used for roofing and boiler insulation prior to 1985 in the United Kingdom, and it also contains asbestos.

Asbestos is a natural substance that is mined and used in the manufacture of insulation boards and other construction materials. As asbestos breaks down or is disturbed by drilling or other means of penetration, it releases a fine dust that is toxic to humans, but the effects of asbestos exposure are not apparent for decades. Asbestos poisoning manifests itself as silicosis, which is a fatal lung disease; therefore, asbestos was banned in Great Britain in 1985 by the Health and Safety Code of Practice and by the Environmental Protection Agency in the United States in the early 1970s. There are other countries throughout the world that still use asbestos in the manufacture of wall board for construction, and construction workers need to be aware that they are being exposed to asbestos dust when they install or demolish wall board (sheetrock).

9.5.6 Miscellaneous Countries

The construction industry in the United States produces over 50 percent of the total hazardous wastes in the country, and prior to the 1970s, hazardous wastes were disposed of in landfills or unmarked sites. The Environmental Protection Agency (EPA) has identified over 500 hazardous waste sites, and the EPA Superfund program is attempting to mitigate the materials in hazardous waste sites. Hazardous wastes should be disposed of by dumping them into protected dump sites that are clearly marked as containing hazardous wastes, but using these sites substantially increases the cost of disposal. The materials in hazardous waste sites have been linked to health issues such as cancer, leukemia, autism, and miscarriages in people who live over former dump sites or close to former dump sites (Meyninger, 1994).

Asphalt and concrete may be recycled in the United States and reused as aggregate for new road construction. Special heavy construction equipment grinds up concrete roadways to form new aggregate. Pressure-treated lumber is being investigated in the United States by the Environmental Protection Agency during the 2000s, because it may be related to health problems.

In the Netherlands, rebates are provided for energy-efficient appliances. In Belgium, Germany, Hungary, and Switzerland, there are "green tariffs," which means that energy generated by renewable sources is purchased by the government at a higher price. Austria, Germany, Belgium, Switzerland, Japan, and Slovenia promote transporting materials by ship or rail to reduce pollutants. Norway and Switzerland charge landfill tariffs if a facility is not sealed to prevent methane gas from escaping from it. Denmark has stabilized its GHG emissions by switching from coal to renewable natural gas. In Japan, energy requirements have been reduced through the redesign of VCRs and DVDs (50 percent reduction), refrigerators (30 percent reduction), and computers (83 percent reduction).

9.6 Summary

To protect the environment, engineers and constructors need to follow international environmental guidelines, regulations, standards, treaties, and laws. This chapter outlined some of the issues of concern to engineers and constructors in the global arena, including the United Nations Framework Convention on Climate Change, the Kyoto Protocol, the Basel Convention, the Rio Declaration, and the Stockholm Convention. In addition, environmental issues that are of concern to the citizens of specific countries were discussed, along with examples of techniques that are being implemented to reduce or minimize hazardous waste by-products from construction and demolition processes and to conserve energy.

DISCUSSION QUESTIONS

1. How does implementation of the Kyoto Protocol affect engineers and constructors?
2. Which of the sources of greenhouse gases that are listed in Table 9.1 are related to construction job sites?
3. What is the goal of the Kyoto Protocol, and how does the UNFCCC plan to reach that goal?

4. Discuss whether allowing countries to trade emission credits helps to reduce greenhouse gasses.
5. How does the Basel Convention affect construction?
6. What additional requirements have to be implemented for countries to be in compliance with the Rio Declaration?
7. How is the Kyoto Protocol being enforced, and who enforces its provisions?
8. Explain why steel continues to be used in the construction industry even though it requires 5 times the energy to produce a beam with the same strength as a wood beam with the same strength.
9. Which of the specific mechanisms for energy conservation for engineering design and construction could be implemented on a commercial office building that will be built in a hot and arid climate.
10. How can global environmental-management issues be incorporated into the processes that are used on construction projects?
11. Why is asbestos dangerous to construction workers and residents of structures that contain it, and why it is not banned everywhere in the world?
12. What is the appropriate agency to contact to investigate environmental issues in (a) Chile, (b) Kenya, (c) Egypt, (d) Iran, (e) the Congo, (f) Russia, (g) Germany?
13. How could construction wastes such as concrete be recycled and used again to meet the 3R's required in Germany?
14. Would requiring jail terms, as India does, for polluting the air and water work in other countries? Explain why or why not.
15. What diseases are linked to hazardous-waste sites?

REFERENCES

Breslau, K. 2006. *New York University website:* www.nyu/edu/library-intl/intenvironmental/html. It can pay to be green, *Newsweek,* May 22.

Dwivedi, D. and A. Jabbra. 1998. *Governmental Response to Environmental Challenges: Global Perspective,* Vol. 6. Geneva, Switzerland: International Organization for Standardization (IOS) Press.

Elliott, L. 1998. *The Global Politics of the Environment.* New York: New York University Press.

Euring, R. C., and A. Ashworth. 1993. *The Construction Industry in Britain.* Cambridge, England: Oxford University Press.

Forester, W. S., and J. H. Skinner. 1987. *International Perspectives on Hazardous Waste Management.* New York: Academic Press.

Gauzin-Miller, D. 2002. *Sustainable Architecture and Urbanism.* Frankfurt, Germany: Kirkhauser.

Garner, R. 2000. *Environmental Politics: Britain, Europe and the Global Environment.* Cambridge. England: Oxford University Press.

Jeong, Ho-won. 2001. *Global Environmental Politics.* New York: Palgrave Publications.

Lavelle, Marianne. 2006. Insurers may cash in on climate change, *U.S. News and World Report,* p. 43.

Meyninger, R. 1994. *The Effects of Toxic Wastes on Humans.* Ph.D. dissertation, Polytechnic University, Brooklyn, New York.

Ministry of the Environment. 2004. *Research about Construction Waste Separation, Recycling and Waste Sources.* Seoul, South Korea: Ministry of the Environment.

Newsweek, New Frontiers, November 21, 2005, pp. 53 and 64; September 20, 2004, pp. E18–20.

Office of Environmental Affairs and Office of Internal Affairs—*U.S. Environmental Protection Agency* (OIA-EPA). 2005. http://www.epa.org.

Solange, E., S. Vinken, and Aoyagi-Usi. 2003. *Culture and Sustainability.* Amsterdam, Netherlands: Dutch University Press.

UNFCCC. 2005a. Essential Background, http://unfcc.int/essential_background/items/2877.php.

UNFCCC. 2005b. Kyoto Mechanisms, http://unfcc.int/kyoto_mechanisms/items/1673.php.

UNFCCC. 2005c. Parties and Observers, http://unfcc.int/parties_and_observers/items/2704.php.

World Health Organization. 2005. http://www.who.org.

Chapter 10

Global Productivity Issues on Construction Projects

10.1 Introduction

When operating in the global engineering and construction (E&C) marketplace, one of the essential components of the bid estimate preparation process is determining labor costs, which requires access to reasonable productivity rates. The cost of construction has been increasing throughout the world in the 2000s because of the rising cost of construction materials, especially steel and timber, and increasing wage rates. The main objectives of construction projects are to complete them on time and under budget, and when productivity improvement techniques are implemented they may help to reduce both time and cost.

The concept of labor productivity improvement has remained highly elusive since labor productivity rates are defined in a variety of different ways, ranging from the value of gross output per worker (referred to as *labor hours* or *work hours*) to careful attempts to measure the physical output of labor taking into account other factors that affect production. External factors that affect the physical output of labor vary from location to location; therefore, the conditions under which labor productivity values are obtained also need to be known when a bid estimate is being prepared by engineers or constructors for a project in a foreign country.

Several work-improvement methods are available for measuring labor productivity and for dividing construction processes into distinct activities. Labor productivity data for each activity are obtained through work-measurement techniques, such as work sampling, time-series techniques, crew balance analysis, and process analysis, all of which were developed to analyze construction processes in developed countries. However, these methods have limited use in developing countries because of external influences beyond the control of project managers. The work-measurement techniques for developing countries should be adapted to local conditions and include appropriate factors for nutrition, education level, motivation, and incentives.

Work-measurement techniques yield data that may be interpreted differently depending on the particular needs of the person or agency that is interpreting the data. Work-measurement techniques are used by contractors or government agencies to measure the impact of a number of parameters on labor productivity. It is important to consider the quality of labor in cross-country comparisons of labor productivity and several dimensions of labor quality including education, training, experience, health, and nutritional status should be examined to determine the primary sources of labor productivity differences. In this chapter, statistical methods are provided that may be used to assess the impact of several factors on productivity rates in developing countries as contrasted with developed countries, using a comparison of the United States and Nigeria as a case study.

Even though productivity rates vary around the world, they are related to the following variables "management (proper planning, realistic scheduling, adequate coordination, and suitable control); labor (union agreements, restrictive work practices, absenteeism, turnover, delays, availability of labor, level of skilled artisans, and use of equipment); government (regulations, social characteristics, environmental rules, climate and political ramifications); contracts (fixed price, unit cost, and cost-plus fixed fee); owner characteristics; and financing. In addition, individual national productivity rates have been found to be dependent upon numerous diverse factors such as national characteristics, climate conditions, local labor practices, and acceptance of equipment" (Koehn and Brown, 1986, p. 299). To improve productivity, the impact of each of these variables on labor productivity may be assessed using statistical methods.

Statistical methods are available that measure the impact of one variable called the dependent variable on another variable called the independent variable. In addition to being able to predict the value of the dependent variable based on information about an independent variable, a measure of the strength of the relationship between these variables may also be determined. One measure of the strength of the relationship between two variables x and y is called the *coefficient of linear correlation* (r), or simply the *correlation coefficient*. Given n pairs of observations (x_i, y_i), the sample correlation coefficient r can be computed as (Yates and Guhathakurta, 1993):

$$r = \frac{Sxy}{\sqrt{S_{xx}S_{yy}}} \tag{10.1}$$

where:

$$S_{xy} = \Sigma xy - \frac{(\Sigma x)(\Sigma y)}{n}$$

$$S_{xx} = \frac{\Sigma x^2 - (\Sigma x)^2}{n}$$

$$S_{yy} = \frac{\Sigma y^2 - (\Sigma y)^2}{n}$$

In order to find the proportion r^2 of the total variability of the y values that are accounted for by the independent variable x, the following equation may be used:

$$r^2 = \frac{S_{xy}^2}{S_{xx}S_{yy}} \tag{10.2}$$

Similarly, $1 - r^2$ represents the proportion of the total variability of the y values that are not accounted for by the x variable. These equations may be used if and only if there is a linear relationship between x and y. Other models become necessary when the relationship between x and y are not linear; that is, y increases or decreases with x but not in a linear fashion. One approach is the *rank correlation coefficient,* which measures the monotonic relationship between y and x; that is, y increases or decreases with x even when the relation between y and x is nonlinear.

The *rank-order correlation coefficient* is calculated by first ranking all x values and y values separately and then calculating the ordinary correlation coefficient for the ranks. The relationship based on the ranks is called the *Spearman's rank-order correlation coefficient r_s.*

10.2 International Comparisons of Labor Productivity

Data obtained through work-measurement techniques yield valuable information for future planning and productivity forecasting. There are several factors that affect labor productivity, and these factors can be broadly categorized as *general, organizational, technical,* and *social factors.* The extent to which these factors affect labor productivity varies from country to country; for example, socioeconomic factors may have a larger impact in developing countries as opposed to developed countries. In global construction, these types of comparisons are useful for planning resources and management methods and selecting appropriate technology.

Although the extent to which each of these factors affects productivity varies from country to country, some type of method is needed to develop international comparisons in order to establish the *relative* importance of the variables for different countries, and a *rank-agreement factor* allows such a comparison.

The rank-agreement factor is a method whereby factors are ranked based on their impact on productivity for each country, and these rankings allow a cross-comparison of the relative importance of variables for different countries. For any two countries, the rank of the ith item for country A is R_i^1, and for country B it is R_i^2. Then the absolute difference D_i between any rankings of the ith item for the countries would be (Yates and Guhathakurta, 1993, p. 16):

$$D_i = |R_{i1} - R_{i2}| \tag{10.3}$$

where $i = 1, 2, \ldots, N$, and there are N items. Define

$$D_{max} = \sum_{i=1}^{N} |R_{i1} - R_{j2}|$$

where $j = N - i + 1$. The formula for the *rank agreement factor* is

$$RA = \sum_{i=1}^{N} \frac{|R_{i1} - R_{i2}|}{N} \tag{10.4}$$

With a maximum RA,

$$RA_{max} = \sum_{i=1}^{N} \frac{|R_{i1} - R_{j2}|}{N}$$

The formula for the percentage disagreement between countries is

$$PD = 100 \times \frac{\displaystyle\sum_{i=1}^{N} |R_{i1} - R_{i2}|}{\displaystyle\sum_{i=1}^{N} |R_{i1} - R_{i2}|} \tag{10.5}$$

The formula for percentage agreement PA is

$$PA = 100 - PD \tag{10.6}$$

A larger percentage agreement would indicate that the ranking of a particular factor under consideration has global significance and is not peculiar to any one of the countries under consideration. A larger percentage disagreement would indicate that the rankings for a particular factor are peculiar only to the countries under consideration. These data have many potential uses in planning for international construction as they may be used to identify areas that have a *relatively* greater affect on productivity of labor in foreign countries as compared to labor productivity rates in someone's native country.

10.3 Case Study: Worker Productivity in Nigeria and the United States

The following issues were identified by workers (artisans) in Nigeria as factors that influence their productivity rates (Olomolaiya et al., 1987, p. 320):

- Lack of materials
- Lack of proper tools
- Repeat-work instruction delays
- Inspection delays
- Absenteeism
- Supervisors' incompetence
- Changing crew members

10.3.1 Lack of Materials

Transporting construction materials was ranked as being the most common problem encountered on each of the five projects in the United States, with an estimated average loss of between 6.40 and 8.40 hours per week per worker. The time lost on Nigerian sites ranged from 2.5 to 5.5 hours per week per worker. The reasons given for higher productivity losses in the United States were (Borcherding et al., 1980)

- Lack of cranes, trucks, or both for transporting construction materials
- Not enough laborers to retrieve material orders from warehouses
- Excessive paperwork required for requisitioning materials
- The nonexistence of certain items at the job site
- Improper materials delivered to the work area
- A lack of proper preplanning by either supervisors or general supervisors

In Nigeria, workers identified outright shortages of certain materials as the main problem. This was traced to incessant cash flow problems experienced by contractors, resulting in suppliers not being paid for materials previously delivered to job sites. Workers also identified lack of proper planning and on-site transportation difficulties as the second and third most frequent sources of problems, and delays were the fourth problem source (Olomolaiye et al., 1987).

10.3.2 Lack of Proper Tools

At U.S. sites, an average of 3.41 to 5.08 hours per worker was lost each week due to a lack of proper tools. In the United States, craftsmen indicated poor quality, improper maintenance, and insufficient tools as the three most frequent causes of problems. Pilferage (stealing) was another significant problem. Pilfering from other crews was condoned and sometimes encouraged by some supervisory personnel. Some craftsmen worked with broken tools, and there were insufficient quantities of power-driven tools. The indifferent attitude toward tools contributed to related problems, according to warehouse officials (Borcherding et al., 1980).

In Nigeria, poor quality tools and poor maintenance of tools and equipment were cited as major problems. At some job sites, steel-bending machines were not available; consequently, steel fixers were forced to bend reinforcing steel by hand, which resulted in productivity losses. Concrete operations would stop for an entire day when concrete vibrators were not working properly, and joiners took turns using power saws because some of the saws were not in working order (Olomolaiye et al., 1987).

10.3.3 Repeat Work

The amount of time spent on rework was between 4.92 and 7.73 hours per week per worker in the United States and 1 to 7 hours per week per worker in Nigeria. In the United States, most of the work redone was a result of change orders, substandard workmanship, poor instructions that resulted from the misinterpretation of drawings and specifications and other regulations imposed by inspectors or supervisors.

A majority of rework was due to poor quality engineering designs. Incomplete drawings, coupled with errors and inconsistencies, were a continuous dilemma for supervisors and tradesmen. In Nigeria, workers blamed most of the rework on design changes. On the other hand, steel fixers claimed that repeat work was caused mainly by poor or inadequate instructions from supervisors.

10.3.4 Inspection Delays

Inspection delays varied between both sites investigated and who was surveyed as to their contribution to lost time and degree of effect. The estimated average weekly loss of hours was between 2.06 and 4.06 hours per worker in the United States and between 1 and 3 hours per worker in Nigeria. A considerable number of workers, in both the United States and Nigeria, indicated that *supervisor incompetence* contributed to low productivity. The workers were of the opinion that a high proportion of inspection delays could have been avoided if there had been clearer instructions (Olomolaiye et al., 1987).

Thus, from a comparison of the two countries, it can be ascertained that worker productivity was affected by similar problems in both countries. The number of hours lost per worker per week was higher in the United States for each of the factors affecting productivity. The amount of time required to build a square meter of finished structure varied from 6.44 to 16.78 worker days in Nigeria compared with 1.53 worker days in the United States. Clearly, other sources of productivity differences had to account for lower productivity in Nigeria.

The next section describes worldwide productivity factors.

10.4 Labor Productivity Variations

Labor productivity factors were compared in a report by the International Labor Organization (ILO) using a norm of 1.0 for the Washington, D.C., area (ILO, 1969, p. 299). The ILO report is useful to construction industry personnel when they are planning global construction projects. Forty-seven labor productivity factors were developed using data from the ILO survey of international contractors and other research projects.

Table 10.1 lists base labor productivity rates for different regions of the world, along with low and high rates. The data in Table 10.1 demonstrate that productivity factors vary substantially throughout the world. This table was developed from a variety of sources (Koehn et al., 1986 p. 299).

One example of how the data in Table 10.1 may be used is the following: The labor factor for Costa Rica is 1.74, and this indicates that 74 percent more work hours are required to complete a project in Costa Rica than in the Washington, D.C. area in the United States. Construction firms that are frequently involved in global contracting develop their own in-house base-productivity factors for locations where they build projects.

Base productivity is defined as the productivity that could be expected on a *standard* project, and it includes the following (International Labor Office, 1969):

- Projects requiring x number of work hours (companies set their own minimum number of hours) of direct labor at a standard location
- Area construction activity at a normal level

Table 10.1 Global Base Labor Productivity Rates

Country	Rate	Country	Rate
Austria	1.6 (1.67–2.10)	Iraq	3.50
Australia	1.20 (0.96–1.45)	Israel	1.23 (0.83–2.44)
Argentina	2.30 (1.30–2.60)	Italy	1.45 (1.10–1.48)
Brazil	2.80 (1.23–3.10)	Jamaica	2.00 (1.49–3.05)
Belgium	1.45 (1.30–2.60)	Japan	1.90 (1.00–2.00)
Canada, East	1.14 (1.08–1.17)	Korea, South	1.23 (1.00–1.67)
Canada, West	1.07 (1.02–1.11)	Kuwait	2.00 (1.32–4.55)
Ceylon	3.50	Lebanon	1.25 (1.00–1.67)
Chile	2.70 (2.00–2.09)	Mexico	2.00 (1.54–3.15)
China	4.00 (2.60–4.50)	Netherlands	1.80 (1.60–3.30)
Costa Rica	1.71 (1.25–3.00)	Nicaragua	2.67
Denmark	1.28 (1.24–1.28)	Norway	1.23
Dominican Republic	2.00 (1.49–3.03)	Oman	1.67 (1.32–4.55)
El Salvador	2.00 (1.50–3.00)	Pakistan	2.64 (1.67–7.14)
England	1.52 (1.11–2.08)	Panama	1.71 (1.43–3.03)
Egypt	2.04 (1.30–5.00)	Philippines	1.80
Finland	1.28 (1.24–1.28)	Portugal	3.00
France	1.33 (9.89–1.54)	Puerto Rico	1.80
Germany	1.28 (1.00–1.33)	Saudi Arabia	1.90 (1.27–4.50)
Ghana	3.5	Singapore	4.00
Greece	1.43 (1.00–1.59)	South Africa	1.58
Guatemala	2.13 (1.50–3.05)	Spain	2.95
Haiti	2.33 (1.49–2.94)	Sweden	1.13 (1.10–1.20)
Honduras	2.22 (1.50–3.03)	Switzerland	1.05 (1.0–1.10)
India	4.50 (2.50–10.0)	Taiwan	1.94 (1.69–3.33)
Iran	4.00	Turkey	2.32 (2.22–2.44)
		Venezuela	1.22

- Normal execution basis
- Normal area weather
- Local work rules and practices
- Normal worker attitudes
- Normal area workweek
- Average contractor performance
- Normal material availability and deliveries for project location
- Sufficient areas for staging, laydown, and work
- Ratio of national to expatriate labor based on historical standards and the number of work hours

Base productivities may be modified using a job-size and activity-correction curve, as well as workweek correction curves. Base productivities may be modified for other criteria such as execution basis, abnormal weather, travel time to site, temporary laydown, work, and staging area, site accessibility, work rules (craft restrictions, breaks, etc.), worker attitudes, contract performance, and other significant considerations.

Construction labor productivity in developing countries is lower than productivity in developed countries because construction operations in developing countries are affected by the nonpayment of wages, work interruptions caused by a lack of proper materials or tools, inadequate management, and various other socioeconomic reasons that affect worker efficiency (Moavenzadah, 1978).

10.5 Labor Productivity Factors

This sections discusses labor productivity variations and the factors that affect them.

10.5.1 Labor Quality

Labor productivity differentials can be explained on the basis of factors such as education, health, nutrition status, and such characteristics as age and education (U.S. Government, 1981). Researching information on the variations in labor productivity rates throughout the world provides insight into the effects of education and health improvement on not only the productivity of workers but also on other resources as well.

10.5.2 Education

Education is a central determinant of productivity because it affects the occupation that someone chooses to pursue as well as productivity within an occupation. Higher levels of education help increase productivity and aid in the adoption of new technology. Institutional vocational training programs do not increase productivity as much as on-the-job training programs, and formal education increases the ability of a worker to gain from on-the-job training.

A study on designer construction knowledge determined that it is beneficial for civil engineering students to take construction courses along with engineering courses in their educational programs and for design engineers to visit job sites to observe construction in addition to visiting job sites to check conformance to the plans and specifications (Yates and Battersby, 2003).

10.5.3 Nutrition

The effects of nutrition on an individual worker's performance level are difficult to quantify because nutrition interacts with several factors, such as disease, education, and motivation, in a complex way. One example that illustrates how conditions affect worker productivity is a road construction project that was built as part of the Pan American Highway that connects the United States to Mexico. Productivity increased up to 500 percent when workers were housed in camps and provided with three meals a day.

The most clear-cut evidence of nutritional impact is for specific nutrients. One study found a correlation in cross section between performance and iron intake in construction workers (International Bank for Reconstruction and Development, 1974). Studies of road construction workers in India and Kenya indicated a correlation between productivity and iron levels in the blood (U.S. Government, 1975; U.S. Government, 1981). The effects of reduced caloric intake are less clear than for iron intake. Studies indicate that productivity is not affected by current calorie intake but rather by indicators of past calorie intake, such as height, weight, and arm circumference.

The effects of reduced nutrition are evident in countries where workers observe the Muslim holiday of Ramadan. For 1 month, workers fast during daylight hours and eat only after the sun has set. The strength of construction laborers is impaired while they are fasting during the day; therefore, some job sites only operate for 4 hours a day during the month of Ramadan, or construction work is shut down during Ramadan.

10.5.4 Motivation

Job content, perceived status of employees, or an increasing sense of participation all significantly affect productivity rates. Poor worker motivation causes low productivity. Frederick Herzberg developed a theory of attitudes toward work that separates out the characteristics associated with any job into *motivation* and *hygiene* characteristics. Motivation factors are factors that give workers satisfaction at work, such as achievement, a sense of responsibility, and pleasure from the work itself. Hygiene factors include salary, working conditions, company policy, and administration, and these are principal sources of dissatisfaction (Herzberg, 1953).

If rewards are primarily intrinsic to the task performed, the provision of extrinsic incentives may have a negative effect on performance. If a task is viewed as requiring someone highly skilled to perform it, and if there is a major emphasis on producing quantities of it, this may lead to a deterioration in quality.

Group motivation influences productivity, and the size of the work group is important because it affects the strength of social relations. Productivity may vary inversely with group size, and workers often set output restrictions for certain piece-rate jobs. *Rate-busters,* or individuals who exceed the worker-determined norms, may be subjected to social sanctions. Output restrictions appear on easy jobs so as to avoid rate cuts by management. Another pattern is *goldbricking,* where workers put forth the lowest possible amount of effort on jobs that do not offer bonuses.

10.5.5 Financial Incentives

The effects of financial incentives on increasing worker productivity were demonstrated on a road construction project in India, where managers discovered that by changing from time rates to piece rates or task rates, worker productivity improved by 50 to 75 percent (U.S. Government, 1975). The ILO conducted studies on road construction projects in Nigeria and Tanzania that confirmed that when workers were paid in a way that made them feel like they were working for themselves, productivity improved on projects (Koehn and Caplan, 1987). In India, changing from a day rate to a finish-and-go rate system raised productivity 45 to 53 percent, and changing to a paid-task rate increased productivity by 60 percent.

10.5.6 Management

Managers affect labor productivity in many different ways, because managers are responsible for setting general goals for firms, as well as supervising day-to-day tasks. Managers make a wide range of allocation and technical decisions, such as choosing construction materials, determining the methods that will be used for construction, and establishing the order in which tasks will be accomplished. Unspecified management factors have been cited as responsible for labor productivity differences with apparently similar techniques, especially on international comparisons, where there are obvious differences in management styles.

Productivity in India or Indonesia on earthwork tasks is as much as four times the standard figures for those countries if there is good management and high incentives (piecework) coupled with longer working days. Conversely, in Chad, a country with no tradition of labor-based construction, in the first months of a project under conditions of low incentives (daily-paid work), fair management, and a 7-hour working day, productivity of one-third the standard figures was recorded. Lower levels of productivity in Africa are the result of poor management rather than the inherent ability of workers. Poor working conditions and lack of financial incentives reduce the motivation of workers in Africa (Coukis, 1983).

10.6 Factors that Affect Productivity on Global Projects

This section discusses specific issues that affect productivity on global projects, except for language barriers, which are discussed in Chapter 2.

10.6.1 Cultural Motivation

While working on global projects, E&C personnel have to manage workers from nations throughout the world who may all speak different languages. Even within one country, several languages may be spoken, and citizens practice different religions. Having a basic understanding of the culture of a host country and knowing something about local religious beliefs are essential when designing methods to improve productivity on construction projects.

In societies based on the Protestant (Quaker) work ethic, such as some parts of the United States, people are taught that hard work will be rewarded by raises and promotions, and individuals are praised for their performance rather than praise being given to groups of individuals.

In Islamic cultures, people accept their position in life, and they do not *live to work;* rather they *work to live.* Since their status in society is determined by birth, their clan, their tribe affiliation, or their place of birth, work is not a means for elevating one's status. Social networks are important to workers, and they may only socialize with people from their own region or clan.

South America and Asian cultures are *paternalistic* societies, where companies are like families and supervisors behave like parents. In addition to supervising work, managers are parent figures, and they are admired at work and treated with reverence by their employees. The group is more important than the individual, and groups are rewarded when objectives are met, not individuals.

The motivation techniques used to increase productivity in areas where there are different religious beliefs vary because workers are motivated by different objectives. If someone is taught to be content with whatever job he or she has rather than attempting to find a better job, he or she is not motivated by anticipated promotions. Productivity rates are influenced by how motivated workers are to perform a task, and motivation is tied to the work ethic of a particular region of the world.

Workers in some Islamic countries may not understand the importance of working quickly because they expect to be rewarded in the afterlife, not the present life. In rural areas of the world, workers may not be familiar with the types of facilities they are constructing, or they may not understand why the facilities are essential to their society because they have lived without certain structures their entire lives. Workers may not comprehend how a facility will personally benefit them when it is complete; therefore, they do not understand the urgency of meeting deadlines during construction. If workers are being compensated on a daily basis, it is to their benefit to work slowly so that they will be paid for additional work.

Work ethics may be affected by whether someone has been working in a Communist society or their society has been influenced by Communism in the past. If the objective of previous projects on which they have worked was to employ large numbers of people, it is difficult for workers to rapidly change their work habits to suit profit-driven projects or for workers to adopt methods designed to increase productivity. Also, if workers are used to being compensated for working overtime, then they may be reluctant to work past normal working hours when overtime is not being paid for extra work.

10.6.2 Process Issues

10.6.2.1 Materials

The location of a project has an effect on productivity rates. In rural areas, productivity declines when there are problems associated with securing and transporting construction materials and equipment to job sites. In some countries, such as parts of Africa, Afghanistan, and Pakistan, materials may still be transported to job sites on waterways or by mules, horses, elephants, camels, water buffalo, or other beasts of burden. One of the beasts of burden used in Southeast Asia to transport materials is a water buffalo, and Figure 10.1 is a photo of a water buffalo. Figure 10.2 shows a sample of a water transportation system in China, and Figure 10.3 demonstrates how items are transported on the tops of heads in some countries.

Figure 10.1 Beast of burden—water buffalo.

Figure 10.2 Waterway transportation system, China.

Figure 10.3 Transporting items by head basket, Indonesia.

Trucks could be available to transport construction materials, but they might be unreliable if there is a lack of trained mechanics or due to the unavailability of spare parts. In some of the less developed regions of the world, there are advanced rail systems and port facilities, but once materials arrive at train terminals or ports, there are no transportation systems to move materials to job sites.

It is also challenging to transport materials to job sites that are located in crowded cities because laydown yards are scarce, so materials may be stored outside a city and brought to job sites on a daily basis. Urban job sites have restricted access because of the existing structures in the surrounding environment.

Another transportation issue is that materials may not be delivered to job sites on their scheduled delivery dates because of delays caused by bureaucratic governments. Without the assistance of *business agents,* and the paying of unofficial *processing fees,* or *gratuities,* officials in the customs office of a host country could delay the delivery of materials. Business agents are familiar with how local systems operate, and they can assist with the expediting of government administrative paperwork and help to obtain the release of materials from customs.

Materials in developing countries may be inferior to the materials used for construction projects in industrialized nations. If materials called for in the specifications are not available locally, or they are illegal to import, substitutions have to be made at the job site, which delays projects or completely stops work. Local contractors will sometimes order materials and have them delivered to job sites without the knowledge of the project manager in charge of a project.

In some cultures, such as India, workers may worship their tools, or they share their tools with workers from different trades, which slows down construction as workers wait for their tools to be returned to them. In some developing countries, if workers lack appropriate tools to work on certain types of construction, they may use inappropriate tools that are available locally to perform work, such as using iron pile rods to dig for electrical and plumbing conduit below grade level during excavation or using wood 2 by 4s instead of hoes or shovels for spreading concrete as it is being placed in formwork.

10.6.2.2 Labor

A business agent may have to be retained before someone can hire labor to work on a construction project. In the United States, and northern European countries, business agents are people who work directly for labor unions; they manage union affairs, they help to settle union disputes, and they operate hiring halls where union workers wait to be hired by construction firms to work on projects. Business agents could be government officials, or they could be unofficial agents who are familiar with local laborers and where to locate workers for projects. In eastern European countries and countries where there are weak or changing governments, there are unofficial agents or members of local *mafias* (illegal criminal organizations) that control materials and the labor supply. In tribal societies, the clan or tribe chief may be the only person who determines who will work on projects.

If a project is being built in a clan or tribal society, workers may slow down their productivity so that additional workers, such as their family or clan members, will be hired to work on the project. Tribal construction workers could refuse to perform work if they are required to work with people who are not from their clan or tribe. Workers might refuse to work for a particular supervisor or take directions from someone who is not from their tribe or clan. If a supervisor is not from the local clan, he or she might have to explain work processes to a clan member, and the clan member, in turn, explains the work to laborers.

In some parts of the Middle East, construction labor is difficult to locate because certain classes of people will not perform manual labor; therefore, workers are imported from Pakistan, the Philippines, Sri Lanka, Thailand, Egypt, and South Korea. Japanese and European firms provide supervisory personnel that work on construction projects in the Middle East.

Unemployment rates are high in the following countries: eastern European countries, where it is 20 to 40 percent; Germany, where it is 18 percent; France, 10 percent; Asian and African countries, 20 to 40 percent; and South and Central America, over 10 percent. In countries with high unemployment rates, workers are desperate to keep their jobs, and they might try to perform work beyond their capabilities. Managers should be cognizant of this problem and not allow workers to perform work that is beyond their capabilities because it leads to accidents and injuries.

Issues of *face* (embarrassment) directly influence productivity at job sites in countries where it is part of the culture such as Asian countries. If a worker loses face, he or she will not return to work, and if the worker who loses face is part of a team, it may affect the entire team. Team members might reduce their output in protest over one of their team members losing face.

Laborers take smoke (Asian countries), tea (British or former British colonies such as India), kava (Fiji Islands), qat (Yemen), coffee (the United States), or other types of breaks during the workday, and Chapter 3 discusses safety issues at construction sites related to different types of breaks. Workers are not as productive right before or right after holidays because preparations are required for holidays that are accomplished after regular work hours. In India, the start of construction could coincide with a religious festival, and certain rituals, called *poojas,* are performed during groundbreaking ceremonies at jobsites.

10.7 Summary

Studying regional sources of productivity variability helps project-management personnel to develop more efficient productivity-improvement programs. The International Labor Office and the International Bank for Reconstruction and Development have conducted a number of studies related to the quality of labor in developing countries, and the quality of labor is improving in nations where there are health and medical care programs and housing and educational programs.

The effects of education and training on productivity are interlinked because education increases the ability of laborers to gain more knowledge from practical on-the-job training sessions. Health and nutrition combine in a complex way that has a synergistic effect on labor productivity; therefore, it is difficult to identify a direct relationship between nutrition and health and productivity rates.

Conducting statistical studies using labor productivity data may yield information that can be used by managers for production planning and forecasting in global contracting. However, when developing cost estimates for global construction projects, labor productivity factors yield more accurate results. Statistical examples presented in this chapter demonstrated how labor productivity factors could be used for planning and forecasting on construction projects.

Managers may use labor productivity factors when they are trying to determine what is appropriate construction technology that may used in developing countries. Various methods have been developed for this purpose, ranging from the use of available productivity data to elaborate mathematical models. When reliable productivity data are not available, simpler techniques suffice for developing cost estimates for comparisons between different techniques for accomplishing a task. More elaborate methods should be used in global contracting because risks are greater, and accurate data are required to develop bid estimates. Education, nutrition, management, work ethics, national transportation techniques, unemployment rates, and religion all affect labor production rates.

DISCUSSION QUESTIONS

1. Of the five primary sources of labor productivity differences, which is the most important, and why?
2. What are the items that cause variations in global productivity rates?
3. How can management help to increase global productivity rates?
4. What are the different factors that influence productivity rates in the United States versus Nigeria related to a lack of materials?
5. What are the labor productivity factors for Iran, Spain, Venezuela, the Philippines, and Sri Lanka? If the labor on a project in Washington, D.C., took 970 hours to complete a task, how many hours would the same task take to complete in these countries?
6. What are the eight items that can be used to adjust base productivity rates?
7. How does nutrition affect construction workers and their productivity rates?
8. How could the difference between Protestant, Islamic, and paternalistic cultures be used to motivate workers on a project where the workers are from all three types of societies?
9. What would be the most efficient way to use the water buffalo shown in Figure 10.1 to transport materials, and which construction materials could be transported using this method?
10. If construction workers are forced to share tools, how could productivity be increased at job sites if no additional tools are available?

REFRERENCES

Borcherding, Sebastian, and Samelson. 1980. Improving motivation and productivity on large projects. *Journal of the Construction Division* 106 (CO1): 73–89.

Coukis, B. 1983. *Labor-Based Construction Programs.* Washington, D.C.: The World Bank.

Herzberg, F. 1965. The Motivation to Work. New York: John Wiley and Sons.

International Bank for Reconstruction and Development (IBRD). 1974. *Study of the Substitution of Labor and Capital in Civil Construction Projects,* staff working paper no. 172. Washington, D.C.: IBRD.

International Labor Office (ILO). 1969. Measuring productivity. *Studies and Reports,* new series no. 75, Geneva, Switzerland: ILO.

Koehn, E., and G. Brown. 1986. International labor productivity factors. *Journal of Construction Engineering and Management* 112 (2): 299–302.

Koehn, E., and S. Caplan. 1987. Work improvement data for small and medium-size contractors. *Journal of Construction Engineering and Management* 113 (2): 327–339.

Moavenzadeh, F. 1978. Construction industry in developing countries. *World Development* 6 (1): 99–116.

Olomolaiye, P. O., K. A. Wahab, and A. D. F. Price. 1987. Problems influencing craftsmen's productivity in Nigeria. *Building and Environment* 22 (4): 317–323.

Tucker, R. L., D. F. Rogge, W. R. Hayes, and F. P. Hendrickson. 1982. Implementation of supervisors-delay surveys. *Journal of the Construction Division* 108 (CO4): 577–591.

U.S. Government. 1975. The Planning of Control and Production, Productivity and Costs in Civil Construction Projects, technical memorandum no. 15. Washington, D.C.: U.S. Government, Library of Congress.

U.S. Government. 1977. The Relationship of Nutrition and Health to Worker Productivity in Kenya, technical memorandum no. 26. Washington, D.C.: U.S. Government, Library of Congress.

U.S. Government. 1981. Labor Productivity: Un Tour d'Horizon, staff working paper no. 497. Washington, D.C.: U.S. Government, Library of Congress.

Yates, J. K., and L. Battersby. 2003. Master builders project delivery system and designer construction knowledge. *Journal of Construction Engineering and Management* 129 (6): 1–10.

Yates, J. K., and S. Guhathakurta. 1993. International labor productivity. *Journal of the American Association of Cost Engineers International* 35 (1): 15–25.

Chapter 11

Global Planning and Construction Delays

11.1 Introduction

A number of factors should be tracked when managing global construction projects in order to determine whether a project is being delayed and the impact of delays. Construction delays may be assessed in terms of cost and time, but concurrent delays are harder to quantify due to the elusive nature of them. Project delays are difficult to predict with any accuracy and each delay causes certain risk factors. Delay factors may be prioritized based on the importance of the activity being delayed and the amount of risk involved in each activity.

Global engineering and construction (E&C) projects are more challenging, and they experience more delays than domestic projects due to:

- The complicated nature of working in a foreign environment
- Working in foreign cultures
- Dealing with foreign government agencies, regulations, and officials
- Adjusting to different concepts of time
- Bribery issues
- Import and export regulations
- Foreign legal systems
- Differing management philosophies
- The absence of safety regulations
- Working with unskilled labor
- Limited quality control
- Inferior materials
- Variations in the educational backgrounds of E&C personnel

The FORMSPEC system of model specifications that was developed in the United States by Project Management Associates provides one definition of construction and project delays (Project Management Associates, 1991, FORMSPEC 89.0710, Section 00780, Glossary):

Acts or events that postpone, extend, or in any other manner alter the schedule or completion of all or any part of the work. Delays include deferral, stop, slowdown, interruption, and extended performance, and all related hindrance, rescheduling, disruption, interference, inefficiency, and productivity and production losses. Delays may result from added work, or without the addition of any work—due to suspension of work, contractor caused delays or delays from any other causes covered under the general conditions.

When contractors assert that delays are for reasons beyond their control, owners may remain unconvinced that the contractor is legitimately entitled to a time extension. Some construction contracts allow owners to recover either liquidated (actual) damages for delays caused by contractors and for contractors to be entitled to extended field and home office overhead costs if an owner is responsible for delays. Thorough construction documentation and an analysis of the indicators of delay help both owners and contractors streamline the processes required to determine the causes of delays (Yates, 1993).

This chapter discusses the types of issues that cause planning and construction delays on global construction projects. It includes information on global construction indices, dogmatism (stubbornness of opinion), scope of work changes, bribery policies, political instability, management issues, regional influences, religious and social factors, labor disputes and strikes, technologic and economic limitations, natural disasters, government restrictions, and global delay factors. Construction documentation is also discussed in terms of its importance for indicating construction project delays.

11.2 Accounting for Global Variations in Construction

11.2.1 Global Construction Indices

Global construction indices are available such as the Hanscomb/Means Construction Cost Index, the Engineering News Record International Construction Cost Data, or Spon's International Construction Index. Hanscomb/Means provides *construction labor purchasing parities* that list a range of values for labor in each currency along with a *parity index* for each country (with 100 being the benchmark cost of labor in Chicago, Illinois). A parity index compares total cost, excluding contractor overhead and profit. The differences in costs in the parity index are attributable to labor, materials, productivity rates, bidding methods, and market condition factors.

The Hanscomb/Means *global construction cost comparisons* provide comparisons of costs using local currency rates based on a market basket of 26 items representing the most common construction materials. Examples from this cost comparison are 163.6 for Great Britain (the most expensive country) to 50.9 for Malaysia and 56.1 for Shanghai (the least expensive locations) (Wiggins, 2005).

Unfortunately, parity indices ignore other issues that contribute to differences in costs between locations, such as (Wiggins, 2005):

- Codes, regulations, and other legal requirements
- Design variances arising from climatic differences
- Design differences arising from cultural and social standards
- Procurement, design, management, and contractual differences

Parity indices are used to equate costs between countries. In addition to parity indices, information on the gross domestic product (GNP) per capita for countries throughout the world is available through the Word Bank or in the *Time Almanac* (World Bank, 2005; Bruner, 2005). Both of these methods of classification can be used to distinguish the economic stability of various nations, which may be determined by monitoring and observing fluctuations in either the Gross National Product (GNP) of a country and the values or construction parity indices. Monitoring the stability of a nation where a construction project will be built is important because, when personnel are working on overseas construction projects, they must deal with and account for countless risks and numerous types of delays not commonly encountered on domestic projects.

11.2.2 Dogmatism

Cultural influences affect construction operations if people from different cultures are working together on a construction project. In some cultures, direct, rational approaches to engineering design and construction problems are not always an efficient way of operating because *dogmatism* (stubbornness of opinion) tends to not work as well applying culturally sensitive techniques. Understanding the motivations of clients, coworkers, suppliers, contractors, subcontractors, and government representatives helps to avoid cultural conflicts between personnel.

11.2.3 Scope of Work

Being familiar with the scope of work on overseas projects is essential because all requests for changes in scope have to be carefully considered in light of their impacts on not only the project but also on the local environment. Construction operations are conducted without the assistance of lawyers in many countries; therefore, scope changes need to be evaluated from the perspective that there may not be a second chance to negotiate scope changes.

11.3 Bribery Policies

A policy on *bribery* should be established at the beginning of a construction project so that native personnel, suppliers, subcontractors, and government officials know what will and will not be tolerated on a construction project. Bribery is "when something is offered in order to influence a person illegally, or improperly, to act in favor of the giver" (McKechnie, 2005, p. 101).

If a bribery policy is not uniformly implemented, then delays may be caused by government officials, suppliers, or subcontractors while they wait to be paid more money. If an understanding is reached during the early stages of a construction project, then subsequent delays and interruptions may be minimized during construction. Sometimes bribes will be offered in frustration by E&C personnel during construction to expedite processes, and this should be avoided because it sets a bad precedent and raises expectations.

Bribes are not discussed openly, so one person should be designated to deal with bribery issues. It is considered improper to call payments *bribes;* therefore, other phrases that may be used are payments, *baksheesh* (Arab cultures), commissions, tips, money for your trouble, gifts, compensation, money for influence, backhanding, graft, payments under the table, kickbacks (these occur when someone is awarded a contract and he or she pays money to the person who awarded the contract), taking care of sensitive issues, persuasion, funds for expediting, administrative fees, money for paperwork, to ensure that someone is taken care of, and so forth. Bribes may be required when:

- Bid proposals are submitted
- Materials are ordered or delivered
- Personnel first enter a country or travel within a country
- Permits need to be issued
- Inspections and certifications are required
- Skilled labor is needed

Most languages have a word for *bribe,* and some examples are:

- *Hui lui* (Chinese)
- *Bestechen* (German)
- *Corrompre* (French)
- *Sobornar* (Spanish)
- *Corrompere* (Italian)
- *Subornar* (Portugese)
- *Davat' vzyatku* (Russian)
- *Wairo* (Japanese)
- *Noemur* (Korean)
- *Rusvet* (Turkish)
- *Lapowka* (Polish)
- *Kenopenz* (Hungarian)
- *Uplatk* (Chekoslovakian)
- *Omkopen* (Dutch)

Since bribery is not legal in every country, E&C personnel should check with their native government to determine the restrictions on bribing because the laws of some country may govern the bribing of government officials in other countries (such as the U.S. Foreign Corrupt Practices Act). Using a business agent to deal with issues related to bribes is one of the best methods of preventing repercussions from bribery. Chapter 14 provides information on whether a business agent is required in different countries.

11.4 Categories of Global Nontechnical Delays

This section discusses the different types of nontechnical delays that affect global construction projects

11.4.1 Political Instability

Countries with unstable governments, military rulers, widespread poverty, high unemployment rates, or widespread corruption expose E&C personnel to immeasurable risks, ranging from wars, military and nonmilitary coups, acts of terrorism, physical harm, naturalization of facilities, and kidnappings (a detailed discussion on terrorism and kidnapping is provided in Chapter 12). One way for E&C personnel to protect themselves in politically unstable environments is to live and work barricaded behind walls, which makes the accomplishment of their work objectives difficult, if not impossible. Tracking political and economic events provides some measure of the current political climate, and knowing local people provides insight into the emotional climate of a region.

11.4.2 Management Factors

In many of the newly emerging countries (NECs) and developing countries, construction is managed through highly centralized systems, which reduces the efficiency of projects. A centralized management system may be used when there is only unskilled labor. Education and training help to decentralize management if all of the trades participate in educational programs.

A survey in Saudi Arabia identified interface difficulties in design and construction as a major problem, along with drawings not providing sufficient detail, unfamiliarity with local conditions, and buildability (Al-Hammad and Assaf, 1992). In Bahrain, another study found that lack of motivation and not using task drawings and job assignment sheets caused project delays (Al-Hamer, 1990). One manager in Bahrain purchased US$30 worth of materials and built a simple wood model of the project to show to the workers. Having the model substantially increased productivity because workers used it to visualize the finished structure in their minds.

In most countries, there is a dominating influence of direct personal experience rather than attitudes between ages, gender, education, occupation, or experience in constructing projects. If someone is new to a country and he or she is trying to manage unfamiliar workers, that person may assume that workers will be able to perform the work assigned to them simply because they have worked on other construction projects, they are educated, or their title leads someone to believe that they are qualified to perform the assigned task. A better indication of ability is to interview workers and ask them questions about their previous work experience and what was the precise nature of the work they performed on other projects.

11.4.3 Regional Influences on Delays

In every region of the world there are specific factors that delay construction projects. Knowing what types of factors occur in particular regions of the world helps project-management personnel prepare themselves for dealing with these factors. This section provides some examples of how regional factors influence construction projects.

In Australia, the powerful construction labor unions dictate when a project will be completed and at what cost (Hanscomb/Means, 1993a). In one instance, the unions shut down a job because of a leak in the union trailer, and another job was shut down because of aromas coming from a Chinese restaurant near the site. In Turkey, the government promotes and advocates the build-own-transfer (BOT) financing method for infrastructure projects, yet there are no standard forms available for contracts between contractors and owners for these projects (Hanscomb/Means, 1991). In Germany, there is a scarcity of cheap labor; thus labor has to be imported from Poland, Hungary, the former Czech Republic, Slovakia, and East Germany (Hanscomb/Means, 1994).

Government regulations may cause construction delays, especially when there are strict building procedures, as was the case on the Kansai International Airport in Hong Kong (Wang, 1986). Financing terms may cause delays if owners have restrictions on their funds and how they are used, as is the case in countries where there are economic sanctions, such as Iran. Legal issues may lead to delays in countries where contractors and clients rely more on the legal system to settle issues of price or scope changes, as is the case in the United States.

11.4.4 Material Transport Issues

Construction materials may be transported long distances by tractor-trailers, flatbed trucks, railways, or ships, but once they reach a local region, they may be transported by animals (e.g., horses, mules, donkeys, water buffalo, camels, and elephants), animal carts, bulldozers, wheel barrows, head baskets, cloth or canvas slings, or stretchers. When humans or animals are used to transport materials it substantially increases the time required for materials to be delivered to job sites.

11.4.5 Optimistic Schedules

The manner in which construction materials are transported varies throughout the world, and sometimes construction projects experience what appear to be delays, but what is really wrong is that the original schedules were too optimistic. If estimators are not familiar with local conditions that surround a potential project, they may create schedules for bid proposals that include durations that are inappropriate for that particular region (Yates and Audi, 1998).

11.5 Global Engineering and Construction Delays

Since the nature of delays encountered on global projects is different from the types of delays experienced on domestic projects, a survey was used to collect data on the causes of delays on global construction projects. Various factors needed to be accounted for, such as political and social unrest, religious and social beliefs, technologic and economic limitations, and government restrictions, regulations, and limitations. The following subsections provide examples of the types of issues that delay construction projects.

11.5.1 Political and Social Unrest

Certain regions of the world are undergoing drastic and sometimes violent changes in their political, economic, and social structures. The regimes of most of the countries of the former Soviet Union in eastern Europe, including Russia, are politically and socially volatile. The ruling parties of these countries have become hesitant and afraid to make political decisions or agree on social or economic reforms for fear of jeopardizing the delicate balance prevailing among its various ethnic groups, political parties, and social classes. Other countries, such as Nigeria, Angola, Liberia, Afghanistan, Iraq, Palestine and Colombia, are areas of violent unrest, and foreign contractors have had limited access to these countries.

11.5.2 Religious and Social Factors

Although small in population but rich in oil and mineral resources, citizens of countries such as Saudi Arabia, Kuwait, Qatar, the United Arab Emerites, and Oman adhere faithfully to their religious beliefs. Economic development in these countries, including construction, may be conducted and executed according to Islamic law. As a result, foreign construction companies have to tolerate delays, especially during the holy month of Ramadan and the Al-Haj season when millions of Muslims make a pilgrimage to the holy city of Mecca. Local laborers are hard to hire in these countries, and in certain cases they are not available for both social and economic reasons. Construction firms in these regions have to import foreign labor through manpower agencies, which is a time-consuming process because it includes securing visas and arranging for medical checkups, travel, and accommodations.

In every region of the world, religious holidays are observed throughout the year, causing construction delays when workers take days off or productivity declines during the days prior to and after the holidays as elaborate preparations are made for the holidays. Chapter 14 provides information on the holidays celebrated in different regions of the world.

11.5.3 Labor Disputes and Strikes

A labor dispute, followed by either a short or a long strike, disrupts construction schedules. Delays are inevitable, losses are certain, and the future of construction companies becomes questionable when strikes last for long periods of time. In developing countries, labor strikes may turn violent if governments interfere by using military force. It is difficult to know for certain who is the proper person to be negotiating with when a strike occurs at a job site. If there are no labor unions in a country, project management personnel need to carefully determine who has the authority to negotiate on behalf of the laborers. Different people may try to negotiate on behalf of the workers, and a determination has to be made on who is their legitimate representative.

11.5.4 Technologic and Economic Limitations

Technologic limitations include such factors as constructability issues, design standards, performance standards, quality standards, material availability, testing, inspection, and safety. Economic limitations may include such factors as inflation, escalation, and the availability of cash flow. The structure of wages is important in absolute terms but also in relation to what workers are being paid at other job sites within a country. If workers discover that they are being paid less than other workers for the same type of work, they may intentionally slow down their work.

11.5.5 Natural Disasters

Unpredictable and uncontrollable natural disasters such as earthquakes, tsunamis, floods, hurricanes, volcanic eruptions, tornadoes, and mudslides are major factors that lead to significant construction delays. Italy, Bangladesh, regions in the United States such as the Gulf States and California, India, Iran, Iraq, Mexico, Pakistan, the Philippines, Japan, Indonesia, Thailand, and other Pacific Rim countries (due to the Ring of Fire—volcanoes and the sliding of tectonic plates across each other) all are regions prone to natural disasters.

11.5.6 Government Restrictions

Governmental factors, such as legal restrictions and regulations, and interference by government officials are imposed in some fashion on all contracts in the global construction industry. Government restrictions include acts of protectionism, as is done in Japan, where foreign firms are not allowed to operate in their country. Other restrictions include provisions mandating that a government has the right to determine where something is built in its country. Requiring certain permits and inspections before, during, and at the conclusion of construction is a formal government restriction.

11.5.7 Global Technical Delay Factors

The results obtained from a global survey of construction projects indicated that the primary technical causes of construction delays in the global arena are the following (Aoude, 1996):

- Design modifications
- Weather (climate)
- Material delivery
- Equipment delivery
- Incomplete drawings
- Material quality

Additional technical causes of global construction delays are provided in Table 11.1. Table 11.2 contains a list of mitigation strategies that can be used to reduce

Table 11.1 Technical Delay Causes

Delay Number	Technical Cause	Delay Number	Technical Cause
1	Labor strike	23	Schedule too optimistic
2	Weather delay	24	Work on noncritical tasks
3	Craft labor shortages	25	More work than planned
4	Poor labor supervision	26	Poor productivity
5	Construction methods	27	Owner interference
6	Labor productivity	28	Poor pay
7	Undermanning	29	Late engineering
8	Later than planned start	30	Bankruptcy of subcontractor
9	Design modifications	31	Subcontractor delay
10	Incomplete drawings	32	Fire
11	Regulatory change	33	Lack of testing equipment
12	Tool availability	34	Change orders
13	Equipment delivery	35	Decision delay
14	Equipment breakdown	36	Unforeseen health- and maintenance-management review
15	Inaccurate estimates		
16	Material deliveries	37	Poor scope definition
17	Material quality	38	Project deferred
18	Inspection/quality control	39	Damaged goods
19	Rework	40	Permit approvals
20	Environmental issues	41	Inadequate plans
21	Overmanning	42	Poor management
22	Overcrowded work area	43	Site conditions

Table 11.2 Delay-Reduction Measures

Delay-Reduction Measure Number	Delay-Reduction Measure	Delay-Reduction Measure Number	Delay-Reduction Measure
1	Overtime	27	Quality assurance data
2	Additional equipment	28	Communication with agencies
3	Material management	29	Schedule expansion
4	Productivity measurement	30	Additional subcontractors
5	Inventory reporting	31	Fragments
6	Improved internal reporting	32	Office fabrication
7	Cost-control program	33	Subcontract more work
8	Increased training	34	Change execution strategy
9	Quality assurance program	35	Construction review and practice
10	Resource leveling	36	Task resequencing
11	Task acceleration	37	Slip completion
12	Expediting	38	Separate entrances
13	Increase supervision	39	Equipment maintenance
14	Work on critical tasks	40	Separate gates
15	Increase crew size	41	Timely submittals
16	Design modifications	42	Timely order placement
17	Process modifications	43	Alternate supplier
18	Better planning	44	Change order
19	Better supervision	45	Better scheduling
20	Second shift	46	Organizational modifications
21	Customer hired striking staff	47	Higher pay
22	Settle strike	48	Time extension
23	Additional shifts	49	Hire scabs
24	Expedite owner	50	Negotiations
25	Quality improvement	51	Weekly progress meetings
26	Procurement activity report		

delays, and Table 11.3 contains a matrix of delays along with three possible mitigation strategies for each delay that are commonly used to reduce delays.

The types of construction delays and mitigation strategies listed in Tables 11.1, 11.2, and 11.3 for foreign construction projects are similar to the types of delays and mitigation strategies encountered on domestic construction projects. The most prevalent mitigation strategies used on global construction projects include (Yates and Audi, 1998):

- Establishing a quality assurance (QA) program
- Working overtime
- Working on critical tasks
- Increasing supervision

Delays that are specific to developing countries include:

Table 11.3 Delay Causes versus Delay-Reduction Measures

Delay Cause	Delay Measure 1	Delay Measure 2	Delay Measure 3
Bankruptcy of a subcontractor	Process modification	Financial assistance to subcontractor	Replace subcontractor
Change orders	Quality assurance (QA) program	Work critical task	Increase crew size
Construction methods	Increase supervision	QA program	Design modification
Craft shortage	Overtime	Increase crew size	Resource leveling
Damaged goods	Timely order placement	Expediting	Use a different supplier or transporter
Decision delay	Expedite owner	Work critical task	Organizational modification
Design modification	Overtime	Design modification	Work critical task
Environmental issues	Additional equipment	Improve information reporting	Design modification
Equipment breakdowns	Additional equipment	Overtime	Resource leveling
Equipment delivery	Expediting	Additional equipment	Overtime
Fire	Additional subcontractors	Better planning	Resequencing
Inaccurate estimating	Cost-control program	Improve information reporting	Overtime
Inadequate plans	Increase supervision	Increase training	Process modification
Incomplete drawings	Design modification	Overtime	Task acceleration
Inspection/Quality Control (QC)	QA program	Increase supervision	Increase training
Labor productivity	Increase supervision	Increase training	Increase crew size
Labor strike	Overtime	Task acceleration	Work critical tasks
Lack of testing equipment	Resequencing	Check with rental agencies	Hire testing agency
Late engineering	Overtime	Increase crew size	Work critical task
Later start	Task acceleration	Overtime	Increase crew size
Material delivery	Expediting	Material-management program	Overtime

Cause	Corrective action		
Material quality	QA program	Material-management program	Increase supervision
More work planned	Overtime	Increase crew size	Work critical tasks
Overcrowded work area	Overtime	Resource leveling	Increase supervision
Overmanning	Resource leveling	Overtime	QA program
Owner interference	Overtime	Work critical tasks	Process modification
Permit approvals	Process modification	Timely submittals	Increase training
Poor labor supervision	Increase supervision	QA program	Productivity measure
Poor management	Organizational modification	Improve information reporting	Expediting
Poor pay	Improve information reporting	Process modification	Better planning
Poor productivity	Increase supervision	Productivity measurement	Overtime
Poor scope definition	Redefine scope	Resequencing	Acceleration
Project deferred	Work on other projects	Resequencing	Overtime
Rework	QA program	Increase crew size	Overtime
Regulatory changes	Design modification	Work critical task	Overtime
Schedule too optimistic	Work critical task	Overtime	Increase crew size
Site conditions	Overtime	Design modification	Organizational modification
Subcontractor delay	Overtime	Increase crew size	Additional subcontractors
Tool availability	Additional equipment	Resource leveling	Process modification
Undermanning	Increase crew size	Overtime	Resource leveling
Unforeseen health- and maintenance-management review	Task acceleration	Overtime	Work critical task
Weather delay	Overtime	Increase crew size	Work critical task

225

- Lack of materials
- Lack of tools, tools that are in poor shape, or use of inappropriate tools for a particular task
- Repeat work (if instructions were not clear when they were issued the first time)
- Instruction delays (workers will only take instructions from an authorized supervisor, such as a tribal, clan, or religious leader)
- Inspection delays (waiting for government officials to inspect every stage of a project)
- Absenteeism (due to bureaucratic, personal, religious, or family issues)
- Supervisor incompetence (difficult to locate qualified supervisory personnel)
- Changing crew members (high turnover rates caused by issues related to problems getting to work, family problems, injuries, and different concepts of time)

11.6 Indicators of Project Delays

Information plays a key role in construction project management, and for a construction project to be well-managed, data from past projects, as well as data from current projects, should be readily available. Construction projects use a variety of different types of documentation to determine the causes of project delays. Being familiar with the different formats that might be used for scheduling and claims analysis provides insight into the most efficient methods for determining the causes of delays and for recording historical information.

Data and the accumulation and documentation of data are essential for project planning, control, and decision making. In each of these areas, effective management of information is an integral part of any successful project management system, where the primary objective is completing a project on time. In all three of these areas, the documents that help to indicate the causes of delays are crucial.

Proving that a delay has occurred on global construction projects is difficult; therefore, Table 11.4 provides a list of the different types of documentation that may be used to help determine whether a delay has occurred during construction. Not all projects use the 43 types of documents listed in Table 11.4, but the list may be used as a reference during the implementation of document retrieval systems or to help locate information that documents the causes of delays.

One example of how the information in Table 11.4 may be used is to relate the indicators of project delays to the technical causes of global delays. The most prevalent technical cause of global delays is *design modifications,* and the primary indicators of delays that are used to determine that design modifications have delayed a project are (El-Nagar and Yates, 1997):

- The schedule (S)
- The correspondence log (COR)

Table 11.4 Indicators of Project Delays

0. (O) Other	22. (TS) Time sheets
1. (T) Time sheets	23. (DMR) Design milestone report
2. (IR) Inventory reports	24. (ER) Expediting report
3. (S) Schedule	25. (CRMR) Milestone review
4. (Q) Quality control report	26. (PML) Project manager's log
5. (COR) Correspondence	27. (FO) Field observation
6. (COL) Chance order log	28. (DCN) Design change notice
7. (PM) Productivity reports	29. (BPR) Bid package reviews
8. (DCR) Daily construction report	30. (DR) Design reviews
9. (DL) Drawing log	31. (EXCR) Exception report
10. (SDD) Superintendent's log	32. (DCM) Daily contractor meeting
11. (JL) Job log	33. (VO) Visual observation
12. (CE) Cost estimates	34. (DS) Delayed start
13. (PC) Progress curves	35. (DP) Delayed permit
14. (VA) Variance analysis	36. (MO) Manual observation
15. (PR) Procurement report	37. (LCO) Labor cost report
16. (MR) Manpower report	38. (MCR) Material cost report
17. (FLL) Field labor report	39. (SN) Strike notice
18. (SL) Submittal log	40. (EUR) Equipment usage report
19. (RFI) Request for information log	41. (TR) Test reports
20. (E) Experience	42. (NA) Network analysis
21. (PP) Permit process	43. (WL) Weather log

- The change order log (COL)
- The request for information log (RFI)
- Daily construction reports (DCR)
- Progress curves (PC)
- The drawing log (DL)

Another technical cause of delays is due to *incomplete drawings,* which uses the following indicators of project delays to determine whether incomplete drawings have caused a delay (El-Nagar and Yates, 1997):

- The request for information log (RFI)
- The schedule (S)
- The correspondence log (COR)
- The change order log (COL)
- Progress curves (PC)
- Daily construction reports (DCR)
- The drawing log (DL)

For these two examples, the top seven indicators of delay for both technical causes were the same; however, the order in which the indicators were used is different.

11.7 Global Project Planning Delays

Private projects require at least 3 years to develop, and public projects are in the planning stage for over 5 years because they involve legal, institutional, political, environmental, and funding issues, all of which are time consuming (Goodman, 1994). If a project is going to be built in a country other than where it is being designed, or if the owner lives in another country, the potential for planning delays increases exponentially.

Typically, several issues affect the development of projects, and two issues that are difficult to quantify are that (1) there is no major cost impact assessment until design and construction actually start and (2) owners are not aware of actual cost impacts because there are no studies or analyses performed at the planning stage. Often actual cost impacts are determined by the loss of profit or income that occurs when the completed project is not in full service. Owners have to refer to previous projects (historical data) or rely on their previous experience with similar projects (expert opinion).

Project budgets contribute to project delays because they are established early in the planning process with insufficiently detailed information. Project budgets are derived from initial construction-cost estimates of major construction elements, such as concrete, as a base and adding amounts or multiplying by percentages to estimate the cost of other activities. The construction-cost estimates that are developed during the project planning stage are approximately 65 to 70 percent of the total project budget cost.

The longer it takes to develop a global project, the more expensive it becomes for the following reasons:

- Additional time requirements for project team members to research global issues and foreign government requirements
- Changing interest rates in the host and native countries
- Changing currency-exchange rates
- Worldwide inflation
- Cost of construction extras caused by design omissions as a result of insufficient design time if global issues interfere with a project
- Regime changes
- Personnel changes because of circumstances that prevent workers from traveling to work on a project (changes in visa requirements)
- Government restrictions on the importing and exporting of materials changing during the planning process
- Changes in the environment surrounding a project (especially in newly emerging and developing countries)
- Changes in local and national laws

The following subsections outline some of the frequent causes of global project planning delays that were determined from a survey of E&C industry professionals and provide recommendations on how to mitigate delays (Eskander, 1990; Yates and Eskander, 2002).

11.7.1 Changes in Project Requirements

One of the most frequent causes of planning delays is constant changes in project requirements. No matter where a project is going to be built, owners are always changing their mind during the planning stage. Therefore, there really is no definitive method for preventing this type of a delay. Engineers may help owners to clearly define their objectives and processes and provide proper information and documentation to owners. When owners are developing multiple projects at the same time, it causes delays to projects that are less important when other projects take precedence. Owners need to monitor multiple projects effectively and prioritize projects.

11.7.2 Lack of Communication between Various Divisions

Communication disconnects within an organization that is planning and designing a project can cause multiple delays during project planning, especially if it is a firm that has offices in different countries. Having an effective project manager with a global perspective who understands how to properly document processes helps to reduce delays, along with:

- Formalizing the planning process
- Scheduling routine (biweekly or monthly) team meetings for reviews by users
- Assigning the project to an experienced project manager
- Using the project manager as a single point of communication
- Empowering the project manager to act as a conduit for communication

11.7.3 Assigning Projects to the Wrong Person

When the wrong person is assigned to monitor a project, it is a frequent cause of project planning delays. Companies sometimes assign one project manager to the planning process who works in the main office of a firm that is experienced in planning projects and then assigns a different project manager to oversee the construction process in the foreign country. While this provides appropriate expertise in each of these stages, it is more difficult for the second project manager, when he or she takes over the project, because the project loses the institutional memory of how issues have been addressed previously or why decisions were made by the first project manager.

11.7.4 Project Funding and Financing

If multiple currencies are involved in a project, this may cause delays, and recommendations for preventing financial problems emphasize initial financial planning and alternative financial resources. Other methods for preventing delays include:

- Keeping a project team current on costs
- Increasing the quality of initial estimates
- Reusing the scope of a project
- Providing help with creative alternative financing ideas
- Assisting clients in making partial budget requests

11.7.5 Slow Decision-Making Processes

When decision makers do not make decisions in a timely fashion during the planning stage of projects, it may delay projects. Therefore, scheduling decisions is an important aspect of the decision-making process, and project schedules should contain decision-making milestones.

11.7.6 Vague Project Scope

When the project scope is vague, the person in charge of the project should emphasize that clear documentation of the scope of work is essential, because it is not always possible for E&C personnel to meet personally with the owner. Communication issues are magnified when planning global projects because it is frustrating to obtain clarification through e-mails or over the telephone.

11.7.7 Limited Resources

Not having enough resources to identify project needs (scope, budget, resource, financing, etc.) may be mitigated by obtaining additional personnel or resources or stopping work until the proper resources are obtained for the planning process. Some companies do not allocate a budget for the planning phase of a project; therefore, expenditures have to be monitored carefully during the planning process.

11.7.8 Unclear Economic Analysis

The economic analysis for a project may not be clear, and then it requires several reviews and revisions. This type of delay factor requires either rework or a competent and experienced project manager and project-management team in order to work within unclear parameters.

11.7.9 Environmental Impact Studies and Archeologic Surveys

Environmental impact studies and archaeologic surveys cause delays because of various unknown factors. Some governments require environmental impact studies

(EIS) that outline how a project will affect the surrounding environment. Environmental impact studies are only used as a reference by decision makers. Governments may require that an archaeologist investigate construction sites to identify whether there are any historical artifacts, and if they are identified, all work has to stop while the site is excavated by experienced archaeologists.

11.7.10 Field Operations Personnel Introducing New Demands

Delays caused by field personnel introducing new demands may be reduced if field personnel are involved early in the planning process and there is proper documentation of their suggestions and resulting impacts on the project.

11.7.11 Political Pressures and Political Requirements

Owners are the ones who need to be more involved in trying to reduce delays resulting from political issues. Other ways of dealing with political pressures and requirements are to:

- Handle these communications and involvement with care and delicacy
- Request that owners take the initiative to alleviate any delays or conflicts of interest
- Assist the client in seeking resolutions
- Review priorities with members of the government and users to reevaluate priorities, cost, and scope

11.7.12 Incomplete Project Requirements

Sometimes written project requirements are not actually what the owner wants, and clarifications and detailed written explanations are primary mitigation strategies suggested for these types of delays.

11.7.13 Multiple Approval Levels

Too many levels of approval are required before any work or studies may be commissioned and begun. Reducing the number of approval levels and delegating authority for approvals, along with an approval schedule, are two methods for dealing with approval delays.

11.7.14 Lack of Project Management

When no one is assigned to be in charge of a project during the planning stage, such as a project manager or an appropriate substitute, projects are neglected by the rest of the staff. Not only does assigning a project manager help to satisfy owner requirements but also provides a point source for documenting the progress of a project toward achieving the objective of generating project plans and specifications.

11.7.15 Delays by Government Officials

For public projects, the requirements set by public officials, their involvement in project reviews, and the assignment of special panels cause numerous delays. Working closely with local public officials, involving the public more in project planning, and trying to anticipate the concerns of the public help to prevent delays on public projects.

11.7.16 Historic Assessments

Historic assessments and evaluations are time consuming. In order to deal with these, a historical inventory library should be created to bypass the entire process.

11.7.17 No One Identifies Project Needs

If there is no clear direction on who is responsible for identifying project needs, someone needs to:

- Establish who is responsible
- Set up a project team
- Stop work until project requirements are identified
- Assign responsibility to a project manager

Using an in-house designated scope-development team helps to prevent delays when a proposal is first being developed for a project.

11.7.18 No Ownership of Project at the Planning Stage

If no one claims ownership of a project at the planning stage (the project is no one's responsibility), delays cannot be avoided, and owners become frustrated when they have to deal with different people.

11.7.19 Site Selection and Acquisition

Site-selection and acquisition problems may be prevented by having a master plan and qualified personnel to help expedite these processes.

11.7.20 Social Impact Studies

Social impact studies and town meetings may cause the planning phase to be extended. Anticipating issues and ensuring that potential issues are addressed properly in planning documents are useful methods for preventing delays related to social impact studies.

11.7.21 Miscellaneous Planning Delays

Other factors that cause global project planning delays include the following:

- Planning is iterative, and federal studies require identification of national economic plans (federal loans) as well as locally preferred plans.
- Lack of knowledge about planning processes and strategic planning.
- Hidden agendas and strong individuals pushing their ideas.
- Client cost-sharing negotiations.
- Engineers are not accustomed to thinking like planners.
- Lack of political and governmental support.
- Planning projects in occupied buildings in many countries (historical buildings cannot be demolished) is more difficult and time consuming and requires multiple-construction phasing that, in turn, requires more preplanning.

11.8 Summary

Preventing delays is more efficient than mitigating them after they have occurred, but in the global arena it is impossible to predict what will cause delays because there are so many unknown factors related to political and social unrest, religious and social factors, labor disputes and strikes, technologic and economic limitations, natural disasters, and government restrictions. The main causes of construction delays in the global marketplace are design modifications, weather, material delivery, equipment delivery, incomplete drawings, and material quality. The most common mitigation strategies used on global projects are establishing a quality assurance program, working overtime, working on critical tasks, and increasing supervision.

Many other delay factors affect global construction projects, and specific delays may be unique. Unique project delays challenge project-management personnel to develop mitigation strategies that they would not normally resort to on domestic construction projects.

This chapter discussed the types of delays that occur on global construction projects, as well as the delays that occur during the planning stage of projects. Items that were discussed include accounting for global variations in construction using global construction indices, issues of dogmatism, bribery policies, and categories of nontechnical delays. Nontechnical delays may be caused by political instability, management factors, regional influences, material-transport issues, and optimistic schedules. Information also was provided on global engineering and construction delays that are caused by political and social unrest, religious and social factors, labor disputes and strikes, technologic and economic limitations, natural disasters, and government restrictions. A list of global delay factors was included in this chapter, along with a list of mitigation strategies that were developed from a survey of global E&C professionals. Project documentation was discussed in relation to how it is used to indicate projects delays, and global planning issues were explained in terms of their causes and possible mitigation strategies to provide insight into how global E&C personnel address planning delays.

DISCUSSION QUESTIONS

1. If the global construction labor purchasing parity is 163.6 for Great Britain and 56.1 in China, what would be the contract bid price in those countries if a project cost US$2,463,000 in Chicago, Illinois?

2. If a government wage is 200 rupias per year in Indonesia, what would be an appropriate bribe for a government official to allow a foreigner through customs into Indonesia? Explain why the amount of the bribe would be appropriate for Indonesia using the information in Chapter 14.

3. If a government wage is 15,000 convertible marks per year in Bosnia and Herzegovina, what would be an appropriate bribe for a government official to allow a foreigner through customs into Bosnia and Herzegovina? Explain why the amount of the bribe would be appropriate for Bosnia and Herzegovina, using the information in Chapter 14.

4. In addition to task drawings and job assignment sheets, what are three other ways of communicating what needs to be done to construction workers who do not speak the same language as the project management team?

5. What would be six appropriate questions to use in an interview guide while interviewing workers at a foreign job site to find out about their previous work experience on other construction projects?

6. What steps could a project manager take to get Australian workers back on a construction project and working again when they go on strike because of the aromas from a nearby Chinese restaurant?

7. In Africa, what three types of beasts of burden are used to transport construction materials? Explain how the beasts of burden could be used to transport construction materials.

8. During the feast of sacrifice in Islamic countries, animals are ritually slaughtered (the throats of animals are cut, and their blood is drained out) in public squares. How might a project manager persuade expatriate workers to return to work after they have walked off a construction job site in protest of the animal slaughters, and it is delaying the project?

9. If it is illegal to form unions, who should project-management personnel negotiate with to get workers to return to work when they walk off a construction job site? Explain why the person selected would be the appropriate person to negotiate with to end a strike.

10. What are mitigation strategies that could be used to reduce construction delays due to design modifications in a developing country?

11. What are four mitigation strategies that could be used to reduce construction delays when there are equipment- and material-delivery problems?

12. What are four mitigation strategies that could be used to reduce construction delays caused by poor-quality materials?

13. What would be an effective strategy that would reduce the amount of time required to wait for a construction-project inspection approval by government officials in Russia (refer to Chapter 14)?

14. Which four types of construction documentation listed in Table 11.4 could be used to determine whether a delay has occurred on a construction project that is related to change orders?

15. Of all the global project–planning delays listed in this chapter, what are the four most difficult ones to solve and why?

REFERENCES

Al-Hamer, B. 1990. *Construction Productivity in Bahrain,* unpublished report. University of Colorado, Boulder.

Al-Hammad, A., and S. Assaf. 1992. Design-construction interface problems in Saudi Arabia. *Building Research and Information* 20 (1): 60–63.

Aoude, Hisham. 1996. *A Computerized Construction Delay Management System.* Ph.D. dissertation, Polytechnic University, Brooklyn, New York. Ann Arbor, Mich.: University Microfilms.

Bruner, B., ed. 2005. *Time Almanac and Information Please.* Needham, Mass.: Pearson Education, Inc.

El-Nagar, H. and J. K. Yates. 1997. Construction documentation used as indicators of delays. *Journal of the American Association of Cost Engineers International* 39 (8): 31–37.

Eskander, A. 1990. *Total Project Management Delay Analysis Method (TPM-DAM) with Emphasis on Project Planning.* Ph.D. dissertation, Polytechnic University, Brooklyn, N.Y. Ann Arbor, Mich.: University Microfilms.

Goodman, A. 1994. *Analysis of Public Works Projects.* Englewood Cliffs, N.J.: Prentice-Hall.

Hanscomb/Means. 1994. Germany: Special In-Country Profile. In *Hanscomb/Means Report.* Atlanta, Ga.: Hanscomb, Faithful, and Gould.

Hanscomb/Means. 1993a. Australia: Special In-Country Profile, 1993. In *Hanscomb/Means Report.* Atlanta, Ga.: Hanscomb, Faithful, and Gould.

Hanscomb/Means. 1993b. Project Delivery Approaches: U.S.–Europe. In *Hanscomb/Means Report.* Atlanta, Ga.: Hanscomb, Faithful, and Gould.

Hanscomb/Means. 1991. Turkey: Special In-Country Profile. In *Hanscomb/Means Report.* Atlanta, Ga.: Hanscomb, Faithful, and Gould.

McKechnie, J. 2005. *Webster's Unabridged Dictionary.* New York: Simon and Schuster.

Project Management Associates. 1991. FORMSPEC 89 Section 00780, Glossary, West Palm Beach, Fla.: Project Management Associates.

Wang, P. 1986. The dogfight over Osaka Bay: Another trade dispute. *Newsweek* 108 (8): 67.

Wiggins, T. 2005a. Construction Cost Purchasing Parties. In *Hanscomb/Means Report.* Atlanta, Ga.: Hanscomb, Faithful, and Gould.

Wiggins, T. 2005b. Global Construction Cost Comparisons. In *Hanscomb/Means Report.* Atlanta, Ga.: Hanscomb, Faithful, and Gould.

World Bank. 2005. *The World Bank Development Report.* Washington, D.C.: World Bank.

Yates, J. K. 1993. Construction decision support system for delay analysis. *Construction Engineering and Management Journal* 119 (2): 226–244.

Yates, J. K., and J. Audi. 1998. Using decision support systems for delay analysis: The project management perception program. *Journal of the International Project Management Institute* 29 (2): 29–38.

Yates, J. K., and J. Audi. 1995. Construction delay communication system. *Project Management Institute Symposium,* 147–154. New Orleans, La.

Yates, J. K., and A. Eskander. 2002. Construction total project management planning issues and delay. *Journal of the International Project Management Institute* 33 (1): 37–48.

Chapter 12

Global Terrorism: Kidnapping and Design Considerations

12.1 Introduction

This chapter provides information about terrorism and kidnapping issues that affect members of the global engineering and construction (E&C) industry. The political and social dynamics taking place throughout the world are demonstrating that terrorism is a serious problem that affects E&C personnel. Every country in the world deals with internal and external aggression, but the E&C industry has survived, and sometimes thrived, through wars and terrorist acts. Terrorism is used to spread fear, and the bombing of major structures is an effective way to disrupt societies.

When structures are destroyed, E&C personnel repair them, restore basic services, and build new structures. Engineering and construction professionals are part of the restoration process after aggressive destruction. During wars, the military targets buildings that are economic centers or functional institutions. When these buildings are attacked and they collapse, it harms a large number of people and has a negative effect on the local economy.

Throughout history there have always been terrorist attacks that have caused either severe damage to buildings and other structures or the total collapse of structures; but in the last few decades there have been more widespread destruction due to terrorists. More people are now aware of terrorist acts because they have access to world events through the media and the World Wide Web (the Internet). Terrorist acts that have received the most media coverage include the federal building in Oklahoma City; the twin towers of the World Trade Center and the Pentagon in New York City and Washington, D.C. (United States); the *U.S.S. Cole* in Yemen, the U.S. embassies in Kenya and Tanzania, Africa; the subways in Madrid, Spain, and London, England; the United Nations building in Iraq; nightclubs in Europe and Bali, Indonesia; military residences in Saudi Arabia; hotels in the Middle East, buses and businesses in Israel; fast food restaurants in Pakistan, and movie theaters in India. In some countries; such as India and Pakistan, terrorist acts are a daily occurrence, and thousands of people are killed every year.

Engineering and construction personnel have had to reexamine the strategies they use to design and build structures to try and provide structures that can withstand terrorist attacks or that minimize injuries and deaths to building occupants. New regulations and codes are being developed for structures, but there is no global enforcement of the new codes; therefore, terrorism will continue to cause widespread destruction to structures, injuries to people, and the loss of life. The E&C industry has only been a target of terrorist organizations a few times (although individuals are kidnapped), but terrorist attacks harm or destroy the final products of the E&C industry such as buildings, rail systems, bridges, utility systems, water treatment facilities, irrigation systems, port facilities, highways, industrial plants, and dams.

How terrorism affects the E&C industry can be divided into three main categories:

1. Designing and building safer structures that can withstand or resist terrorist attacks
2. Retrofitting existing structures that have been damaged during terrorist attacks
3. Protecting laborers during construction against terrorism, sabotage, and kidnapping

Dealing with these three categories requires more time and staff for designing and building structures, better quality materials, larger construction elements, and the implementation of new technologies, all of which increase the cost of construction. New designs and regulations are being created and implemented on global construction projects, but since there is no global enforcement of codes or standards, there is no guarantee that all newly built structures will conform to the new standards. The issues that influence the design of structures to withstand terrorist attacks are similar to the issues that influence the design of structures to resist earthquakes.

Previously, structural designs only took into consideration gravitational forces, but after new information was discovered about earthquake forces, seismic design became an integral part of structural design. Another similarity between seismic hazards and terrorism is that in both cases there is no way to predict exactly when or where an earthquake or terrorist act will occur or how devastating it will be to a structure. Engineers can only hope to provide designs that minimize the impact on structures and that help to protect the inhabitants of those structures when they fail. In regions where earthquakes are prevalent, engineers may perform seismic calculations for the structure they design. Guidelines for terrorism-resistant designs are being developed and implemented in some countries, but it will take years for these guidelines to become industry standards. The *Unified Facilities Criteria* (UFC) (MIL-STD 3007) published by the United States military are standards and codes that provide design regulations for all Department of Defense (DOD) projects (Office of the Deputy Undersecretary of Defense, 2005). The U.S. Federal Emergency Management Agency (FEMA) has developed some guidelines for protecting structures and for designing safer structures: *FEMA 426: Reference Manual to Mitigate Potential Terrorist Attacks Against Buildings* (FEMA, 2003). They are in the process of developing additional guidelines.

This chapter provides information on terrorist organizations to help engineers and constructors deal with the threats posed by members of terrorist groups. Information is included on how to avoid potentially dangerous situations and to protect job sites and offices. It is impossible to provide recommendations that are applicable to all situations, but having some basic knowledge of the behavior of terrorists and how to react when kidnapped is useful to anyone working in the E&C industry.

The topics covered in this chapter include definitions of terrorism, terrorist behavior, financial support for terrorism, terrorist strategy and profit, dealing with terrorism, kidnapping issues, preventing terrorism, and suggestions for protecting labor. This chapter also includes a section on building protection that provides information on new design guidelines, methods for protecting existing structures, and how to design buildings to withstand terrorist attacks.

12.2 Definitions of Terrorism

The word *terrorism* originated in revolutionary France, and the first definition to appear in the *Oxford English Dictionary* refers to "government by intimidation, as carried out by the party in power in France during the Revolution of 1789–1797" (Stern, 1999, p. 11). The aim of terrorism is to intimidate people and to cause fear, and the impact of terrorism is more psychological than physical. Even though more people are injured or killed annually in car accidents than in terrorist acts, the impact of terrorist attacks is more widespread because of its political nature. Acts of terrorism always have been a part of civilized societies, and members of ancient societies used terrorist acts in the same manner as they are being used today—for coercing people to make certain political decisions or for money.

One definition of *terrorism* is "the threat or use of violence for political purposes by individuals or groups, whether acting for, or in opposition to, established governmental authority, when such actions are intended to shock, stun or intimidate a target group wider than the immediate victims. Terrorism has involved groups seeking to overthrow specific regimes, to rectify perceived national or group grievances, or to undermine international political order as an end in itself" (Long, 1990, p. 9). Another definition of *terrorism* is "an act or threat of violence against noncombatants with the objective of exacting revenge, intimidating, or otherwise influencing an audience" (Stern, 1999, p. 11). *Terrorism* has also been defined as "a criminal act that has planning, the use of a technology and a socio-political cause, and is designed to foster fear and dread outside the immediate impact area" (Clarke, 2005, p. 1).

Acts of terrorism may be direct or indirect. Direct actions are the acts themselves, whereas indirect acts of terrorism consist of illegal activities for acquiring money to support terrorism or activities by which new members are persuaded to join terrorist organizations. The most common goal of terrorists is political, and the immediate objectives of terrorist organizations are to create fear, not necessarily to destroy structures. The reason terrorists destroy structures is to create fear, and then they use that fear to obtain their objectives; but sometimes the actions of terrorists are merely for revenge. Since some terrorists seek maximum public exposure for their actions, public buildings are their targets.

Terrorist actions are increasing because of the relatively inexpensive cost and increased availability of materials that are used in explosive devices and chemical weapons. Citizens of any country that has petrochemical plants or fertilizer factories would be able to produce chemical weapons. It only takes a few weeks or months to turn a commercial plant into a military plant that makes nerve agents. For example, the chemical thiodiglycol, used as a solvent in making ink for ballpoint pens, lubricant additives, plastics, and photographic solution, is one of the ingredients that is used to produce mustard gas. During World War I, alcohol, bleaching powder, and sodium sulfite were used to produce thousands of tons of mustard gas (Levine, 2000). After World War I, members of governments in industrialized nations banned the use of mustard gas in warfare because of its horrible consequences when it is used on human beings.

During the 1970s, there were 8114 terrorist incidents reported throughout the world that resulted in 4798 deaths and 6902 injuries. During the 1980s, the number of incidents increased to 31,426, with 70,59 deaths and 47,849 injuries. From 1990 to 1996, there were 27,087 incidents that caused 51,797 deaths and 58,814 injuries (Stern, 1999). In 2004, there were 651 major acts of terrorism (ones that caused widespread destruction) that resulted in 9321 casualties (*U.S. News and World Report*, May 9, 2005).

12.3 Terrorist Behavior

Charles Russell and Bowman Miller, who are part of the United States Air Force (USAF), have tried to provide a profile of terrorists based on captured or known terrorists, and their description of terrorists includes "male, in his early twenties, single, from a middle or upper class family, well educated, with some university training, although he may be a university dropout who often joined, or was recruited, into the group while at a university" (Long, 1990, p. 17). The Unabomber (Ted Kasinsky), who was an active terrorist in the United States during the 1990s, did not match the USAF terrorist profile because he was highly educated (an undergraduate degree from Harvard University and a Ph.D. in mathematics) and had been an assistant professor at the University of California at Berkeley for two years. He acted alone rather than through a terrorist organization, and he targeted people in the engineering and technology areas. Because he was opposed to industrial society and modern civilization, he killed 3 people and injured 22 others by sending letter bombs through the mail that went off when someone tried to open them. The Unabomber example demonstrates that psychological profiles are not always useful when trying to determine who might become a terrorist.

Two other traits that might be found in some terrorists are low self-esteem and a willingness to take risks. Terrorists may be people who have experienced psychological damage during their childhood, but this characteristic seems to be more valid for leaders rather than followers. Terrorists are people who are attracted by situations that involve risk, and they see taking risks as a method of self-affirmation. Surprisingly, an attraction to violence is not always one of the main characteristics of terrorists. In fact, there have been terrorists who were terrified by weapons, such as Ulrike Meinhof, the leader of the Red Army Faction in Germany, who did not like guns.

Some terrorist leaders view the taking of hostages as a way of shifting the responsibility for killing somebody to someone else (Long, 1990). Terrorist leaders seem to be more attracted to the group of terrorists rather than to the ideas the terrorist organization supports. To be part of a terrorist organization someone has to be a good team member, they have to be able to keep secrets inside their network, and they have to obey rules (Long, 1990). Some people join terrorist organizations in the same manner that others join gangs—to belong to a group. But once they have joined a terrorist organization, they may not be able to leave it because they may be threatened or their families may be threatened by other members of the terrorist organization (kneecapping or breaking kneecaps, is a common threat) (Aviv, 2003).

An exact description of how people become terrorists and why is impossible to determine because there are too many psychological, economic, social, religious, and political factors that contribute to the formation of terrorists. What is clear is that terrorists are real people committing real atrocities in a real world. They were not born terrorists, but something caused them to become terrorists.

In the book *Engineering and Terrorism* it states that "engineers that come from societies that are in themselves under threat from internal and external influences, and where alternate (and legal) means of expression are either banned or methodically suppressed, will have the third terrorism necessity—sociopolitical cause and will be recruited by (or found offering their services to) terrorist organizations" (Clarke, 2005, p. 2).

Since there is no unanimously accepted definition for terrorism, some people argue that there is no difference between defending a country and terrorism. From the perspective of the members of terrorist organizations, their members are defending their country, religion, or political interest. They do not call themselves terrorists or criminals, and some terrorists believe that they are fighting for freedom. "Freedom fighting refers to a political goal, whereas terrorism refers to a tactic used to attain that goal; terrorist tactics can be, and often are, adopted by organizations and groups fighting in the name of freedom" (Long, 1990, p. 29).

12.4 Financial Support for Terrorism

Terrorist organizations usually operate around a specific purpose, and they need financial support in order to exist, and some nations may be supporting terrorist organizations but identifying countries that support this form of aggression is difficult because no nation admits to sponsoring terrorism (Long, 1990).

Terrorist groups may have connections with other illegal organizations in order to obtain funds. Since they are illegal organizations, they are forced to obtain money for their organization by dealing with other illicit groups such as drugs dealers, black marketeers, and identity thieves. However, some terrorist groups are financed through legitimate business ventures, such as construction companies. In order to survive and to be able to train their members, terrorist organizations require funds. Estimates of the cost of the attacks on the twin towers of the World Trade Center and the Pentagon in the United States are that it cost over half a million dollars to plan the attacks (Aviv, 2003). Terrorist organizations need to obtain enough money to purchase weapons or other materials from the black market and to pay their members.

One way of securing funds for terrorism is by selling drugs, such as opium. Many of the countries with high rates of drug trafficking are also on the list of countries that support terrorism.

Advances in technology that have reduced the cost of technology and that have provided easier access to information and communication techniques may be aiding terrorist groups. While democratic countries guarantee freedom, access to information, rights of people, and free speech and private communication, terrorists also use these rights to benefit their organizations. Sometimes terrorists use the knowledge they have gained while living in another country against members of that country. In the case of the attack on the World Trade Center twin towers in New York City in 2001, terrorists learned to fly airplanes by taking flying lessons in the United States, and then they hijacked U.S. commercial airplanes and used them to attack two 110-story skyscrapers and the Pentagon.

12.5 Terrorist Strategies and Profiles

The most common types of terrorism are "armed attacks, including bombing, arson murders, and physical injury along with hijacking and kidnapping. The targets can be individuals or properties or both. Usually terrorist acts are done in crowded areas, to maximize the number of casualties. Thus, public buildings are often good targets for meeting this goal. Physical targets include military installations, government offices, public transportation, public utilities, and private properties with symbolic value" (Long, 1990, p. 124).

Some reasons for terrorist acts include (Long, 1990, p. 124):

- To instill a sense of fear and helplessness among civilians.
- To instill a sense of impotence among government officials or to intimidate them.
- To badly affect the national economy by discouraging foreign investments and tourists. Economic targets frequently include key installations, key infrastruc-

ture, public carriers, and transportation networks such as railroads and high-ways.
- To create international incidents to publicize their cause.

There are over 600 terrorist organizations that operate throughout the world. The U.S. Department of State provides descriptions of some of the main terrorist organizations on its Web site at www.state.gov/s/ct/rls/c14812.htm. Information is available in *The Anatomy of Terrorism,* by David Long, and *The Complete Terrorism Survival Guide,* by Juval Aviv, about many of the terrorist groups that operate around the world, including their names, country of origin, year founded, preferred locations, methods of acting, number of members, names of the leaders of organizations, their ideologies, and their targets (Long, 1990; Aviv, 2003). Engineers and constructors should be familiar with the terrorist organizations that operate close to where their projects are located in order to have a basic understanding of how they operate and the risks associated with working in that location.

12.6 Kidnapping Issues

Another form of terrorism is the kidnapping of E&C personnel while they are working on projects overseas. The threat of being kidnapped has always been present whether an engineer or construction professional works in a developed or developing country. Unfortunately, kidnapping professionals for ransom has become a major industry in some Central and South American countries such as Colombia.

12.6.1 Kidnapping Risks

Collecting money is not the only reason for kidnapping, as some organizations kidnap foreigners for political ransom. Since the liability of E&C firms has increased due to more kidnappings, some multinational companies have had to adopt drastic measures for protecting their labor, while other firms have obtained kidnapping and ransom (K&R) insurance. The political risk level of a country is one factor that members of E&C firms consider when they are trying to decide whether to invest in a project even if the project has a potentially high rate of profitability.

Previously, E&C personnel were taught how to deal with uncertainty through risk management, but now they also have to be trained to deal with terrorism threats. When potential projects are being evaluated in terms of political and economic risk, they also must be evaluated based on the potential risk for terrorism or the possibility that personnel will be kidnapped, which requires data that are next to impossible to determine with any accuracy. As is done with other risks, the only way to compensate for unknown hazards is to provide additional funds for contingencies. However, the unknown contingencies involved with terrorism and kidnapping include the loss of lives, and no amount of money compensates family members.

One way to prepare for these types of risks is for companies to train personnel on how to deal with the dangers related to acts of terrorism and what to do if someone is trying to kidnap them (both of which substantially increase the cost of projects). When there are terrorism risks, compensation packages for employees increase, and sometimes companies are paying three to five times more than domestic salaries in order to be able to hire personnel that are willing to work in high-risk areas. Even laborers and support service personnel have to be compensated at rates higher than the rates paid on projects being constructed in low-risk zones.

The globalization of economies, which creates an environment where more people from different countries interact and depend on each other, has resulted in more individuals working for multinational companies, global nongovernmental agencies, charities, aid agencies, and governments around the world. With increasing globalization, the risks associated with terrorism and kidnapping are increasing because there are now more foreign targets.

12.6.2 Political and Economic Kidnapping

For ELC personnel kidnapping issues could be divided into four main areas:

1. Preventing kidnappings
2. How companies can reduce their risk
3. What companies should do when a kidnapping occurs
4. What companies should do once a kidnapping case is solved through negotiations

Engineering and construction personnel should know that 96 percent of all kidnappings are resolved by the hostages being released; therefore, if hostages are able to stay alive and withstand the kidnapping, they have a good change of eventually being released by their captors (Ariv, 2003). Professional negotiators have helped to increase the release rate for hostages, along with having funds available for negotiations through kidnapping and ransom insurance.

There are many different reasons for terrorist acts, but there are only two types of kidnapping: *economic kidnapping* and *political kidnapping* (Briggs, 2001). Economic kidnapping (the *kidnapping business*) takes place when a financial demand is made in exchange for releasing hostages. Political kidnapping takes place when kidnappers make political demands in return for the release of hostages, such as having laws changed, prisoner exchanges, foreigners leaving their native country, or changes in government policies. It is estimated that globally kidnappers receive approximately US$500 million each year in ransom payments and that there are approximately 10,000 kidnappings each year (Briggs, 2001).

The highest number of kidnappings takes place in Columbia, where in 2000 the National Police recorded 3162 cases of kidnapping (Briggs, 2001). The Columbian kidnapping groups are also active in Venezuela and Ecuador, and these three countries are the top 3 countries in the top 10 list of where the highest number of kidnappings occur, that includes Brazil, Colombia, Ecuador, the former Soviet Union, India, Venezuela, Nigeria, Mexico, the Philippines, and South Africa.

A U.S. Department of State study indicated that U.S. business executives are the most frequent targets of global terrorism. If kidnappers are trying to gain financially from a kidnapping, it is easier do it in the private sector, where companies have more money. In the U.S. Department of State study, 58 percent of kidnapping victims were from the business sector, and 18 percent were diplomatic staff (U.S. Department of State, 2005). The E&C industry is one of the largest industries in the world, and it requires a vast amount of personnel, which increases the chances of kidnappers targeting personnel from the E&C industry. Top executives are not always the main targets of kidnappers; operations level personnel are easier to kidnap than top executives, because they travel more frequently and work exposed at job sites.

12.6.3 Effects of Kidnapping

In kidnapping cases, there are three main issues:

1. The physical and psychological effects on hostages
2. The psychological trauma on family and friends of hostages
3. Harm to employers

Many former hostages admit that their families suffered more during their captivity than they did while they were being held as hostages because they had a better idea of their chances of being released than their family members (Briggs, 2001). The employer of a hostage is also affected because the employer usually is the one that negotiates for the release of a hostage, which is expensive in terms of time and money. When firms fail to get their employees released from captivity, it is publicized widely, and it harms the reputation of a firm, which is an added incentive for executives to successfully negotiate for the release of their employees.

12.6.4 Kidnapping Prevention and Kidnapping and Ransom Insurance

Members of E&C firms should be aware of kidnapping prevention-techniques, and know how to cope and behave if they are kidnapped while working overseas. Companies should provide training for their employees before sending them to work in foreign countries, but it is also the duty of employees to obey guidelines set by their employer and to try to minimize their exposure to risks. It is difficult for a company or an institution to dictate what employees can or cannot do in their free time, and kidnappings may occur while employees are driving between home and work or during their leisure time.

Kidnapping and ransom (K&R) insurance policies cover losses related to a kidnap, such as the ransom payment, any travel or living expenses for family members, the wages of the hostages while they are being held captive, any counseling costs, and the cost of a negotiator who provides advice on how to manage the case. Unfortunately, K&R insurance does not minimize the risk of a kidnapping occurring, nor does it always guarantee that someone will be released after they are kidnapped and held hostage, because K&R insurance only allows firms to negotiate for the release of a hostage up to the *limits of the insurance policy*.

A better strategy for dealing with kidnapping threats is initiatives such as those provided by the Overseas Security Advisory Council (OSAC), set up by the U.S. State Department in 1984. This institution assists members of U.S. companies with security issues such as kidnappings. According to its official Web site, the OSAC is a federal advisory committee that promotes security worldwide. The OSAC has over 2700 member organizations, 600 associates, and 5 U.S. government agencies that participate in the council: (1) the U.S. State Department, (2) the U.S. Agency for International Development, (3) the U.S. Chamber of Commerce, (4) the U.S. Department of Commerce, and (5) the U.S. Department of Treasury. For additional information, see the OSAC Web site at www.ds-osac.org.

Every firm should have a crisis plan that is activated when an employee is kidnapped, but not many companies have them. In a survey of 5000 companies, only 45 percent had a business continuity or management plan in place (Business Continuity Institute, 2005). One way to increase the use of crisis plans would be for owners to require them as part of bid proposals or before signing a contract, or insurance companies could require a crisis plan before selling insurance policies to a firm.

When trying to protect their employees companies need access to reliable information and companies should be able to acquire information from as many sources as possible and to exchange information with other firms. Sometimes this is not possible because of restrictions created by confidentiality agreements. The method used for communicating information to employees also contributes to the effectiveness of the information. The E&C industry employs people from many different cultures, and the same information could be interpreted differently depending on the country of origin of an employee. Advice might conflict with religious practices or cultural norms or be offensive in some cultures.

Communications should be as clear and easy to access as possible. For example, using a Web site to provide emergency instructions could be ineffective if construction teams are working in remote locations where there is no access to the Internet or cell phones. Laborers may not have access to computers to look up information; therefore, in remote locations, project management personnel need to communicate information directly to their subordinates and ensure that their employees follow company policies and guidelines. Individuals respond much more positively to advice when they are presented with information that demonstrates the relevance of the advice (Briggs, 2001).

12.7 Preventing Terrorism

Terrorist acts are hard to prevent unless people increase their awareness of their surroundings. Engineering and construction professionals should be trained to be aware of safety issues on construction projects. Moreover, their heightened awareness should extend beyond job sites to items that could be related to terrorist acts.

12.7.1 Awareness Issues

Examples of situations that could be suspicious include (De Becker, 2002, p. 26):

- Payment of a large amount in cash
- Impatience, nervousness, or uncommunicative behavior
- Buying a product that can be part of a bomb and having a desire to take the product right away, with no interest in having it delivered to them
- A person's lack of familiarity with a product he or she is buying
- People asking information about how to get onto the roof of the building or into the basement or other areas (information about access to different parts of the building that are not usually used by the public)
- People asking for information about the air-conditioning or heating, system of a building
- Someone hurrying away from the building after putting down a suitcase, bag, or other large object
- Someone wearing an official uniform (police uniform) that is not a police officer
- Someone who gets nervous when asked to complete a form with personal information when buying something that is considered to be a hazardous material

Other things that should cause suspicion include:

- When objects in an office or job site have been disturbed while no one was suppose to be there
- Being followed by the same person or vehicle
- Signs of being watched by someone
- Vehicles parked in a neighborhood for long periods of time with people sitting in them
- People who visit an office or home without identifying themselves
- People wearing jackets or coats in hot weather
- People carrying heavy objects that are concealed from view
- People wearing dark glasses inside buildings
- People concealing their hands inside their jackets or shirts, not their pockets
- People wearing clothing that is out of place in that region of the world
- A person who is not known by anyone asking detailed questions about an office or a job site and the mechanics of facilities
- If a company Web site has an unusually high number of visits during a particular day or week
- Someone (or the same vehicle) being seen close to an office or job site for several days

12.7.2 Suggestions for Protecting Labor

This subsection provides suggestions for reducing the chance of terrorist attacks on engineering offices or construction job sites located in foreign countries.

In some parts of the world the targets of terrorist groups are people from developed countries; therefore, when a project is being built by personnel from developed countries, it is helpful to hire predominantly local laborers because that may reduce the probability of a terrorist attack, because it is less likely that terrorists will attack people from their own country.

Displaying the flag of a foreign country at a job site, an office, or on personal belongings attracts the attention of terrorists. It is better for members of foreign companies to remain as discrete as possible in terms of displaying their name and country flag.

Access to job sites should be limited to authorized personnel only, and job sites should be equipped with bomb detectors and video cameras for monitoring every person and every vehicle that enters and exits. Adding more emergency stairs or ladders than are required by local building codes to buildings and temporary structures provides more exits for evacuation of employees during a crisis.

Try to blend in with people from the local area by adopting local customs, by wearing the same style of dress, and behaving the same as local people. Global risk consultants offer self-protection seminars about how to blend in, how to detect surveillance, and how to travel globally.

Additional suggestions for protection include learning some basic words in the local language including emergency phrases for embassies, employer's telephone numbers, and words for automobile, medical facilities, police, fire department, military installations, towing services, and pharmacies (Aviv, 2005). Having these words written on the back of a business card that may be handed to someone in an emergency helps to save time during a crisis. While living in a foreign country, ask local people about risks in that country. Clubs and bars are places where unofficial information is shared by local people.

12.7.2.1 Job Site and Vehicle Access Issues

Employees should drive the same type of car as those driven by other people in the area. At some job sites, everyone drives the same type and color of vehicle, and the only way to tell them apart is by the numbers on the vehicles. Vehicles should be examined before they are started and driven. Items to be investigated include checking fluid levels, especially the brake fluid and steering fluid, and making sure that no critical lines have been cut, such as fuel lines or oil lines. Also look for foreign objects that have been attached to the sides of the vehicle or under the vehicle. Any lose wire, missing parts, or changes in the location of parts should be closely examined because it could be an indication that someone has tampered with the vehicle. Other suggestions for automobiles include the following:

- Do not put personalized license plates on vehicles.
- All vehicles used at a job site or office should have their engines and mechanical parts checked frequently.

- Some bombs used in vehicles are set to detonate when the ignition is engaged, but vehicles can be equipped with devices that start the engine remotely so that the driver is not inside the vehicle when the engine is started.
- A thin film of dust left on the car will show whether anyone has touched the vehicle (dust can be added by using talcum powder).
- Do not park in underground or self-park garages.

12.7.2.2 Food and Medical Considerations

When working overseas in areas that are less developed, and far away from urban areas, or where a rescue crew cannot arrive in a short period of time (such as oil refineries, nuclear power plants, fertilizer plants, and water treatment facilities), workers should be vaccinated, and/or they should carry with them vaccines and antibiotics that can neutralize viruses or spores in case of a biologic attack (ciprofloxin for anthrax and other broad spectrum antibiotics, such as tetracycline). Bottled water may not be safe to drink because it may be taken directly out of municipal water systems. Tablets may be purchased that kill some bacteria found in water supplies, but they do not neutralize chemicals. Only a small percentage of people in the world have safe drinking water; therefore, water should be boiled and strained through cloth before drinking (boiling does not eliminate chemicals, only bacteria).

Job sites should have reliable food suppliers, and if anything appears suspicious, have it tested by a reputable laboratory. There should be a system in place that sets certain standards for food and its preparation. Food should be checked not only visually but also by smelling and testing it before large groups are fed. Find out when the food was prepared and by whom to reduce the risk of contamination.

Medications that should be available at job sites include medicines for preventing vomiting (Emetrol), antinausea liquid to induce vomiting (ipecac syrup), antidiarrheals (Imodium liquid or pills), and ciprofloxin or other antibiotics (for anthrax exposure). The first three items are sold without a prescription in pharmacies.

Foreign companies should not rent space in buildings with underground garages, because explosives may be hidden in vehicles and used to damage the foundation of the building. Workers should receive training similar to what the military uses on how to survive during a crisis. Military personnel are taught how to survive in critical conditions using only natural resources from the environment.

12.7.2.3 Security Guidelines and Emergency Numbers

Additional information related to protection issues is published by the U.S. Department of State Overseas Security Advisory Council in the publications *Personal Security Guidelines for the American Business Traveler* and *Security Guidelines for American Enterprises Abroad,* which can be found on the Web site at www.ds.state.gov/about/publications/osac/personal/html.

For U.S. citizens, the phone number for the U.S. Department of State is 202-647-6225, and the phone numbers for the Office of Overseas Citizen's Service are 888-407-4747 and 317-472-2328. The after-hours emergency number is 202-647-

4000. Similar organizations are located in countries throughout the world, or embassies may be contacted when there is a crisis.

12.7.2.4 Reacting to Bomb Blasts

If a bomb is detonated, always be prepared for a second blast. After the first explosion, quickly crawl into an interior room or walkway, because that provides an extra wall for protection against a possible second blast. Rooms should be rearranged to minimize injuries in offices and homes. Move desks, tables, beds, and seating areas away from exterior walls. If a bedroom has an exterior wall, move beds into other living spaces, such as the dining room. If an office has an exterior wall, move desks into interior rooms, such as reception areas, lunchrooms, conference rooms, or storage rooms.

12.7.2.5 Suggestions for Avoiding Kidnappings

This subsection contains information on procedures that may be followed to reduce the risk of having personnel kidnapped while working overseas.

- Do not leave maps on car seats or carry them around because it advertises that someone is not from the local area. Plan trips so that travel is completed before dark.
- Stay at major worldwide hotel chains, monitor global radio stations such as the Voice of America, the British Broadcasting Corp, Univision, etc.
- Do not travel on the same route to work every day (be unpredictable). If possible, use different vehicles and different hours to go to work. Do not park vehicles in marked parking places.
- Never go anywhere alone, and make sure that someone in the office knows where personnel are at all times. Have contingency plans that include coded signals such as hand signals or code words.
- Do not label belongings with names or company names.
- Do not drive with the windows of the car open or with unlocked doors.
- A large dog discourages and sometimes may stop a kidnapper. In Islamic countries dogs are considered to be unclean; if the dog touches someone who is Muslim they will have to wash their hands or bodies seven times.
- Only carry firearms if they are legal, obtain a license to use them, and learn how to use them. The penalties for carrying firearms may be severe in some countries, including lengthy jail sentences (jails are primitive in developing countries).
- Use satellite cell phones because they function in most regions of the world by using satellite networks not local networks, therefore, communication cannot be disrupted even if terrorists attack local communication systems. Cell phones in the United States are on different radio frequencies, and they use different wireless technology than cell phones in the rest of the world. Seventy percent of the world, 162 countries, use a digital platform called Global System for Mobile (GSM) Communications, but this system is not used in the United States. In the United States, Voicestream and Cingular sell and rent world

phones, but Ericsson and Nokia phones will not work in Japan or Latin America. Check with cell phone providers to see if their service works in other countries and make sure their phones have international roaming (Ariv, 2005).

- A coded chip may be inserted underneath the skin that sends information on the location of an individual (coordinates). This technology is already in use to monitor the locations of wild animals that are used in scientific research and to identify lost pets through a national registry. The chip is injected under the skin with a small needle.
- Never let anyone know if your firm has kidnap and ransom insurance because it might void the policy or give someone the idea to kidnap employees of the company.
- When working overseas, find out if the government has any type of program to reward people who provide tips for locating known terrorists, such as the *Rewards for Justice Program,* which is a counterterrorism program run by the State Department Bureau of Diplomatic Service in Colombia, or other types of counterterrorism centers.

If someone is kidnapped, the following suggestions are useful for surviving the kidnapping:

- If someone is trying to kidnap someone, yell "fire," because it gets the attention of more people.
- In case of a kidnapping, try to stay as calm as possible (or pretend to be calm).
- While being abducted, look for landmarks that will help someone locate you. Information on someone's location can be transmitted using code words when verifying that they are still alive during negotiations.
- Never look a kidnapper in the eye because it can be interpreted as a sign of hostility or that someone is a troublemaker (if someone is going to be killed early on, it will be troublemakers). Act as passive as possible.
- If kidnappers keep their faces covered, especially if they are alone with the victim, it is a good sign; it may mean that the kidnappers intend to release the victim, and they do not want their faces to be identified.
- Try to find out the names of the kidnappers, and use their names when speaking to them. People are more responsive and empathetic when they hear their own names. In the subconscious part of their mind, kidnappers form a bond with their victims, and the probability of using violence against them decreases.
- Try to talk to the kidnappers, and pretend to understand their situation. It is harder for terrorists to injure or kill, someone who seems to understand them or who is sympathetic to their situation.
- Do not make any promises or tell kidnappers that there is money somewhere.
- Engineering and construction professionals should have some understanding of psychology or personality types so that they can try to figure out the per-

sonality type of the kidnapper (which helps someone to negotiate with him or her more effectively if code words are used to describe him or her).

- Someone convinced them to be terrorists, so there is a chance that they can be persuaded not to be terrorists.
- Never accuse a terrorist of anything or argue with him or her because he or she might become unpredictable, angry, or act violently. Talk in a soft and soothing voice to your captors.

12.7.2.6 Being a Kidnap Victim

If kidnapped, follow a set routine every day, and do everything slowly and deliberately to fill the time. Kidnap victims should also do the following to help maintain their sanity:

- Figure out a method for keeping track of time.
- Exercise every day, even if it is merely walking in place or moving extremities while lying down. It is important to stay healthy as that prevents the immune system from breaking down.
- Eat whatever is offered because it helps to maintain strength.
- Try to mark the days to prevent disorientation.
- Recite memorized passages or poems, or sing to keep the mind sharp and to prevent feelings of isolation.
- Use whatever is available to create puzzles or to play simple games (sticks, rocks, dirt, pillow stuffing, scraps of paper to play tic-tac-toe hangman, word puzzles, etc.) to pass the time.
- Do not sleep 24 hours a day as the body will break down quickly (along with the mind).

12.8 Building Protection

Protecting structures from terrorist attacks requires the coordination of several agencies within a country. Since terrorists may hijack airplanes and use them as weapons, buildings and other structures are now more vulnerable to major explosions. Car bombs are being used throughout the world to damage or destroy buildings and other structures. Structures are designed to withstand live loads (e.g., human beings, equipment, furniture, and other objects that add weight to floors) and dead loads (i.e., the weight of the structure), both of which exert pressure on the top of floors. When a structure is subjected to an explosion that originates at ground level (such as a car bomb) or below a floor, individual floors experience load reversals, and they have to withstand a net uplift equal to the dead load plus one-half of the live load. Most existing structures were not analyzed for load reversals.

Many of the buildings that have been bombed by terrorists using car bombs have been newer, lighter structures, not older, heavy masonry structures. Designing structures to withstand explosions would require massive structures with no points of vulnerability (unreinforced areas or areas with a limited number of column supports or redundant systems). Terrorist organizations have members who are engineers and who participate in designing bombing schemes, and if they have access to the plans for a structure or they are able to view the inside of a structure, they may be able to determine points of vulnerability. To counteract this possibility, the engineers responsible for designing structures need to do a *vulnerability analysis* when they are designing structures or when someone is doing a constructability review (designs being reviewed by construction personnel during the design process). Designs should be altered accordingly to eliminate points of vulnerability by adding redundant systems or additional structural components.

12.8.1 Design and Safety Tradeoffs

Clients need to realize that the structures they desire might have to be altered to prevent floors or structures from collapsing during explosions. Engineers have a responsibility to make clients aware of the risks associated with certain designs, because the first obligation of engineers is to protect the public (*public safety*) as set forth by the American Society of Civil Engineers (ASCE) in the code of ethics. The Institution of Engineers Code of Ethics states that "the primary responsibility of engineers is to the community, members shall at all times place their responsibility for the welfare, health, and safety of the community before their responsibility to sectional or private interests, or to other members" (Clarke, 2005, p. 4). The Institution of Engineers is the professional society of engineers throughout the world.

If a client does not want to alter the proposed structure, then design engineers may use the environment surrounding the structure to help protect it from explosions. Minimum standoff distances (the distance from the controlled perimeter to the closest point on the building exterior or the inhabited portion of the building) are 33 feet, but the effective standoff distance is 82 feet. Structures may be protected by concrete planters, concrete or steel artwork, or other barriers. Concrete planters may be used for hiding bombs; therefore, other types of concrete barriers are safer. Concrete barriers need to be 3 to 5 feet thick to be effective (or two barriers may be used), especially if vehicles such as concrete trucks are filled with explosives and detonated near a structure. If concrete barriers are located close to buildings, the explosion creates a shockwave that rises vertically and damages the upper floors. The slabs on concrete bridges are held in place by the dead weight of each slab. If an explosion takes place under the slab, it may rise vertically and be dislodged from its supports. During major flooding, concrete bridge slabs are lifted off of their supports by bouyancy forces.

Since the surface area of a sphere is proportional to the square of its radius, doubling the distance between the barrier and the building will reduce a bomb's shockwave intensity by a factor of 4, tripling the distance reduces bomb shockwave intensity by a factor of 9, and so forth. Therefore, concrete barriers should be located as far away from buildings as possible, preferably at the minimum standoff distance of 33 feet.

Locating building entrances and windows away from roadways helps to prevent shattering glass from injuring building occupants and pedestrians walking in front of the building. Buildings where most of the exterior wall surface area is covered by glass or buildings with large windows are the most dangerous to occupants because of flying glass and debris, and their support columns are located further apart to accommodate large windows. Reducing the number of windows and the size of windows helps to reduce flying debris, and antishatter film may be installed on windows to prevent shattered glass from injuring building occupants or people below the building. If antishatter film is not available, then clear packing tape may be used as a temporary measure to keep glass from shattering during an explosion. It is difficult to design buildings that are completely *bombproof* without substantially increasing their cost, but they may be designed to be *bomb resistant* at an additional cost of 3 to 5 percent.

Some structures are designed to withstand the impact of an airplane without collapsing but if buildings are designed with multiple redundencies they would not be able to have large open spaces on each floor. The World Trade Center twin towers in New York City were designed to withstand the impact of a Boeing 707 airplane—the largest plane available at the time the buildings were designed—but the designers did not anticipate the heat generated by the fireballs created by the burning fuel from the Boeing 747 airplanes that was fed by vast amounts of paper and other flammable materials. The floors of the twin towers of the World Trade Center were only braced at the exterior edge and in the center area surrounding the elevator shafts to create open floor spaces. The floor to column connections were designed to hold the weight of one floor plus the live load of each floor.

When the ninety-fifth to ninety-sixth floors in the North Tower and the seventy-ninth through eighty-third floors in the South Tower were compromised by the two airplanes that hit both structures and their resulting fireballs (which melted the fireproofing on the steel members), the beam-to-column connections failed when the steel was heated past its melting temperature, and the weight of each of the floors that were hit was transmitted to the floors below them, which instantaneously doubled the load. Then, when the floor beneath those two floors failed, the load was tripled on the next floor, and so forth, causing each subsequent floor to fail more quickly. Structures are designed with larger columns in the lower floors and more slender columns in the upper floors to compensate for more weight on the lower floor columns, but the connections from the columns to the beams are usually the same throughout buildings.

If the areas of the building closest to roadways are used for offices or residences, more people will be injured if the building is bombed from a roadway. These areas could be used for low-occupancy activities to reduce the number of injuries or deaths to building occupants.

12.8.2 Increasing Building Safety

The British Home Office publishes, *Bombs: Protecting People and Property—A Handbook for Managers,* which is available on the Web at www.homeoffice.gov.uk

/atoz/terrorist.htm. This handbook describes techniques and methods that may be used to help prevent harm to the occupants of structures.

Buildings may be designed to include *protected spaces,* as is done in Israel where such spaces have been required by law since 1992. Apartment or floor protected spaces (FPS) include "a blast door that opens outward, a filtered ventilation system, and emergency lighting, plus a telephone and connections for TV and radio reception." The rooms are stacked one above the other, extending the full height of a building, so that they form one contiguous tower of secure rooms. Sealed openings in the room's floor and ceiling interconnect the FPS to facilitate escape or rescue (Aviv, 2003). More information on protected spaces is available at www.idf.il/English/organization/homefront/homefront2.stm.

Since structures are targets of terrorists, firms that rent space in buildings should investigate methods for protecting their employees and occupants of the building by including new forms of composite materials that make it possible to replace ordinary window glass with clear substitutes that are shatterproof, bulletproof, and bombproof (Davidson and Rees-Moog, 1993). Bulletproof glass has been developed by ArmorVision Plastics and Glass and tested in the United States, Israel, Hungry, and Venezuela. This bulletproof glass does not shatter when it is hit by a bullet on one side but someone could fire a shot through the glass in the other direction because it is one-way bulletproof glass. Any automobile may be equipped with this glass, with an increase in weight of 200 pounds. Cars may be retrofitted with protective materials and devices (such as has been done to Lincoln Town cars), but it increases the cost of the vehicle to over US$200,000.

The ventilation systems of buildings are vulnerability points in terms of terrorism. The larger the building, the more impact there would be if toxic substances were put into the ventilation system. Some large buildings have only a main ventilation system that serves the entire building, and once a terrorist agent is introduced into the system, it spreads to the entire building.

Information on the name, symptoms, and level of mortality for some terrorist chemical agents and viruses are listed in *The Ultimate Terrorists* and *The Complete Terrorism Survival Guide* (Stern, 1999; Aviv, 2003). Some of the antipersonnel biological warfare agents are *Bacillus anthracis* (anthrax), *Yersinia pestis* (pneumonic plague), *Francisella tularensis* (tularemia), Filoviridae (ebola virus), *Clostridium botulinum* (botulism), smallpox, etc. (Aviv, 2003). All these agents have a rate of mortality that exeeds 90 percent.

The following are some techniques for preventing or reducing air contamination that have been proposed by the U.S. Office of Occupational and Radiological Health (*HVAC-Building Vulnerability Assessment Tool,* 2004):

- Use individual air-conditioning units installed in walls or windows or free-standing units
- Pressurize stairways with 100 percent outside air to provide safer evacuation in case of a terrorist threat (this method is not effective in structures where firefighters use the stairwells to access floors to fight fires).

- An emergency air-handling unit (AHU) shutdown plan should be available.
- The air intakes should be as inaccessible as possible; they should be placed at the highest possible level and secured with access limited to authorized personnel.
- Increase the air-filtering efficiency of ventilation systems.
- Provide *shelter in place rooms* that have minimum air infiltration and where the occupants can stay in case a toxic agent is released in the building.
- Mailrooms or delivery areas should be ventilated separately than the rest of the building in case toxic agents are sent through the mail.

Engineers need to consider other issues related to terrorism when they are designing and operating facilities, such as the following:

- Safety features and backup systems to protect nuclear power plants, coal-fired power plants, and other critical infrastructure systems and facilities against terrorist attacks including backup systems for the supervisory control and data-acquisition systems (SCADA)
- Safety features to prevent disruption of utility systems and natural gas and gasoline pipelines by terrorists
- Safety features to protect buildings and structures from excessive damage during explosions

One example of a government agency that provides standards to enhance the protection of structures during a terrorist attack is the U.S. Office of the Deputy Undersecretary of Defense, which published a document in September 2002 called the *Anti-Terrorism Standard for Buildings* that was updated and approved by the Department of Defense (DOD) Engineering Senior Executive Panel (ESEP) in October 2003. The *Anti-Terrorism Standards for Buildings* includes (U.S. Office of the Deputy Undersecretary of Defense, 2002):

- Standards that were developed by the Department of Defense Security Engineer Working Group and published as a *Unified Facilities Criteria* (UFC 4-010-01 under Mil-Std-3007).
- Designing building resistance to terrorist attacks into all DOD buildings that are built after 2004.
- Any building, including leased facilities, occupied by DOD personnel, regardless of the funding source, must meet antiterrorism requirements provided in the code.
- The standards used by the DOD are not based on a specific type of threat, but they provide methods for minimizing injuries and fatalities in the event of a terrorist attack.

The main strategy of the DOD standards is to (U.S. Office of the Deputy Undersecretary of Defense, 2002):

- Maximize standoff distance (the distance between the building and the closest access point for vehicles)
- Avoid the progressive collapse of structures (where floors collapse onto floors below them)
- Reduce flying debris hazards during terrorist attacks
- Limit airborne contamination within a structure
- Provide mass notification in the event of a terrorist attack

12.8.3 Antiterrorism Standards

Requirements for antiterrorism standards for the DOD construction are included in the construction-programming document *DOD Form 1391: Military Construction Project Data*. This document includes the estimated additional costs for antiterrorism requirements and such items as the following considerations (U.S. DOD, 2005):

- Incorporating antiterrorism standards into new buildings where required standoff distances are available, which increases construction costs by 3 to 5 percent.
- Costs vary considerably where sufficient standoff distances are not already available for upgrading existing buildings.
- Site and interior building layouts are a low- or minimal-cost method to meet antiterrorism requirements.
- Items that increase costs are upgraded windows, structural detailing for the prevention of progressive collapses, and modifications of the building interior to minimize hazardous flying debris.

The antiterrorism standards for buildings provide information on (U.S. Office of the Deputy Undersecretary of Defense, 2002):

- Level of protection
- Critical facilities
- Explosive safety standards
- Design strategies
- Maximum standoff distance
- Preventing building collapses
- Minimizing flying hazardous debris
- Limiting airborne contamination
- Minimum parking distances
- Recommendations for additional antiterrorism measures for old and new buildings

A complete version of the antiterrorism standard may be downloaded from the Web site: www.acq.osd.mil/ie/irm/irm_library/UFC4_010_01.pdf.

The U.S. Federal Emergency Management Agency (FEMA) is another institution that provides publications containing information on reducing physical damage to structural and nonstructural components of buildings and related infrastructure during bombings and biological, chemical, or radiological attacks. The FEMA publications provides design criteria to strengthen buildings to resist forces that may be created during a terrorist assault.

Printed copies of FEMA publications are available at the Government Printing Office (GPO), Washington, D.C. (phone: 202-512-1800; nationwide: 866-512-1800; fax: 202-512-2250) or at www.bookstore.gpo.gov, or they can be downloaded from www.fema.gov/fima/rmsp.shtm. The following paragraphs include brief descriptions of some of the FEMA publications. Additional information on FEMA publications can be found at the FEMA Web site at www.fema.gov/fima/rmsp. shtm.

FEMA Document 426: *Reference Manual to Mitigate Potential Terrorist Attacks Against Buildings.* Document 426 provides guidance on reducing physical damage to buildings, related infrastructures, and people caused by terrorist assaults. The document provides incremental approaches that may be implemented over time to decrease the vulnerability of buildings to terrorist threats. FEMA Document 426 contains information from FEMA, the Department of Commerce, the Department of Defense, the Department of Justice, the General Services Administration, the Department of Veterans Affairs, the Centers for Disease Control and Prevention, the National Institute for Occupational Safety and Health, and other sources.

FEMA Document 426 describes a *threat-assessment methodology* and presents a *building-vulnerability assessment checklist* to support the assessment process. It also discusses architectural and engineering design considerations, standoff distances, explosions, and chemical, biological, and radiological (CBR) information. The appendices in Document 426 include a glossary of chemical, biological, and radiological (CBR) definitions, as well as general definitions of terminology. The appendices also describe design considerations for electronic security systems and provide a list of associations and organizations in the building-security area.

FEMA Document 427: *Primer for Design of Commercial Buildings to Mitigate Terrorist Attacks.* Document 427 introduces a series of concepts that are useful to building designers, owners, and members of state and local governments who are trying to mitigate the threat of hazards resulting from terrorist attacks on new buildings. FEMA Document 427 addresses four high-population private sector building types: (1) commercial office, (2) retail, (3) multifamily residential, and (4) light industrial. This manual contains qualitative design guidance for limiting or mitigating the effects of terrorist attacks, and it focuses mainly on explosions. However, it also addresses chemical, biological, and radiological attacks.

FEMA Document 429: *Insurance, Finance, and Regulation Primer for Terrorism Risk Management in Buildings.* Document 429 introduces the building insurance, finance, and regulatory communities to the issue of terrorism risk management in buildings and the tools currently available to manage risks. Document 429 provides information related to insurance and the Terrorism Risk Insurance Act of 2002, and it highlights current building regulations related to terrorism risk and vulnerability. The manual also includes a *building-security checklist* categorized by data collection, attack-delivery methods, and attack mechanisms.

FEMA Document 452: *Methodology for Preparing Threat Assessments for Commercial Buildings.* Document 452 outlines methods for identifying critical areas and functions within buildings, and it helps to determine threats to the areas and vulnerabilities associated with specific threats. Document 452 provides methods to assess risks and to make decisions about how to mitigate them by reducing physical damage to structural and nonstructural components of buildings and related infrastructure and reducing casualties during conventional bomb attacks, as well as attacks involving chemical, biological, and radiological (CBR) agents.

E155 Document: *Building Design for Homeland Security.* A course is offered by FEMA at its training center in Emmitsburg, Maryland, and at Virginia Polytechnic Institute and State University (Virginia Tech) that teaches building design and assessment methodologies to help identify the relative level of risk for various threats, including explosions. The primary target audience for this course includes engineers, architects, and building officials.

FEMA is in the process of developing four additional publications in these areas:

- Document 430: Primer for Incorporating Building Security Components into Architectural Designs
- Document 453: Multihazard Shelter Designs (Safe Havens)
- Document 455: Rapid Visual Screening for Building Safety
- Document 459: Incremental Rehabilitation to Improve Building Security

12.9 Web Sites with Additional Information on Terrorism

Sources that provide additional information related to terrorist issues and specific information related to terrorist organizations may be found at the following Web sites:

List of the most-wanted terrorists, including pictures, http://www.ccmostwanted.com/

Full and complete annual report on terrorism, http://www.state.gov/s/ct/rls/c14812.htm

Federal Emergency Management Agency publications, http://www.fema.gov/fima/rmsp.shtm

Links to the top 25 newspapers in the world, http://www.eclecticesoterica.com/news.html

Links to national and international publications, news, and government organizations, http://www.assignmenteditor.com/default.aspx

Reuters news (a global information company) on war, aid, and disasters, http://www.alertnet.org/

All international government links, http://www.firstgov.gov/

U.S. Office of Homeland Security, http://www.whitehouse.gov/infocus/homeland/index.html

U.S. Federal Bureau of Investigation information about terrorists, http://www.fbi.gov/

Private research institute that focuses on issues of national and international security, http://www.stimson.org/home.cfm

U.S. Department of Health and Human Services, how to get help if affected by biologic terrorism, http://www.hhs.gov/

General recommendations on how terrorism affects people psychologically, the warning signs of stress, and how to cope with trauma, http://www.helping.apa.org/

Safety data and advice for traveling on passenger airplanes, including accident reports, http://www.airsafe.com/

U.S. Department of Defense anthrax information, http://www.anthrax.osd.mil/

U.S. Centers for Disease Control and Prevention, http://www.cdc.gov/

Middle East Media Research Institute, articles about terrorism, interviews, and other news, http://www.memri.org/index.html

Terrorism knowledge base, information about terrorist incidents around the globe by location and date; names of organizations, their ideologies, and location of operations, http://www.tkb.org/Home.jsp

Advice in case of any disaster, http://www.ready.gov/

National Memorial Institute for the Prevention of Terrorism, http://www.mipt.org/

Descriptions of terrorist groups and state sponsored terrorism, http://www.terrorismanswers.org/terrorism/types.html

Disaster Recovery Journal Bookstore, books related to terrorism, http://www.drj.com/bookstore/drj683.htm

12.10 Summary

This chapter discussed terrorism and kidnapping issues that E&C professionals should consider when they are working on global projects. Engineering and construction professionals are responsible for minimizing risks to building occupants by designing structures that are capable of resisting forces generated by terrorist attacks. Terrorism and kidnapping issues are becoming more complicated due to globalization, which creates situations where people from different nations are working together on foreign projects. In most countries, there is no antiterrorism legislation or building codes, but technical professionals should routinely inspect their designs for points of vulnerability, even if it is not required by law.

No one can predict when terrorist attacks or kidnappings will occur because each terrorist attack is unique. This chapter provided background information on terrorist organizations and dealing with terrorism and kidnappings. Suggestions were provided on how to protect laborers and to avoid kidnapping and for designing buildings for protection in order to help E&C professionals understand the problems encountered while working overseas and the issues that should be considered when designing structures to withstand explosions or other terrorist attacks. This chapter also discussed how to locate information related to terrorism and kidnapping and included information on the availability of U.S. government publications related to designing and analyzing structures to withstand terrorist attacks.

DISCUSSION QUESTIONS

1. Write a definition of terrorism that could be used by E&C companies when they are training their personnel to work overseas.
2. Write a profile of a typical terrorist that could be used by E&C companies when they are training their personnel to work overseas.
3. Discuss four methods that terrorists use to raise funds for their organizations.
4. Discuss the difference between political and economic kidnapping, and cite examples of both of them.
5. Discuss what kidnapping and ransom insurance covers for employees and employers and what it does not cover if an employee is kidnapped while working at a foreign company job site.
6. Write a kidnapping plan of action that includes six steps that should be activated if an E&C employee is kidnapped while working at a job site.
7. Create a ten item list that could be used at construction job sites that includes brief (five or six words per item) warnings on the 10 most important ways to help identify a potential terrorist.
8. In addition to the words and phrases listed in Section 12.7.2, there are other words and phrases that an expatriate E&C professional should know how to say in a local language during a crisis. What are six additional words or phrases to use in a crisis?
9. What are seven items that should be checked before turning an automobile ignition on in a foreign country?
10. Create a 10-item list that could be used at construction job sites of brief suggestions (five or six words per item) on how to avoid being kidnapped while living in a foreign country.
11. Discuss how a kidnapping victim may occupy his or her mind and body while being held captive.
12. Explain the effects of a bomb blast on structures.
13. Develop a list of what should be checked during a structural vulnerability analysis.
14. What are three other methods for protecting structures from bomb blasts besides using concrete planters, concrete or steel artwork, and concrete barriers?
15. What are six methods that may be used to help protect the occupants of buildings from injuries when a bomb is detonated next to a structure?

16. What are five methods that would help to protect the occupants of a building if a toxic agent is released into the air circulation-system?
17. List five resources that may be used as references when an engineer is either designing a structure to be bomb resistant or performing a vulnerability analysis on an existing structure?
18. Explain what *protected spaces* are and how they are used in structures to provide protection to building occupants.
19. Explain *minimum standoff distances* and what could be done in place of them in crowded urban areas to protect building occupants from sustaining injuries during a bomb blast.
20. What are the five most essential items that E&C personnel should know about kidnappings before accepting overseas work assignments, and why should they know about these items?

REFERENCES

Aviv, Juval. 2003. *The Complete Terrorism Survival Guide: How to Travel, Work, and Live in Safety.* Huntington, N.Y.: Juris Publications.

Briggs, R. 2001. *The Kidnapping Business.* London: Foreign Policy Centre.

Briggs, R. 2003. *Keeping Your People Safe: The Legal and Policy Framework for Duty of Care.* London: Foreign Policy Centre.

Briggs, R. 2005. Spring tackling the risk of kidnapping. In *Guild of Security Controllers Newsletter.* London:

Business Continuity Institute. 2005. http://www.thebci.org/.

Clarke, M. C. 2005. *Engineering and Terrorism: Their Interrelationship.* Brisbane, Australia: Griffith University Press.

Davidson, J. D., and Lord W. Rees-Moog. 1993. *The Great Reckoning.* New York, N.Y.: Touchstone.

De Becker, G. 2002. *Fear Less: Real Truth about Risk, Safety, and Security in a Time of Terrorism.* Boston, Mass.: Little, Brown.

Federal Emergency Management Administration. 2005. *U.S. Government—Federal Emergency Management Administration.* Washington, D.C., http://www.fema.gov.

Federal Emergency Management Administration (FEMA). 2003. *FEMA426: Reference Manual to Mitigate Potential Terrorist Attacks Against Buildings.* Washington, D.C.: FEMA. http://www.wbdg.org/pdfs/fema426.pdf.

Federal Emergency Management Agency (FEMA). 2002. *Islam and Terrorism.* Lake Mary, Fl.: Charisma House. http://www.fema.gov/fima/rmsp.shtm.

Hoffman, B. 1954. *Inside Terrorism.* New York: Columbia University Press.

Levine, H. M. 2000. *Chemical and Biological Weapons in Our Times.* New York: Franklin Watts.

Long, D. E. 1990. *The Anatomy of Terrorism.* New York: Free Press.

Nash, J. R. 1998. *Terrorism in the 20th Century: A Narrative Encyclopedia from the Anarchists, Through the Weathermen, to the Unabomer.* New York: M. Evans and Co.

National Memorial Institute for the Prevention of Terrorism. 2005. http://www.mipt.org/Patterns-of-Global-Terrorism.asp, http://www.tkb.org/Home.jsp, http://www.mipt.org/, and http://www.mipt.org/pdf/CJCS-Guide-5260.pdf.

Office of Deputy Undersecretary of Defense. 2005. http://www.acq.osd.mil/ie/index.htm and http://www.acq.osd.mil/ie/irm/HomepageFiles/anti_terrorism_standards.html #background.

Overseas Security Advisory Council. 2005. *http://www.ds-osac.org.*

Office of Occupational and Radiological Health. 2005. http://www.health.ri.gov/environment/bvat.pdf.

Official Web Site of Indianapolis and Marion County, Indiana. 2005. Http://library7.municode.com/gateway.dll/in/indiana/635/956/957.

Stern, J. 1999. *The Ultimate Terrorists.* Boston, Mass.: Harvard University Press.

U.S. Department of Defense (DOD). 2005. *Form 1391: Military Construction Project Data.* Washington, D.C.: DOD.

U.S. News and World Report. 2005. Terrorism's latest report card. *U.S. News and World Report.* New York, May 9, p. 16.

U.S. Occupational Health and Safety Administration, Office of Occupational and Radiological Health. 2004. Heating, ventilation, and air conditioning vulnerability assessment tool. *U.S. Office of Occupational and Radiological Health.* Washington, D.C.

U.S. Office of the Deputy Undersecretary of Defense. 2002. *Anti-Terrorism Standards for Buildings.* Washington, D.C.: U.S. Department of Defense.

U.S. Department of State. 2005. Http://www.state.gov and http://www.state.gov/s/ct/rls/pgtrpt/2003/31644.htm.

Chapter 13

Preparing Engineers and Constructors to Work Globally

13.1 Introduction

In order to work in the global arena engineering and construction (E&C) professionals need to be able to adjust to living in foreign environments or to working with foreign E&C professionals. The preparations for moving overseas are similar to planning any E&C project—they require planning, organizing, directing, and controlling. The most crucial phase of any project is the planning phase. Having a well-researched and viable plan helps make the transition from living in a familiar environment to moving to a foreign country easier. This chapter provides information that may be used to create a template for a plan for mobilizing an employee and his or her family and for preparing expatriates to live in a foreign environment.

Culture and technology transfers are defined in the first section of this chapter so that E&C personnel have an understanding of what they mean in the context of working in foreign countries. The second section discusses what should be included in compensation packages and how to negotiate for a viable package. The subjects of securing passports, visas, and work permits are only briefly mentioned in this chapter because most companies provide assistance to their employees in obtaining official government documents to work overseas, and Chapter 2 has a section on government information and how to locate it.

The third section of this chapter provides material on what personnel should know before they move overseas, including information on vaccinations, shipping household items, language training, educational considerations, investigating entertainment, short- and long-term housing, food and water supplies, transportation issues, and work issues. The fourth section defines culture shock and how to recognize it, and the fifth section explains what it means when someone "goes native." The last section provides suggestions on how to avoid foreign jails.

13.2 Definition of Culture

In reference to the E&C industry, culture is used in this chapter as a "training of the mind, manners, and tastes or the results of such training. It is said that culture accounts for everything about the other person that we do not understand" (Lucas, 1986, p. 61).

The things that make people different from each other include: "communication (language), dress and appearance, food and eating habits, time and time considerations, rewards and recognition, relationships, values and norms, sense of self space, mental processes and learning, beliefs and attitudes" (Lucas, 1986, p. 58). Sometimes traditional methods for achieving technical solutions do not work in other cultures; therefore, recognizing cultural differences is a governing factor in structuring global communications (Lucas, 1986).

13.3 Cultural Differences

When dealing with other cultures, sometimes people develop a false sense of understanding about the way another person thinks by judging him or her on their outward appearance. When someone has the same job title or college degree, it does not necessarily mean that they share the same values. Having limited knowledge about different cultures causes misunderstandings, and there is no room for misunderstandings in a world that is becoming increasingly globally interdependent. If people are provided with accurate information about different cultures, some of the stress between foreign and native coworkers is reduced or eliminated completely.

13.4 Language Differences

Although cultural differences are the primary focus of this chapter, language and how it affects communication between native and foreign coworkers are also important concerns for E&C personnel. One major reason language training is important for members of global E&C firms is that the strength of the world economy is shifting eastward to India, Japan, the People's Republic of China, Hong Kong, Singapore, and South Korea.

English is accepted as one of the business languages in the global marketplace and French and Mandarin are two others, but once a deal is closed, it is the expatriates at the engineering or operations level who build projects. Therefore, for people working at the operations level, knowledge of local languages facilitates effective communication. Outside the business realm, most of the local people only know how to speak their native language. Chapter 2 includes a discussion on language translation issues, and this chapter discusses formats for learning foreign languages.

13.5 Technology Transfer

The training of expatriate E&C personnel should include information on how to effectively transfer technology to workers in other nations. One definition of *technology transfer* is: "It describes a characteristic human property: the capacity to store and transmit to people the accumulated experiences of others; a technology transfer takes place when a group of people, usually belonging to one body, become capable of performing one or several functions attached to a specific technique in satisfactory condition" (Seurat, 1979, p. 16).

Some of the issues related to transferring technology include issues that are closely related to cultural differences:

> The difference is far more than a technical gap. The know-how gap spreads into relations with other people, their cultures, their attitudes to problems, and even their professional and social values. The transfer relies heavily on the existence of training executives to link sender and receivers; executives must learn the technical, cultural, and organizational situation of the receiver and the receiver executives must learn the facts of the industrial techniques to be transmitted. (Seurat, 1979, p. 25).

13.6 The Importance of Cross-Cultural Training Programs

Cross-cultural training programs may be used to help teach potential expatriate employees how to work and live in foreign environments. This section discusses the importance of having cross-cultural training programs available for employees.

13.6.1 Benefits of Training Programs

One obstacle to cross-cultural awareness and exchanges is the difficulty in locating information on foreign cultures. Employees are fortunate if they receive cursory training in foreign languages or if they are provided with information pertaining to the technical aspects of doing business in foreign countries before they move overseas. It would be beneficial to employees if their company provided information on housing, obtaining passports and visas, job assignments, school options for children, safety concerns, job expectations, length of assignment, or other basic job-related information.

Cross-cultural awareness is not a new concept. State department officials have been receiving cross-cultural training since World War II. Other agencies that provide cross-cultural training are embassies, the military, the Peace Corps, the Red Cross, government agencies that operate in the global arena, and some multinational E&C corporations. Engineering and construction personnel should have a good understanding of cross-cultural relations before they agree to work on a foreign project rather than undertaking a foreign project and then investigating cross-cultural relations (Lucas, 1986). A lack of culture self-awareness may cause people to commit errors that they are not even aware of while they are working overseas or with foreign personnel.

13.6.2 Criteria for Cross-Cultural Training

In order to operate more effectively in foreign environments, global E&C companies could include the following topics in their cross-cultural training programs (the chapter that covers a particular topic is listed next to the topic):

- Forms of government (Chapter 4)
- Acceptable social standards throughout the world. (Chapter 14)
- Working with, being supervised by, or supervising foreign personnel (Chapter 13)
- Political issues that affect E&C project personnel (Chapter 7)
- Legal issues of concern and the difference between native and host-country legal systems (Chapter 7)
- Salary differentials between native and host countries and the problems it causes at job sites (Chapter 13)
- The consequences of a lack of adaptability of E&C personnel when they are working in foreign countries (Chapter 13)
- Cultural differences between host and native countries (Chapters 2 and 14).
- Quality of work in host countries (Chapter 8)
- Technological differences (Chapter 2)
- Environmental regulations in host nations (Chapter 9)
- Setting and achieving project objectives and managing projects overseas (Chapter 3)
- Safety issues in host countries (Chapter 3)
- Concerns related to construction failures and how to investigate construction failures (Chapter 3)
- Competitive influences on overseas projects (Chapter 4)
- The importance of forming and maintaining international alliances at the project and personal level (Chapters 5 and 14)
- How to pay local subcontractors and suppliers when working overseas (Chapter 6)
- International technical standards and the ISO 9000 and ISO 14000 series of standards (Chapter 9)
- Global productivity issues on construction projects (Chapter 10)
- Where to locate information on specific countries (Chapters 2 and 14)
- What types of delays are encountered most frequently on global projects and which mitigation strategies are used on global projects (Chapter 11)
- Treating foreigners with respect (Chapters 2 and 14)
- Using technology appropriate to a particular country (Chapter 2)

- Understanding foreign company structures and management styles and systems (Chapter 3)
- The importance of socializing with foreign business associates (Chapters 2 and 14)
- Dealing with the issue of bribery (Chapters 11 and 14)

13.7 Developing Cross-Cultural Training Programs

The benefits of developing and implementing cross-cultural training programs for E&C personnel are not fully realized on individual projects because the long-term effects on overseas personnel and members of host countries are more evident on subsequent projects and whether a foreign firm is able to secure additional projects in the host nation. When personnel are sent to work on overseas projects without receiving any cross-cultural training, it may strain the relationships between host country workers and expatriates. In most cultures, people will not express their feelings of contempt verbally, but they will express them when future projects are not awarded to expatriate engineers or constructors.

13.7.1 Materials for Cross-Cultural Training Programs

When developing cross-cultural training programs, trainers should be aware that some of the material written about foreign countries is written by foreign nationals who have merely visited foreign countries or conducted library research by reviewing other publications. Material that is written by host country nationals or expatriates who have lived in a country for more than a year is more reliable. Other sources of information about foreign countries that capture the essence of foreign countries are biographies written by native authors.

Instead of exclusively using company personnel to train E&C professionals, it is more beneficial to bring in speakers from the countries where E&C personnel will be working. Sources for locating foreign nationals who are able address the concerns of E&C personnel are professional societies (both national and international), trade organizations, universities that have foreign professors, and multinational E&C companies. Providing an outline and time restrictions for each topic to guest speakers prior to their presentations keeps speakers focused on topics of interest to E&C personnel.

If guest speakers participate in cross-cultural training programs, having long periods scheduled for questions and answers allows employees to express their concerns about their future overseas work assignments. If employees are not provided with accurate information on foreign countries their employers may have to deal with the early repatriation of their employees from foreign countries.

13.7.2 Areas of Concern to Expatriates

The areas that cause the most frequent problems for expatriates are unrealistic housing expectations and compensation packages not covering all contingencies. A third area of concern is the unavailablity of adequate educational institutions for their children. If high-quality educational institutions are not available close to job sites, expatriate employees may be hesitant about sending their children to boarding schools in other locations or abroad (and they may turn down overseas assignments for this reason alone).

Employees should be aware that they could be assigned to temporary housing until long-term housing becomes available and that their household items probably will not arrive until several months after they move to another country. Therefore, they should plan accordingly and bring whatever is necessary to accommodate their families during the period before their household items arrive from overseas. There are weight restrictions and limits on the amount of luggage that may be checked as baggage on airlines, and if employees change airlines in transit, luggage limits may change.

If an employee has never traveled, or lived overseas, he or she may not be familiar with what items need to be brought overseas. This chapter provides suggestions, but employers should investigate the availability of household items at job-site locations and provide their employees with either a list of items that are available at job-site locations or a list of items that cannot be obtained overseas.

Employers should provide assistance to employees on securing proper travel documents and making the move as easy as possible for their employees by verifying travel document requirements and any requirements for maintaining visas and work permits while employees are working overseas. Embassies are the best sources of accurate information on the requirements for visas, passports, and work permits.

In the current digital era, E&C personnel and their families rely on electronic devices for communication and entertainment, and they should be aware that the country they are assigned to work in may not have digital access outside major cities. If employees are provided with accurate information on the unavailability of digital access before they move overseas, it provides them with additional time to make alternate arrangements to communicate with their families in other parts of the world.

The following sections provide additional information that may be incorporated into cross-cultural training programs.

13.8 What to Know Before Working Overseas

Given to the nature of the E&C profession, personnel sometimes are sent to work overseas on short notice, yet moving overseas requires a great deal of preparation and foresight. If an employee will be moving overseas with his or her family, moving preparations become even more cumbersome. Locating accurate information about a foreign country helps to make the preparation process go more smoothly because it provides insight into what to bring overseas, how to prepare to live in a new country, and what is required to adapt to the foreign culture. Concerns about living overseas may be categorized into the following areas: travel documents, compensation packages, vacations, work-related issues, family, cultural issues, shipping household items, language and educational considerations, short- and long-term housing, food and water supplies, and transportation. Each of these areas are discussed in the following subsections.

Before accepting an assignment to work in a foreign country, E&C professionals should (1) make sure that they have a thorough understanding of their job assignment and responsibilities, (2) negotiate an adequate compensation package that is summarized in a written contract signed by their employers, and (3) prepare adequately for the move overseas.

13.8.1 Negotiating Compensation

The section covers specific areas that should be included in employment contracts that are used for overseas employment.

13.8.1.1 Uplifted Salaries

Before accepting an assignment to work overseas, employees should inquire as to whether the work assignment includes an *uplifted salary* (*uplifts*). An uplifted salary means that an employee receives his or her regular salary plus an additional percentage of that salary while working in another country. Once the foreign job assignment is completed, the employee once again is paid his or her regular salary. An uplifted salary compensates employees for cost of living differentials and the inconvenience of moving overseas. Employees may be paid in their native currency, and their checks are deposited directly into their bank accounts in their native country. An additional consideration is that it is difficult to cash checks overseas; therefore, employers may have to provide arrangements for cashing checks for their employees.

13.8.1.2 Employment Contracts

Managers might ask their employees to sign employment contracts that state that the employees will work on a foreign project for a set period of time (employment contracts are usually for 1 year at a time). The only way an employment contract may be legally enforced is if it has a clause that states that a bonus will be paid to an employee if he or she stays for the duration of the contact because there is no legal recourse for a firm if an employee quits his or her job before the time period stated in the contract. If an employee returns to his or her native country before the end of the employment contract, he or she is not entitled to a bonus. Employers cannot legally bind their employees to live in a particular location for a set duration of time.

One reason people work overseas for a minimum of 1 year is tax laws. If expatriate employees return to their native country before the end of one year, they might be taxed in both their native country and their host country. Before employees accept an overseas assignment, they should investigate the tax laws of both the host country and their native country. Employment contracts might include a clause that states that an employer will pay all foreign and/or domestic taxes of their employees while they are working overseas.

Other issues that should be clarified before employees accept overseas work assignments are whether workers' compensation insurance will cover employees while they are working outside of their native country, as well as whether life, disability, and medical insurance policies are valid overseas.

Overtime is another consideration when working on foreign projects because overtime may be required to complete projects on time if they are delayed by unforeseen circumstances. Companies may not pay for overtime for management personnel, but *compensation time* of one day off for every 8 hours of overtime worked may be negotiated before accepting an overseas assignment. An 8-hour day and 5-day workweeks are not a standard work schedule in every country. Therefore, employees should verify what is considered overtime in the country where they are assigned to work. A job description should be provided for overseas work assignments that outlines job duties and responsibilities. Without a job description, overtime work may be considered to be part of a regular job assignment.

Companies usually provide rest and relaxation (R&R) for their employees, which is time away from the job site in another country or trips back home. A common practice is to provide employees and their family with R&R every 3 months if a project is on an accelerated schedule and one trip home per year. Employers may pay for plane tickets for employees and their family members, a per diem for short R&Rs, and plane tickets for one trip back to the home country per year.

All the compensation that an employee receives might be taxed, which means that moving expenses, housing expenses, education expenses, transportation expenses, vehicles, vacations, gas allowances, and anything else that an employer pays may be considered to be taxable income. If all the benefits are taxed, employees might have to pay more in taxes than they received in actual wages for a year.

13.8.2 Moving Expenses

Moving expenses may be included in compensation packages for overseas work assignments, and they may include travel expenses for all family members, including airfare, hotels in transit, meals, surface transportation, airport taxes, and incidentals. A firm may provide a travel allowance for each family member in lieu of paying for exact expenses. There are moving expenses associated with shipping household items overseas *and back*. Since people acquire other household items or souvenirs while they are working overseas, there should be an allowance for additional weight when household items are shipped back to an employee's native country.

13.8.3 Native Comparisons

One area that causes problems between E&C professionals from developing countries and E&C professionals from industrialized nations on overseas assignments is salary differentials. Expatriate E&C professionals from developed countries may be paid at a rate that is over US$100,000 per year when they are working outside their native country, whereas their foreign counterparts might be paid less than a thousands dollars per year.

This disparity in salaries causes friction on projects, especially if foreign nationals are also provided with housing and other forms of compensation. Expatriates usually are allowed to bring their families overseas with them, whereas native professionals may be living in dormitories on a single-status basis. Realizing that friction exists because of salary and compensation differentials helps to explain why some expatriate E&C professionals have a difficult time establishing viable working relationships with native workers and why there is a lack of camaraderie between expatriates and native personnel.

13.8.4 Moving Preparations

This subsection provides information on how to prepare for an overseas work assignment.

13.8.4.1 Travel Documents, Passports, Visas, and Work Permits

Every family member who will be moving overseas with an employee has to have a valid passport and appropriate visas including work permits for employees. Employers should assist their employees with obtaining passports, visas, and work permits, but employees are responsible for providing appropriate documentation including birth certificates, copies of driver's licenses, multiple copies of passport photos, and other official documents. Visa requirements vary throughout the world, and host-country governments determine the amount of time a person is allowed to stay in a particular country with only a visa. Western European countries and the United States do not require visas for travel between them. Even if employees have a contract for a 1-year overseas assignment they may have to leave the country to have their visas or work permits renewed during the year. Biometric passports are being issued in western Europe, and the United States will start issuing them in 2007. Biometric passports have a computer chip with either a retinal scan (an eye scan) or fingerprints or both encoded in the computer chip.

Copies of official documents, including passports, visas, and work permits, should be kept in several different locations, and one copy of all travel documents should be left with an employer so that the documents will be available in an emergency. There are fraudulent or deceptive schemes used by customs officials, including stating that a passport photo is not the person who is traveling with the passport in order to obtain passports so they may be sold on the black market for thousands of dollars.

13.8.4.2 Vaccinations

The World Health Organization (WHO) provides information on which vaccines are required for each country and whether preventatives such as quinine, or hydrochloroquindine, which is a synthetic quinine, are recommended for malaria prevention (www.who.org).

13.8.4.3 Shipping Household Items

Before shipping household items overseas, it is helpful to know what is available locally and what items may be legally shipped into a country. Employers establish weight limitations for household shipments; therefore, only the essentials should be included in shipments, including items that cannot be bought locally. Charges for exceeding weight limitations for household shipments are paid by employees. Luggage size and weight restrictions change from country to country and from airline to airline, and each airline will charge overweight and oversize fees if luggage exceeds its allowable size and weight.

Knowing the exact location of a job site before moving overseas is helpful because a job site may be listed as being in the closest major city, but it actually may be located in a remote region that does not have a name. The availability of household items and food is more limited in remote regions compared with large cities, so everything that is required for basic needs will have to be imported or purchased during R&Rs.

Cell phones do not operate in remote locations, but satellite phones will operate anywhere in the world. Remote job sites may not have telephone service; television reception, cable, or satellite service; newspapers; libraries; or regular mail service. Figure 13.1. shows a housing complex in China, where the occupants each have an individual antenna for their televisions. Mail may be sent via the home office of a firm and delivered to employees when someone from the main office visits the job site. Seventy percent of the cell phones in the world use the Global System for Mobile Communications (known as GSM); GSM is used in 162 countries but not in the United States' cell phone system. Cell phone users should check with their service provider before they move overseas to see if it supports GSM.

Food supplies might be limited if food shipments are the only means for obtaining food in remote locations. Custom's officials may not release food shipments immediately after they arrive in a country so food shipments may not arrive on schedule. In one developing country, custom's officials would sometimes hold food shipments on the docks for 6 months while they were waiting for an appropriate bribe to be paid. Expatriate E&C personnel learn to be resourceful when it comes to obtaining food. At one job site that was located in a developing country, many of the children and some of the adults that lived in the camp had braces on their teeth because having braces guaranteed a trip out of the country to a major city once a month to have their braces tightened, and fresh food was brought back to the camp in suitcases packed with dry ice.

Common medications and prescription drugs should be hand carried overseas, and they should be in their original bottles. A letter from the prescribing doctor should be carried, with prescriptions that state why the drug is being prescribed

Figure 13.1 Antennas for television reception, China.

to the patient, because custom's officials might ask to see a letter from the prescribing doctor. Prescription drugs may be difficult to obtain overseas, so enough pills should be brought overseas to last 1 year.

Over the counter medications such as Tylenol, Motrin, Aleve, aspirin, cough syrup, antacids, gas relievers, Alka-Seltzer (sodium bicarbonate), antinausea liquid (Emetrol), Imodium (antidiarrheal agent), antibiotic cremes, Cortaid (cortisone), Lanacaine (topical relief of pain), bandages, gauze pads, medical tape, an eye wash, ipecac syrup (to induce vomiting), and rubbing alcohol and pads (for disinfecting wounds) should be included in first-aid kits in case of an emergency. Adequate medical facilities are not available everywhere in the world, and when there is a major medical emergency, patients may have to airlifted to a large city or to another country.

Expatriates should find out whether there are any religious or government restrictions on what may be brought into a country. In Islamic countries, alcohol is strictly forbidden, and women may be required to wear clothing that covers their arms and legs when they are in public. There are restrictions on the importation of cell phones and computers in Russia. Computers may be imported, but computer software could be confiscated when computers are brought out of Russia if they contain encryption programs or global positioning systems (GIS) software.

Household shipments take several months to arrive at their destination, so anything that is required during the first few months overseas should be brought in luggage. Some employers provide household items for their employees, but employees may not know what their employer will supply before they move overseas. Clearing airport security is more arduous with household items because some security systems are more sensitive or more invasive than others. Items that clear security in one country may not clear it in other countries. Suitcases may have to be opened before they are put on planes for a wand to be passed through them that "sniffs" for plastic explosives. In remote regions of developing countries, suitcases might be thrown on the tarmac, and bomb-sniffing dogs circulate among the suitcases. After the search is concluded, travelers may be required to locate their luggage on the tarmac so having a distinguishing mark on suitcases or other means for identifying personal luggage quickly is beneficial to travelers.

Household shipments should include rat and roach poison, insect repellants, and insect sprays because insect sprays that are available in developing countries contain substances such as DDT and malathion that are banned in industrialized countries because they are carcinogenic (cause cancer). Most countries are infected with ants, cockroaches, mosquitoes, flies, spiders, lizards, snakes, cicadas (flying insects), and other insects. In remote jungle regions, monkeys or tiny apes swing from the eves of buildings or play in children's wading pools. However, do not be tempted to purchase wild animals or birds while living overseas (especially orangutans) because many species of monkeys may not be exported, and if they are abandoned after being domesticated, they will not survive in the wild. It takes at least 3 years to train an orangutan to return to the wild, and there is only one place in the world where this is done—the island of Borneo in Indonesia. Many countries require that animals be quarantined for 6 months or a year if they are imported from a foreign country.

Wallets and valuables should be carried in front pockets or in pockets or purses inside clothing because back pockets may be slashed quickly with knives. In Italy, groups of gypsies pet foreigners while one of them is picking a pocket or removing a wallet from a purse. They may carry cardboard over their hands to disguise the hands of the person picking the pocket. Wallets, passports, a list of possessions, all legal documents and other valuables should be kept in hotel safes at the front desk, not in private rooms, because rooms may be searched when no one is in them.

13.9 Language Training

Digital language translators are available that translate and pronounce words through the Lingo Company, Quicktionary, or the Franklin Company. Alestron manufactures tablets that may be attached to computers and used to input Mandarin or Cantonese characters, and the software program Trans Star, which is also sold by Alestron, translates the characters into other languages. Languages may be learned by using cassette tapes (for children, music tapes with translations), CD-ROMS, videos, DVDs, local music, local television stations, books, and flashcards.

13.10 Educational Considerations

Options for educating children overseas include employer-sponsored schools if there are enough children of expatriates to justify having international schools where the teachers are predominately from the United States, native schools, and boarding schools either in the host country or in another country. If children will be attending boarding schools or international schools, employees should negotiate for tuition and trips to visit the children at school when employees are negotiating their compensation package.

13.11 Entertainment Issues

Employees should investigate whether their will be local options for entertainment where they will be living overseas. Learning about what makes a country interesting—art, museums, landmarks, historical places, unusual geography, recreation areas, national holidays and celebrations—and knowing which areas are safe to visit helps family members to adjust to living overseas. Figures 13.2 and 13.3 show two indigenous forms of entertainment in Morocco (camel rides) and Indonesia (a Balinese play). At remote job sites, companies may provide recrea-

Figure 13.2 Local entertainment: camel rides, Morocco.

Figure 13.3 Local entertainment: Balinese play, Bali, Indonesia.

tional facilities such as swimming pools, running tracks, basketball and tennis courts, a recreation center where movies are shown, facilities for religious services, or other facilities for family activities.

Electronic devices from one country may not work in other countries without adapter plugs and transformers. Different types of plugs are used throughout the world, and they require adapters (there are 14 permutations of adapters), which may be ordered with portable transformers at www.magellans.com. Telephones also require adapter plugs, and there are 34 different adapter plugs available for telephones. Heavy-duty transformers are heavy, so only small electronic devices may be used with portable transformers. Electronic devices designed for 60 hertz will run on 50-hertz systems, but motors will run slower. Using electronic devices with incompatible electronic systems for long periods of time may damage them.

13.12 Short- and Long-Term Housing

Knowing what type of housing will be provided by employers before moving overseas is helpful, but most companies do not assign housing until employees arrive overseas. Temporary housing may be provided for a short period of time (a few months) while permanent housing is arranged for families. Figure 13.4 is a photo

Figure 13.4 Expatriates job-site housing, Borneo, Indonesia.

of expatriate housing in Indonesia and Figure 13.5 shows Indonesian indigenous housing. Figure 13.6 shows a sample of island housing in the Fiji Islands.

Living arrangements vary depending on whether someone is overseas on *single* or *family status.* Houses or apartments are provided for families, but an employee on single status could be assigned to live in a small apartment or a dormitory. Houses and apartments may not have modern conveniences such as air-conditioning, central heat, dishwashers, washing machines, clothes dryers, or garbage disposals, and ovens may have numbers, instead of temperatures listed on the stove dial. Figure 13.7 shows a housing complex where bamboo poles are hung from apartment windows to dry laundry.

Housing compounds are residences that have high walls surrounding them instead of fences and they are common residences for expatriates in developing countries because they provide privacy and security. Housing compounds may consist of one home and supporting facilities or several homes. Expatriates from a particular country either all live in one location, or housing facilities could be scattered throughout a city. Being able to live immersed in another culture is a unique experience, but for some people, the familiarity of living with people from their native country makes the transition easier for them. If safety is a concern in particular neighborhoods, employers may provide guards. In countries where there are high unemployment rates, it is common for families to have domestic help, including cooks, housekeepers, chauffeurs, gardeners, maids, guards, and nannies because the cost of labor is inexpensive. Domestic help may cost less than US$20 a month per person.

Figure 13.5 Indigenous housing, Borneo, Indonesia.

In expatriate communities, some homes may have large gravel or paved areas next to them. If a company owner, an important government official, or upper-management personnel live in homes next to large gravel or paved lots, the lots are probably helicopter landing pads that are used in an emergency to evacuate executives and their family members.

Employers provide pictures of *potential* housing, but there is no guarantee that the housing shown to an employee will be where they will live overseas. Employees should ensure that adequate housing will be available when they arrive overseas for both them and their families, and they should have a contingency plan for alternative housing if their assigned housing is not available when they arrive overseas. Companies may require employees to move overseas alone and then have their family members follow them at a later date, or family members may have to wait in another city or country until housing becomes available at a job site.

13.13 Food and Water Supplies

Less than 20 percent of the world has safe drinking water, and the water that is sold in bottles could be from municipal water systems (tap water). Figures 13.8 shows what tap water looks like in some developing countries. Boiling water a minimum of 10 minutes and straining it through a fine material such as cheese cloth, thin towels, coffee filters, or plain paper helps to remove particulates and

Figure 13.6 Indigenous housing, Fiji Islands.

kills some bacteria, but it does not remove chemicals. Tea and coffee are safer to drink than plain water, because they are boiled before they are served to anyone.

Cholera is a common problem in countries where local water systems, such as rivers, are used for all household purposes. Cholera is an infection of the small intestine, caused by bacteria in water or food, that starts suddenly 1 to 5 days after ingestion of tainted food or water. Over a pint of fluid may be lost per hour from diarrhea, so if cholera is contracted, it is essential to maintain hydration by drinking fluids that contain both salt and sugar (fructose or sucrose). Advanced cases of cholera require intravenous fluids to achieve rehydration. Drugs that stop diarrhea (such as Imodium) and antibiotics (tetracycline) help to shorten the period of diarrhea, and cholera vaccines are available. However, the vaccines only provide protection for approximately 6 months.

Large multinational companies may import food for their E&C personnel, but food shipments may not always arrive as scheduled, or they may be delayed by customs officials and be spoiled by the time they arrive at their destination. Food

Figure 13.7 Drying laundry in high-rise apartments, China.

Figure 13.8 Tap water in a developing country.

may have to be purchased in native villages if food shipments do not arrive promptly. Figure 13.9 shows a photo of a market in a native village in Indonesia that is built over the water to provide sanitation services to its citizens.

Food expiration dates on packaged food should be checked, and if the food is past its expiration date, it should not be eaten. When food expires in industrialized nations it might be sold to food resellers in developing countries. Augmenting food shipments with local food provides variety, but fresh vegetables and fruits may cause dysentery, which is severe diarrhea that results in dehydration and abdominal cramping. The symptoms of food poisoning are similar to those of dysentery, and both are treated with fluids and Imodium. Scurvy, which is caused by a vitamin C deficiency, is a problem if fresh milk, citrus fruits, meat, vegetables, or fish are not consumed regularly.

Each country has unique foods, such as dog in South Korea, sea snakes in China, and lobster brains in Indonesia. Toiletry products may be different from what people are used to having in their native country. One example is toilet paper, which is soft and pliable in the United States, but in other countries it may resemble crepe paper as it stretches when it is pulled, it might be a bright color (possibly purple), and the dye may come off during use.

Figure 13.9 Food market in a native village, Borneo, Indonesia.

13.14 Transportation Issues

Employers may supply vehicles for expatriates, but driving in a foreign country is risky because vehicles may be driven on the opposite side of the road from what someone is used to, and the steering wheels could be either on the right or the left side of the vehicle. Having a driver while living in a foreign country helps to prevent legal entanglements that might develop when a foreigner is driving a vehicle that is involved in a traffic accident. Spouses of employees may not be provided with vehicles. It is illegal for females to drive vehicles in some Islamic countries such as Saudi Arabia.

Requirements for government issued driver's licenses may include residency requirements, but foreign nationals may legally drive with an international driver's license in most countries. In the United States, the American Automobile Association (AAA) issues international driver's licenses to licensed drivers.

13.15 Foreign Work Environments

Expatriates may have more responsibilities when they are working overseas than they had in their domestic positions, because the local support staff and foreign coworkers may have limited knowledge about a project or its technical requirements. Issues related to status surface when expatriates are supervising projects, and status is discussed in Chapter 2.

Socializing with coworkers outside of work if it is socially acceptable in a culture improves working relationships. In some countries, expatriates may never meet the family members of coworkers because of cultural restrictions. In some Asian cultures, such as Japan, coworkers socialize without spouses and in other cultures, such as India, spouses are included in business functions. If expatriates only socialize with people from their native country, a barrier is created at work that is difficult to transcend during work hours.

Carefully explaining why things are done in a certain way, what is expected of native coworkers, and why it is expected helps to circumvent cultural misunderstandings. Providing instructions in writing, after explaining them verbally, allows native coworkers more time to absorb information on how a task should be performed or exactly when a task should be completed. Most people understand foreign languages better when they are written rather than spoken.

Workers may follow native traditions, and expatriate supervisors need to determine what they will tolerate at job sites. If supervisors attempt to ban certain traditions from job sites, such as workers wearing a dagger attached to their abdomens in Yemen or workers wearing brass rings hanging from their ears as shown in Figure 13.10, workers may quit in protest.

How foreigners are treated varies throughout the world, and the type of welcome employees and their family members receive in a foreign country depends on historical and governmental influences. Citizens of countries with repressive

Figure 13.10 Brass rings hanging from ear lobes.

regimes (such as Communist countries) or formerly repressive regimes may be afraid to talk to foreigners for fear of reprisal. Public and private conversations may be recorded in Communist countries, and hotel rooms and houses might be searched by the authorities.

Rewards may be given to local citizens when they report information to the authorities about foreigners or about other native citizens having contact with foreigners. Even casual conversations could jeopardize local citizens and their family members. If local citizens do not respond to friendly overtures, it may be because they are afraid of their government. Jail sentences, torture, and being sent to forced labor camps are punishments for conversing with foreigners in some countries. During the Communist regime the government of Russia incarcerated 5 million people who were accused of being dissidents (openly disagreeing with the government), some of whom were merely corresponding with foreigners. Sharing sensitive technical information with foreign coworkers might be dangerous for citizens or result in native coworkers being discharged from work or expatriates being repatriated to their native countries.

13.16 Prejudice

One root of prejudice is fear of the unknown. When people look unfamiliar, act or dress different than the local people, speak a foreign language, speak with an accent that is not familiar to the local people, have darker skin tones or uncommon facial features, or use a unique vocabulary, people might be hesitant about approaching them or being around them. *Prejudice* is defined as "an unreasonable attitude for or against something or someone" (*Webster's Unabridged Dictionary,* 2005, p. 863). In some remote regions of the world, people from an indigenous society may not have ever seen human beings that look any different from local people. It is a natural reaction for some people to fear situations they have not experienced before and to be defensive as a protective mechanism. Unfortunately, this prevents some people from being able to live around or socialize with people from other cultures. The more people know about a culture, the easier it is for them to adapt to foreign environments.

13.17 Culture Shock

Moving away from familiar surroundings or family is difficult for anyone, but moving overseas requires more time to adjust and adapt to local cultures. *Culture shock* results from losing all familiar social cues and symbols, including verbal and facial expressions, as well as gestures. It also results from people having their value system, which they may have previously believed to be absolute, questioned or challenged by new experiences, places, and people:

Symptoms of culture shock include withdrawing from interactions or feeling rejected by the local social system. Some people who are experiencing culture shock will start criticizing the local culture because they feel that the host country is bad because it makes them feel bad. Regression also takes place when there are strong feelings of ethnocentrism (a belief that one's own culture, race, and nation are the center of the world). Another symptom of culture shock is always showing local people how things are done back home. Memories of life at home are filtered to only provide a positive picture of life there.

Physical symptoms of culture shock include:

- Excessive fatigue (depression)
- Compulsive eating or drinking large quantities of alcohol
- Excessive hand washing
- Excessive concern over eating local food and drinking local water
- Fear of physical contact with local people
- Dependence on expatriates
- Increasing frustration over minor incidents
- Refusing to learn the local language
- Excessive fear of being harmed by someone
- Agoraphobia, abnormal fear of open spaces

Culture shock may be alleviated by learning the local language or learning more about the local culture and history, getting to know local people, trying new activities or indigenous hobbies, going on short walks to become familiar with new surroundings, or volunteering to help others.

Culture shock does not start when someone first moves overseas, because their mind is occupied with many different activities during the first few weeks. Family members who do not work experience culture shock more frequently than their employed relatives, especially if they are left alone without a social network or the support of family members.

13.18 Going Native

The opposite of culture shock is when someone "goes native," which is a term used to describe someone who totally immerses themselves in another culture to the exclusion of their home nation culture. Signs of going native include exclusively wearing native clothing and speaking the local language, living in native housing or converting housing into native housing, not associating with people from other countries, and not returning to home nations during scheduled vacations.

13.19 Avoiding Foreign Jails

When someone first enters a country, they are vulnerable because they are not familiar with local laws or the local language. Foreign officials might insinuate that someone's travel documents are not valid and expatriates may not know what is happening to them nor be able to explain their travel documents to foreign government officials. If the exchange between expatriates and government officials degenerates, expatriates may be taken to detention rooms or a local jail until the matter is cleared up, which could take hours, days, or weeks. If travel documents are seized by government officials, the situation becomes critical.

Offering to pay *administrative expenses* (bribes) may prevent detention or it might result in arrest for attempting to bribe a government official. One method to circumvent accusations of bribery is to place a twenty-dollar bill inside a passport between the pages where it is going to be stamped by a government official because this provides an opportunity to deny that a bribe was being offered to government officials (the money was merely mixed in with their passport).

If government officials are paid low wages, they are forced to augment their salaries with bribes. In developing countries, bribes could be for as little as US$10 or US$20, but if someone appears to be affluent, then government officials will expect higher bribes. If a bribe is not accepted immediately, try negotiating for higher and higher amounts, and provide the government officials with a reason for accepting the bribe without being confrontational, such as you know they have other people to deal with and their time is limited or that they may need the money to pay for additional paperwork or to pay for another form or other methods that make the offer look legitimate.

When individual efforts are not successful in preventing detentions, contact embassy officials and employers, but be aware of the fact that local governments may not honor foreign constitutional rights.

Suggestions for avoiding foreign jails include the following (Copeland and Griggs, 2001):

- Be familiar with local laws and customs.
- Never buy anything on the black market (items that have been secured illegally, especially in government-regulated economies such as Communist countries).
- Always have permits and other legal documents that are required by the local government verified by several sources.
- Make sure that someone always knows where you are at all times.
- Avoid driving in foreign countries by using a chauffer, taking a taxi, or riding the transit system.
- Remain calm, and do not argue with government officials for any reason.
- Do not try to hide items in a suitcase that are illegal to import into a country (e.g., guns, drugs, or pornography).
- Do not drink alcohol or use drugs (or transport them) in Islamic countries or other countries where drugs are illegal.

13.20 Summary

Planning for a move overseas is one of the crucial phases of overseas work assignments. Engineering and construction professionals need to investigate foreign cultures in the same manner that is used to conduct technical research projects. The criteria for cross-cultural training and techniques for developing cross-cultural training programs were delineated in this chapter, including a list of topics that should be covered to prepare E&C personnel to operate effectively in foreign environments.

This chapter provided information on the types of issues that E&C personnel should investigate before accepting overseas work assignments, including technology transfer, criteria for cross-cultural training, negotiating compensation packages and employment contracts, moving preparations, shipping household items, language training, educational considerations, entertainment, short- and long-term housing, water and food supplies, transportation, and workplace concerns.

Culture shock was defined in this chapter and suggestions were provided on how to alleviate it. A discussion on prejudice was included in this chapter to provide insight into the roots of cultural misunderstandings. The last section of the chapter provided suggestions on how to avoid foreign jails.

DISCUSSION QUESTIONS

1. Explain why it is important to know about different cultures when working overseas.
2. Write a simple definition of *technology transfer* that could be used by E&C firms in cross-cultural training sessions.
3. What are the 2 areas that cause the most frequent problems for expatriate employees?
4. What five areas should E&C employees have accurate information about before they accept an overseas assignment?
5. What items should be included in an employment contract between employers and potential expatriate employees?
6. What legal documents do employees need to work overseas?
7. List 10 items that could be included in a first aid kit to be used overseas.
8. What are the most difficult aspects of shipping household items overseas?
9. Explain housing compounds and why some expatriates live in them.
10. Explain why food and water supplies are major concerns for expatriate employees.
11. Explain why citizens of Communist countries may be reluctant to speak to foreigners.
12. Why is it more effective to provide instructions in writing, after explaining them verbally, than only providing verbal instructions to people from cultures that are different from the culture of the person providing instructions?
13. How does ethnocentrism contribute to culture shock?
14. What are some techniques for alleviating culture shock.
15. Explain what it means to "go native."

REFERENCES

Copeland, L. and Griggs, L. 2001. Going International, New York: Random House.

Lucas, C. 1986. *International Construction Business Management: A Guide to Architects, Engineers, and Contractors.* New York, New York: McGraw-Hill.

Seurat, S. 1979. *Technology Transfer.* Houston, Tex.: Gulf Publishing Company.

Webster's Unabridged Dictionary. 2005. New York: Simon and Schuster, p. 863. Copeland and Griggs, 2001, on page 287.

Chapter 14

Country-Specific Information

14.1 Introduction

In order to work effectively in foreign countries or with personnel from foreign countries, it is useful to have information on languages, cultural nuances, local customs, business practices, local religions, climates, traditions, holidays, social standards, whether business agents should be used, and what construction materials are available locally. This chapter provides country-specific information on these topics in alphabetical order by regions of the world and highlights characteristics that may help foster relationships across national boundaries.

Other publications are available that provide information for Europe and Asian countries on meeting and greeting people, names and titles, corporate cultures, body language, dining and entertainment, dress, and gifts, such as the Web site www.windowontheworldinc.com/countryprofile/index.html, which provides excerpts from the series of books called *Put Your Best Foot Forward* (Bosrock, 2005). The book *Going International* (Copeland and Griggs, 2001) provides information on U.S. foreign commercial service posts in other countries, fundamentals of business, sensitivities, forms of address, courtesies, business dos and don'ts, entertainment, and religion. Since there are other publications that provide information on countries in Europe, Asia, and South America, the emphasis in this chapter is on countries that are not covered in other publications. The specific countries highlighted in this chapter are countries where major construction projects are being built or countries where major construction projects are needed to replace deteriorating infrastructure or to replace structures that were destroyed by terrorism, war, police action, political uprisings, natural disasters, such as those that have taken place in Eastern Europe, the Middle East, Africa, and Southeast Asia. In this chapter, each region of the world is discussed and a few countries within each region are used to illustrate what is appropriate for that particular region of the world.

At the beginning of each major section, a description of local cultures and customs is provided, and then specific information is provided for several countries within that region about local languages, religions, national and religious holidays, climate, culture and customs, business practice standards, whether business agents are required or recommended, what construction materials are available locally, and social standards.

In addition to the cultural information provided in this chapter, cultural issues are also discussed in other chapters in relation to how they affect engineering and construction processes and personnel. Religions are specified under each country and Appendix C contains a short description of each religion.

Languages throughout the world are either high-context or low-context languages. Examples of high-context languages are Arabic, Chinese, Japanese, Greek, Spanish, and Italian, and examples of low-context languages are English, French, Scandinavian, German, and Swiss. *Low-context languages* are languages "where information is explicit and words have specific meanings," and *high-context languages* require people to "read between the lines" or to know the context in which a communication is used to understand its meaning (Copeland and Griggs, 2001, p. 107). Some high-context languages, such as Arabic, are not exact languages. Sentences contain many adjectives and information is repeated to demonstrate the importance of the subject. In Chinese, expressions such as *perhaps, maybe,* and *we'll consider it* are common, and "when something is inconvenient, it is mostly likely impossible" (Copeland and Griggs, 2001, p. 106).

14.2 Africa

Africa is one of the most diverse continents in the world, because it has a variety of different ethnic groups, climate zones, cultures, and customs. Parts of Africa have been devastated by droughts that have led to widespread famine. People who used to live off the land have been forced into living in refugee camps that are filled with people who originally were nomads but, due to unrelenting droughts that destroyed all of the vegetation, their herds of animals died off, and they had no other means to support themselves. In several countries, civil wars have torn countries apart. The spread of AIDS (autoimmune deficiency disease) has killed 25.8 million people as of 2005 and left millions of children orphaned and homeless (Newsweek, May 22, 2006).

In addition to AIDS, Africa is home to the Malberg and Ebola viruses, both of which can destroy a human being in a matter of days due to liquefaction of their internal organs (the virus mutates rapidly inside human hosts). Ebola jumped species from monkeys to humans, when human beings started eating monkey meat. The Ebola and Malberg viruses are now able to be spread by contact between humans because they are airborne viruses. There is no cure for either virus and nine out of ten people who contract the Ebola virus die within a few days. If people are quarantined when there is an outbreak of the Ebola or Malberg virus, it helps to prevent the virus from spreading to other humans.

All the preceding issues have limited the potential for engineering and construction projects in parts of Africa, but infrastructure projects are needed to provide basic services throughout Africa. This section provides information about Africa and why each country is unique, and it also includes a brief synopsis of geographic features and examples of tribal differences for several African nations. Detailed information is included for Yemen and Ethiopia that explains tribal societies, along with information about Egypt.

Africa has large metropolitan areas along with deserts, jungles, mountain ranges, and savannahs. The scenery could be scattered palm groves with thatched huts on dunes tapering into the sea or deserts that are hot and desolate with no bushes or even blades of grass for thousands of miles. There may be an occasional oasis that has date palms, irrigation systems, *tahing* holes (watering holes), and mud houses.

In Kenya, women wear modern clothing, colorful sarongs (a cloth wrapped around the lower half of the body), or they might be veiled and wearing black robes. Men may wear long, colored shirts and white caps or modern clothing. Ancient Arabian *throws* (ships) may still be seen along the coastline.

In Tangiers, Morocco, travelers may wander through the *Casbah,* which is shown in Figure 14.1, while drinking mint tea, and across the Atlas Mountains is the desert of Morocco. Cities in Morocco are a mixture of old towns and new cities, medieval and modern. The *Casbah* (pronounced "cazbaa") is a walled city with 12 arched entrances where streets are merely alleyways (too small for automobiles), and the scene is medieval, with donkeys carrying bundles and the air filled with exotic smells. Figures 14.2 and 14.3 show some of the more unusual sites in Marrakech. Figure 14.2 shows goats standing on the branches of dragon trees eating the nuts, and Figure 14.3 is a donkey parking lot in Rissani next to a local village.

In desert areas of Africa, bodies dry out as they give up moisture to the parched atmosphere, and watering holes are only found every 100 miles or more. In the desert regions, people may travel for days without seeing anything move or that is alive because all there is to see is the sun and sand. The nights are cool, and daylight brings out hordes of flies.

In local villages, tribes live in traditional structures, they follow ancient customs, and they wear native costumes. Rural Africans are natural, friendly, and honest. Members of tribes may wear long robes, and some members also wear head wraps.

Figure 14.1 Casbah, Skours, Morocco.

Figure 14.2 Goats in dragon trees, road to Marrakech, Morocco.

Figure 14.3 Donkey parking lot, Rissani, Morocco.

In some tribal villages, local women may stare and giggle if they are spoken to by foreigners.

Tribes in the northern part of Morocco, Algeria, and Niger are Arabs or *Berbers*. *Tuareg,* or *Fulani,* are located in the middle regions, and the *Housa* live in the South. Members of the Twareg tribe are tall, slender, and stately, and they may be seen wearing royal blue robes and veils with black hoods. Around their necks they might wear square silver engraved cases with an inscription from the Koran sealed inside for good luck. The Fulani are a primitive tribe of nomads who sometimes still wear goat skins as clothing and go bare from the waist up. The more primitive Fulami wear their hair in hundreds of braids, their faces are painted orange, their lips are painted dark red, and they wear necklaces and earrings as a sign of wealth.

To celebrate the rising of the new moon at the end of Ramadan, tribes converge in one place, and there will be drumming, chanting, and wild dancing all through-out a town, along with camel races. Each day at dawn, the *Emirs* (native rulers) ride about town with their princes and guardians on bejeweled horses preceded by rows of drummers.

In Nigeria, the deserts give way to cornfields and grazing cattle. Food supplies are more plentiful in Nigeria, including luxury foods such as oranges, pineapples, potatoes, and cold drinks. Nigeria is the most densely populated nation in Africa, and its cities are congested and full of smog. Local citizens advise foreigners to *go slow.*

In Nigeria, women carry things on their heads in baskets or bundles, and native Nigerians may have tribal scars on different parts of their bodies. Tribal rather than individual consciousness controls societies. People are friendly to foreigners and stop whatever they are doing to help out a foreigner. At night, tables appear along streets with items for sale. People sit outside in the evenings, and laughter is a familiar sound. Natives yell out *Ovimbo* ("white man") in a friendly fashion when foreigners pass by. In the cities that are on islands, there is intense heat and humidity.

Cameroon and the Central African Republic are countries with sloping hills rather than flat terrain. At nightfall, in the cities, people fill bottles with hundreds of green grasshoppers that congregate under the streetlights, because the grasshoppers are a local delicacy. The landscape changes again in the Republic of the Congo, where it becomes dense jungle, and only one-tenth of the roads are paved with asphalt. Citizens of the Republic of the Congo have experienced years of colonial bloodshed, and parts of the country are cut off from the rest of the world. French is spoken in some locations in the Congo and other African countries.

In the Republic of the Congo the forests are still populated by hunters carrying quivers of poisoned arrows and *pygmies* (people of small stature), and in rural areas drumming and animal sounds are common. In rural areas, monkeys and baboons swing from the trees and run across the roads. In government animal preserves (called *irunga*), there are elephants, gazelles, and wild boar that roam the countryside. In local villages the natives may swim out to greet foreigners when they are traveling by on the rivers and attempt to trade dried fish or dead monkeys for manufactured items. Others play in the wake of boats or dive off of them.

The eastern frontiers of the Congo have the largest mountain range in Africa, which stretches from Uganda to Zambia and from the Congo to Tanzania, and it contains the active volcano Nirangano.

Rwanda is the home of the tall *Watusi*, who wear dark red robes and produce handmade baskets with intricate designs. Some of the Watusui are shy and may be unaware of the rest of the world except for the intertribal strife of their native land.

Tanzania is on the eastern side of the mountains, and it has large *masetas* (savannahs) with animals such as giraffes, elephants, wild bears, lions, hippopotamus, *dik diks* (a medium-size roaming animal), hyenas, vultures, storks, rhinoceroses, monkeys, baboons, and zebras. The Nagarangora crater is in the Serengeti National Park, and it is populated by thousands of pink flamingos.

Kenya is where snow-covered Mount Kilamenjaro rises out of the plains at the edge of the Ambrosela National Park, and some local people believe the mountain is a stronghold of the *Masai* gods. Nairobi is a modern, Westernized, cosmopolitan city that is 6000 feet above sea level.

Mozambique is one country where Portuguese is spoken rather than French, and it has been an independent country only since the 1980s. It is also a country where few tourists venture.

In addition to the countries mentioned previously, the continent of Africa includes Egypt, Libya, Sudan, Ethiopia, Somalia, Uganda, Mozambique, Zimbabwe, Botswana, Namibia, Angola, Gabon, Benin, Ghana, Côte d'Ivoire (or the Ivory Coast), Liberia, Sierra Leone, Gambia, Senegal, Guinea, Mauritiania, Western Sahara, Tunisia, Mali, Burkina Faso, Niger, Swaziland, Malawi, Burundi, Eritrea,

Yemen, Chad, Equatorial Guinea, São Tomé, and Princripe, and Madagascar (an island country). Each of these countries has a unique culture and different customs.

The following subsections provide detailed information about Egypt, Ethiopia, and Yemen to help illustrate some of the customs and cultures in Africa.

14.2.1 Egypt

Languages spoken: The official language is Arabic, but some people speak French and English.

Religions: Islam and Christianity.

Holidays: Islamic holidays are celebrated (see Yemen).

Climate: Egypt has a Mediterranean climate, which is hot and humid in the summer months and mild in the winter.

Culture and customs: Egyptians address people using the titles of Mr., Mrs., Miss, or Dr., along with a first name only. Titles are used if someone has earned advanced college degrees. People in the country are courteous and social etiquette is practiced in personal and professional situations. It is not appropriate to eat, drink, or smoke in front of people who are observing Ramadan when they are fasting during daylight hours. The Muslim Brotherhood has been active in Egypt since the 1930s.

Business practice standards: Business cards should be used in Egypt and everyone should be greeted with handshakes. Hands should also be shaken again when leaving business situations. Meetings might be interrupted by other people, phone calls, or tea breaks. Business progresses at a leisurely pace and interpersonal interactions are valued in business transactions. People socialize with each other before they conduct business. A person has to persuade his or her business associates to have confidence in him or her before business is conducted by either party.

Use of business agents: Use a business agent who has good connections within the country.

Construction materials available in the country: Egypt has a thriving construction industry and construction materials are available in the country or imported from other countries. Native construction techniques take advantage of locally produced materials.

Social standards: Do not discuss politics or religion. Lunch is the main meal and it takes place between 2 and 4 in the afternoon, but lunch meetings may be held at noon. Egypt is a country of contrasts because there are large cities and small villages. Tourism generated by the pyramids and other Egyptian artifacts has introduced many Egyptians to foreigners, but their interactions with foreigners may be limited to having only dealt with tourists, not foreign business associates.

14.2.2 Ethiopia

Languages spoken: There are 83 languages spoken and 200 dialects in Ethiopia. The largest language and ethnic groups are the Oromo, who live in the central and southern regions, and the Somali, who live in the southeast section of the country. The Amhara and Tigrayan live in the northern plateau region and speak Semitic languages (languages that belong to the Afro-Asian language family, which also includes Arabic and Hebrew), such as Amharic and Tigrinya. The official language of Ethiopia is Amharic, and three other main languages are Cushitic, Omotic, and Nilo-Saharan.

Religions: As of 2005, approximately 40 percent of Ethiopians were Christian (the Ethiopian Orthodox Church), 45 percent Muslim, 12 percent animist, and 3 percent other religions, such as ancient forms of Judaism.

Holidays: Christmas—January 7; Epiphany—January 19; President's Day—February 18; Arafa—February 1; Id Al Adaha—February 22; Victory of Adwa—March 2; May Day and International Labor Day—May 1; Mawlid—May 2; Good Friday—May 3; Easter/Patriot's Victory Day—May 5; Birthday of the Prophet Mohammed—May 24; Downfall of the Dergue—May 28; New Year's Day—September 1; Meskal (the finding of the true cross)—September 27; and Id Al Fetir the end of Ramadan (which varies each year).

Climate: Temperate, moderate, or even cool, but not tropical. Highest daytime temperatures are around 70°F for most of the year, and there are two rainy seasons—from late January to early March and from June until mid-September.

Culture and customs: A common greeting is bowing heads, and the greeting process is longer among close friends. Greeting someone by kissing their cheek three times is also acceptable in Ethiopia. Appropriate titles are *ato* for a man and *weiziro* for a woman. Titles are not used among close friends. Some elders use their titles in the evenings (which causes confusion as to what is the exact time when the title is once again being used). Leaders are respected for their education and for promoting teamwork.

The *coffee ceremony* is an integral part of the Ethiopian culture and the ceremony may take up to 3 hours if it is an elaborate version. Coffee beans are roasted by hand, then ground in a special way, prepared in a pot, and poured into ceremonial cups.

Business practice standards: The engineering and construction industry is referred to as the architectural, engineering, and construction (AEC) industry in Ethiopia. As of the early 2000s, there were no environmental regulations in Ethiopia. The primary trading partners are Germany, Italy, Japan, Saudi Arabia, and the United States.

Construction equipment and spare parts are exempt from custom's duties, sales tax, and excise taxes, but if expatriates are used on a project there is a 40 percent tax on their salaries. Local construction firms specialize in building roads and bridges.

A bachelor of science degree is all that is required to design construction projects. If an engineer has graduated from an approved university and has 5 years of experience, he or she may obtain an engineering license (no examinations are required for licensing in Ethiopia). Government regulations restrict engineers and contractors to performing only work permitted by their license. Licensing provisions may require firms to have certain types of construction equipment to maintain their license, even if they do not need it for their projects. There are labor shortages in masonry and plasterwork.

There are currently no alliance requirements in Ethiopia, but contractors want local participation requirements to be set by the government. A lot of construction projects are funded by the World Bank, and the World Bank has prequalification requirements for contractors (in 2005, the requirements were for a firm to have a capital base of US$3 million and extensive international experience). Obtaining funds for construction projects is difficult because letters of credit and collateral are required for loans, and banks have a difficult time establishing who is the true owner of construction equipment. The largest construction firm in Ethiopia is MIDROC.

A construction company may be held liable for any type of construction delay, even if a delay is caused by government interference through their agencies, such as the custom's authority. Contractors are responsible for the cost of delays, including fines and damages, because the legal process is too expensive for contractors to use as a legal means for assigning blame for construction delays. Nonpayment is a common problem, even on government contracts. Ethiopia has land-ownership regulations that limit development within the country.

Use of business agents: Not required but useful.

Construction materials available in the country: Coal, iron ore, natural gas, petroleum, copper, lead, potash, silver, sulfur, and zinc. Wood and concrete may be obtained locally, as well as some steel products. Diesel fuel, explosives, steel panels, welding parts, composite materials, lumber, and plywood have to be imported from Europe or Saudi Arabia.

Social standards: Ethiopia is located in northeastern Africa, and it is a landlocked nation of rugged mountains and a high plateau. The term *Ethiopia* means "burned faces" in Greek. Agriculture is the main economic activity; and it employs 80 percent of the population, only 5 percent are in manufacturing, and 10 percent are in service industries (construction is a minor economic activity). The majority of roads in Ethiopia are unpaved, which makes it difficult to transport construction equipment and machinery.

The Oromo organize their citizens into groups of either seven or eleven, and each person assumes a different responsibility in society every 8 years (for additional information, see the web site www.oromia.org). Ethiopians are a proud culture, and their pride stems from their sense of self (the people are dignified). The term the *power of the day* is used in Ethiopia to refer to whomever is currently in charge, and the availability of financial and human resources are affected by who is in power. Some Ethiopians respect and sometimes fear authority figures, and they may consent to do something even if they do not agree with it; therefore, acceptance of instructions does not necessarily mean approval.

Social status is derived from a history of family wealth, achievements in school, or the workplace, and links to people with political and economic influence. Foreigners are given immediate respect in the workplace as a result of the tradition of showing hospitality to visitors. In the workplace, people may hide their true feelings or express their feelings to other people by making cryptic remarks to ensure that no one is offended by what they say. Maintaining the status quo is a preoccupation, especially when it is related to maintaining one's position in an organization.

14.2.3 Yemen

Languages spoken: Arabic is the official language, but English is also spoken. Construction contracts, drawings, and specifications are written in English.

Religion: Islam.

Holidays: Ramadan is in the tenth month of the Islamic calendar and it lasts for one month. Qat parties (which are defined later in this subsection) during Ramadan may last until 4 A.M., and many construction sites do not operate during Ramadan or they only operate on half-day schedules. Official business hours during Ramadan are 11 A.M. to 3 P.M. The first four days of the month after Ramadan are called *Shawwal,* and they mark the breaking of the Ramadan fast. Muslims do not eat or drink during daylight hours during Ramadan. Most construction work restarts 10 days after the end of Ramadan and it starts on a Saturday because Saturday is the beginning of the week in Yemen and other Islamic countries. Two months and 10 days after the end of Ramadan is the Id-Aladha, the Feast of Sacrifice, and it begins on the tenth day of the month of the Hajj, which is the month where Muslims make their pilgrimage to Mecca. It last for 6 to 10 days.

Secular holidays include: Labor Day—in May; Day of the National University—May 22; Victory Day—July 7; Revolution Day—September 26; National Day for South Yemen—October 14; and Independence Day for South Yemen—November 30.

Climate: Yemen has a coastal-plain region, a high-mountain region, a mountainous basin, a plateau area, and a desert region. The climate is hot because of the influence of the desert, but it also has rainy monsoons from the Indian Ocean in the south and southwest in the summer. In winter, dry monsoons blow from the north and northeast. Traveling along the mountains during the rainy season is treacherous because of violent, torrential rains.

Culture and customs: Yemen is still a tribal society. Being in a tribe means that someone is part of a social and political organization. Tribal affiliations are stronger in the north and they diminish in the southern regions of Yemen. Each tribe has its own wells, market towns, friends and enemies, history, and leadership traditions. Conflicts are solved at the lowest possible level of a tribal unit. If someone kills a member of another tribe, everyone from the tribe has to pay compensation. Every tribe elects a sheikh, who will resolve conflicts or raise an army if conflicts cannot be resolved peacefully.

First names are used in Yemen, along with a last name related to someone's tribe or province. It is impolite to ever refuse tea when it is offered and it is served in business negotiations. When a Yemeni starts a conversation with foreigners, the foreigners should converse with him or her even if they do not understand anything that is being said in the conversation.

Visitors to homes should take off their shoes when they enter a home, and they are not allowed to enter a room reserved for women. Foreigners should never try to socialize with Yemeni women, because they are off-limits to foreigners. Yemenis will gather around a foreigner to greet him or her, and they are genuinely interested in assisting foreigners. The main meal of the day is lunch and people eat with their fingers.

After lunch there may be *qat* chewing, which means taking the smaller leaves of the qat plant, mixing them with water or soda, and chewing them for hours, which releases a volatile oil that provides a mild high. Qat leaves are not swallowed, but they are pushed to one side of the mouth in a large bulge. Qat may also be chewed during work. Qat parties are attended at least once a week and daily if someone can afford to do it. Qat parties are spontaneous, and although many people use them to gossip, business may also be conducted during them.

Yemeni males wear a dagger, called a *jambiyyah,* attached to their belt in the front and center of the body after they have transitioned into manhood. If someone is charged with a crime, he forfeits his dagger, and the dagger is used as a bond that guarantees that he will return for his trial if he is released from jail. It is unacceptable for someone to unsheathe his dagger to threaten someone, and if he does, his tribe will punish him by making him pay a fine. Yemen has more guns per capita than any other country in the world, because guns are used for warfare as well as during celebrations such as weddings. If a man does not have a gun he is not worthy of respect from others.

Business practice standards: The construction industry is monitored by the Ministry of Public Works (MPW), which designs and supervises the construction of public buildings, highways, and rural water-supply systems. Other agencies that are involved in the construction industry include the Central Planning Organization (CPO) (which allocates financial resources to projects and executes projects) and the Ministry of Power and Water (which overseas construction of the national electric grid, the power supply networks for cities, and water supply networks).

The Ministry of Construction, Housing, and Urban Planning is responsible for the construction of roads and bridges and for issuing licenses, overseeing food and safety, and designing urban plans. Payments to contractors by the government might be delayed for 2 to 4 months; therefore, construction is interrupted and delays are a normal occurrence on job sites.

Construction employs 7 percent of the population in Yemen. Power in Yemen is allocated by the Public Electricity Corporation, but the government supplies only 45 percent of households with water, and the rest is provided by the private sector. There are only approximately 500,000 telephone lines for a population of 20 million people (2 lines per 100 people).

Construction labor generally is unskilled labor, and the quality of the products they produce is below the standards obtained in industrialized countries. Plumbers, carpenters, electricians, and plasterers do not receive adequate training. Foreigners are employed by various aid donors and private Yemeni contractors, and foreign engineers and architects design and supervise most private sector structures.

Workweeks start on Saturday and end Thursday evening. In the construction industry, workdays start at 7 A.M. and end at 5 P.M. with a half-hour break for breakfast and an hour break for lunch. There are no unions in Yemen, no paid holidays, and no paid vacations. In the public sector, workdays start at 8 A.M. and end at noon and workweeks are from Saturday until Wednesday evening.

Construction foremen, supervisors, and laborers usually are related to each other, and they may have a family relationship with the owners. Each tribe specializes in one aspect of construction. Carpenters and masons come from the middle provinces (Ta'izz and Ibb), and steel fixers (iron workers) are from the west (Zabid and Bajil). Projects managers cannot make any decisions without first consulting with owners.

There is no official contract law in Yemen, and contract disputes are settled on the basis of each individual contract and agreement. Disputes are settled by the Yemeni Center for Conciliation and Arbitration in cooperation with the Association of Banks and the Federation of Chambers of Commerce. Government arbitration is used to solve disputes between state agencies and the government. Arbitrations are conducted by the Public Authorities and the Ministry of Parliamentary and Legal Affairs, and their decisions are final. Seventy percent of disputes are solved through tribal arbitration, because most Yemenis do not trust the court system and their first loyalty is to their tribe.

Use of business agents: The government owns most of the industrial sector and private companies are encouraged to participate in joint ventures with the government. Some private sector firms are owned by government officials, which creates large scale corruption including kickbacks (where a contract is awarded to a contractor if he or she will pay part of the money back to government officials) from contractors. Since public sector wages are nine times lower than private sector wages, bribery has become a common practice in Yemen. Salaries and promotions are based on relationships and personal connections rather than on ability.

Bid documents in Yemen usually are incomplete, and specifications do not provide sufficient details. Contractors are not allowed much time to review contracts before bidding on them. Without the oil industry only 4 percent of the gross domestic product (GDP) per capita is generated by industries. The government owns most of the oil and gas facilities through the following organizations: the Yemen Petroleum Company, the General Corporation for Oil and Mineral Resources, the Yemen Refining Company, the General Department of Crude Oil Marketing, the Yemen Corporation (which produces cement), and the Petroleum Exploration and Production Authority.

Construction materials available in the country: Yemen lacks appropriate infrastructure for transporting heavy construction equipment and large construction materials (only 9 percent of the roads are paved), and it is difficult to import materials through customs. Construction materials that are available locally include gas, cement, colored stone, marble, flagstones, concrete blocks, sand, gravel, wooden panels, pipes, and plastic tubing. Bricks still are made by hand, and mud bricks are common in south Yemen. Steel and lumber are imported from other countries.

Structures usually are six stories high, and the first three stories are built of thick stone (50 cm at the base) to provide insulation from the heat. The upper three stories are made of brick because it is lighter, easier to elevate, and less costly. Windows are large on the upper floors, and they are surrounded by intricate plaster decorations. Arches are made of gypsum, and they are called *gumaria.*

Social standards: The official name Yemen came into being in 1990 as a result of a merger between North Yemen (the Yemen Arab Republic) and South Yemen (the People's Democratic Republic of Yemen). Yemenis have an average family size of 7.4 people, and an average of 3.1 people share one room. Only 37 percent of the population is literate above the age of 15.

Architecture in Yemen is unique in the highlands, where multistory (five- or six-story) structures are made from stone, brick, or mud. Extended families live in the same house. In the plateaus, hard-dressed stone is a common building material, but mechanical methods of cutting stone are replacing hand-cut stone due to labor shortages.

14.3 Asia

In Asia, harmony is stressed, and people are judged by the manner in which they perform a task and how they perform a task rather than on what they achieve as individuals. Since group achievements and performance are rewarded rather than individual performance, people strive to work well in groups. The way someone gets along with other people is more important in business than his or her expertise or the profits they generate for a company.

Honor, dignity, and family are protected at all costs, and people are modest and self-deprecating in Asian cultures. If a compliment is given to someone, he or she may deny it or minimize his or her abilities. People sit up straight with their feet on the floor in front of them during meetings because slouching is not acceptable in public. A slow, measured pace is used when talking to maintain harmony.

A quote that captures the essence of the work environment in Asia is that "more important than winning is the grace of the runner" (Copeland and Griggs, 2001, p. 187). People may see forces that are beyond human control, such as fate, God, or people with influence, as controlling their lives; therefore, they view hard work as futile, especially in countries where people are rewarded for currying favors with the right people. In many countries in this region, seniority systems determine who is promoted or who receives raises.

In Buddhist cultures, people are taught that suffering is caused by people's desire for possessions and selfish enjoyment. The practice of Confucianism is based on the life and work of Kong Fu Ze (479–551 B.C.), and it stresses "harmony with nature and reciprocal relationships such as emperor/minister, husband/wife, or father/son with moral imperatives of filial piety, fidelity, and paternalism, personalism, and a sense of systemic insecurity" (Martin and Westwood, 1997, p. 188).

Paternalism within companies causes firms to have centralized directive management, and decisions are made based on the intuition of owners and experience rather than group consensus (Japan is an exception to this philosophy). Trust and personal relationships dominate Asian societies such as Hong Kong where business opportunities are created through personal referrals. There is little difference between personal and professional relationships in Chinese societies because they are usually one and the same.

Major developers hold more than 10 percent of the stocks listed on the Hong Kong Stock Exchange, they own most of the major corporations, and they dominate the Hong Kong stock market, which is called the *Hang Seng Index*. Hong Kong is a contrast in cultures, as is shown in Figures 14.4 and 14.5, where there are high-rise buildings next to Chinese architecture and slums along the hillsides.

Members of Communist societies, such as China or North Korea, always feel insecure about the future because of unstable political systems and how easily everything can be taken away from them when there is a regime change. Insecurity leads to citizens working harder in order to accumulate wealth to protect their families in the future or to purchase immigration opportunities (when Hong Kong reunited with China, citizens of Hong Kong were able to purchase visas to live in the United States and Canada for US$2 million per person.

People in Hong Kong identify more with Western societies than Eastern societies, and managers have learned to adopt organizational-management styles that are used in Western societies. However, these practices are merely a mask because the internal structure of firms is Chinese. Employees in Hong Kong are expected to work long hours (9 A.M. to 8 P.M.) and on Saturday morning and to be punctual (in contrast to mainland China where citizens work regular business hours). A common greeting in Hong Kong is "Have you eaten yet?" rather than "How are you?"

Indigenous teaching methods affect the overall ability of engineering and construction (E&C) personnel, because they may be taught to never question their superiors and not to speak up in class, and they may not be required to participate in any group projects during their educational programs. Students are required to

Figure 14.4 Contrasting architecture, Hong Kong.

Figure 14.5 Slum housing, Hong Kong.

stand or bow when their teachers enter a room, and they show respect for their educators at all times so there are no casual discussions. Respect for authority is paramount and education comes second. In China, the purpose of studying is to learn something useful in order to make a contribution to society, and government campaigns are used to teach people to act or behave in a manner that is acceptable to the government.

Job applicants may include photos in their applications for work, which is illegal in some other parts of the world. Honor, family, and dignity are the most important aspects of Asian societies. In Confucian cultures, harmony and paternalism are stressed and units within companies are run like families (Copeland and Griggs, 2001). In Japan, consensus decisions require more time, but once a decision is made it is easer to implement than in Western cultures where decisions are made and then employees are convinced to follow the decisions. In Asian cultures, if a business unit looks good then the boss looks good, so there is no competition within business units.

Dignity is revered throughout Asian societies, and if someone loses his or her dignity, it dishonors his or her family. Humility, respect, job security, a good personal life, social acceptance, and the achievement of power are what people strive for in Asia.

14.3.1 Japan

Languages spoken: Kanji is the official language, and it is a system of pictographs including 50,000 different characters. People rely on *haragei*, which translates to "visceral communication" or "belly language"—people determine the reaction of another by observing the facial expressions of the other person, the length and timing of their silences, and the sounds they use when listening to someone else.

Religions: Confucianism with traces of Shintoism and Buddhism.

Culture and customs: The proper way to bow in Japan is at a 15-degree angle with hands at the side. A more formal bow that shows respect is 30 degrees. *Bun* is a word that refers to being in a high place in the social order.

Festivals that revere nature are an integral part of the Japanese culture and people participate in "flower viewing" or "moon viewing." *Mah-jong* (a game that is played with tiles) and *pachinko* (electronic gambling) are national pastimes, although gambling is illegal (promissory notes are used to pay mah-jong winnings; therefore, cash does not change hands). Children play *go* or *shogi*, a form of Japanese chess. Executives play golf, archery, or practice swordsmanship. There are 1500 VHF and 10,000 UHF television stations in Japan, most of which are operated by the Nippon Hoso Kyokai (NHK), which is a public broadcasting system funded by user fees.

In Japan, homes and apartments are small; therefore, everything in them, including the appliances and furniture, is smaller than in Western countries. People may sleep on mats on the floor that are rolled up during the day and stored in closets. Apartments and many homes only have a few rooms so all activities are conducted in the same room.

Business practice standards: The process of making decisions by consensus that is used in Japanese firms is called *nemawashi* ("root binding"). Before a decision is made everyone is consulted and everyone in the group has to accept the decision before it is implemented, which may delay decisions for years.

Lawsuits are not common in Japan, because the legal system takes from 2 to 20 years to settle cases. The main purpose of lawsuits is to force the opposing party to take moral responsibility for his or her actions. Contracts are regarded with suspicion by the Japanese, because they feel that relationships should change with the circumstances. Even though elaborate contracts are not used, the Japanese are honorable in their business dealings and use the term *seig,* which means "right principles," in reference to legal matters.

The Ministry of International Trade and Industry regulates the economy, and government central planning includes a consensus process whereby private enterprises may provide input into the planning process. Being a civil servant is a prestigious job in Japan.

The government of Japan is a bicameral parliament that includes a diet of two houses—an upper house that performs ceremonial functions and a lower house. Prime ministers are selected by the parliament rather than by popular vote, and there is a monarchy.

If someone is assigned to sit close to a window in an office, it may mean that they are close to retirement. If someone says that something is inconvenient that means that it is impossible to do. Looking directly into the eyes of someone who is a superior is considered to be a hostile gesture, or it is interpreted as judgmental or punitive. *Chigu* means that someone is different, and it would be used in context to foreigners.

Use of business agents: Candidates for public office receive "back money," which is money acquired through large, undeclared contributions. Corporations do not pay money directly to candidates because they use a *kuromaku* (a "black curtain man"). The Yakuza is an organized-crime group that operates in many different sectors of Japan.

Construction materials available in the country: Most construction materials and energy products have to be imported into Japan, because there are no natural resources available in Japan.

Social standards: The Japanese race descended from a proto-Caucasian race called *Ainu.* The Japanese culture was influenced by the tribal values of feudal systems, and the Japanese still emphasize hierarchy and loyalty to the group. Japan is a homogeneous society, but there is a group called the *Burakumin* (the "people of the hamlet") or *Eta* ("full of filth"), and they live in the ghettos of Japan.

Arranged marriages still exist in parts of Japan, and the divorce rate is less than 10 percent due to the fact that Japanese courts award custody of children to fathers not to mothers. Extended families live together, and the elderly are cared for by their children. Men have close relationships with their mothers, who are called *amae* (mother-in-laws may be referred to as *mamagon,* which means "dragonlike"). Wives make household decisions and control the finances in Japan.

Teachers are called *sensei,* and their students revere them. Prestigious universities, such as Todai (Tokyo University), are the ones that students need to attend to be offered elite jobs. Education is a means to an end—to obtain a house, which may mean an apartment or a small plot of land with a home on it.

Commonly misunderstood Japanese terms are *kamikazi* and *geishas. Kamikaze* means "the divine wind." *Geishas* have been replaced by bar maids. There is a high suicide rate in Japan, and one form of suicide that is used in Tokyo is called *tobikomi,* which means "diving in." *Tobikomi* takes place when someone throws themselves onto the tracks of computer trains.

14.3.2 People's Republic of China and Taiwan (The Republic of China)

Languages spoken: In China, the languages spoken include Mandarin (*Putonghua*), Cantonese (*Yue*), Shanghaiese (*Wu*), Minbei (*Fuzhou*), and Minnan (*Hokklen*); in Taiwan, *Xiang, Gan,* and *Hakka,* as well as hundreds of local dialects. A system for phonetically representing Mandarin characters, called *Pinyin,* is used by foreigners to learn Mandarin. Taiwanese indigenous groups include the Atayal, Saisiyat, Bunun, Tsou, Thao, Paiwan, Rukai, Puyuma, Amis, Yami, and Kavalan. Mandarin is the official language in Taiwan, but there are 14 major dialects.

Religions: In the People's Republic of China, the official government position is atheism, but people practice Daoism, Buddhism, Islam, and Christianity. Taiwan is polytheistic and syncretistic, and the religion is a blend of Taoism, Buddhism, and folk beliefs.

Climate: In China, the climate varies throughout the country from the cold-temperate north ($-30°C$) to the tropical south ($28°C$), with wet monsoons in the summer and dry monsoons in the winter. Taiwan has a subtropical climate.

Holidays: Months are referred to as month 1, month 2, and so forth, because names are not used for months. In China, holidays include New Year—January 1; a 3-day Spring Festival—New Year by the lunar calendar; International Working Women's Day—March 18; Tree Planting Day—March 12; a national 3-day International Labor Day—starting May 1; Chinese Youth Festival—May 4; International Children's Day—June 1; Anniversary of the Founding of the People's Liberation Army—August 1; Teacher's Day—September 1; and a 3-day National Holiday—starting October 1. Traditional festivals include the Spring Festival, the Lantern Festival, Pure Brightness Day, the Dragon Boat Festival, and the Mid-Autumn Festival. Ethnic minorities have retained some of their festivals, including the Water Sprinkling Festival (Dai people), the Nadam fair

(Mongolian), the Torch Festival (Yao), the Third-Month Fair (Bai), the Antiphonal Singing Day (Zhuang), New Year and Onghor Festival (Tibetan), and the Jumping Flower Festival (Miao). Holidays in Taiwan are Founding Day of the Republic of China—January 20; Lunar New Year—February 8–11; Memorial Day—April 5; the Dragon Boat Festival—June 11; Armed Forces Day—September 3; the Mid-Autumn Festival—September 18; and National Day—October 10.

Culture and customs: Family names are the first name. People are used to waiting for everything. Introductions are formal. Banquet etiquette requires that guests not eat until the main host indicates that everyone may start eating (meals include several courses), the seating arrangements at tables are determined by status, the location of particular individuals in a room is determined before the banquet, and speeches will be given during the banquet.

Banfa is a term used that means "circumventing bureaucracy rules and regulations, a way out of problems, and resourcefulness through regulations" (Copeland and Griggs, 2001, p. 167).

Wall posters, called *dazibao,* are posted on walls that are located adjacent to areas that have extensive pedestrian and vehicular traffic and they are used to express grievances, political views, and frustrations that are related to current events. *Dazibao* are posted quickly when an area is busy so that the authors remain anonymous to avoid the authorities. The *dazibao* are an informal means of communicating with a large audience, and hundreds or possibly thousands of people read them before government officials remove them. Space is left on wall posters for people to respond to them.

Business practice standards: Sometimes members of foreign firms run into road blocks called "internal documents" that refer to secret laws that protect the interests of the government. Construction is the third largest business sector behind manufacturing and agriculture. Construction firms may either be foreign-invested construction enterprises (FICEs), which include Sino–foreign equity construction joint ventures and Sino–foreign cooperative construction firms, or wholly foreign-invested construction enterprises (WFICE). For a firm to be classified as a FICE, its overseas personnel must reside in the People's Republic of China for 3 months each year. Members of FICEs who want to upgrade their business classification have to meet government requirements that are based on a minimum number of years performing work in China, previous work experience, registered capital, technical capabilities, and number of personnel.

There are three classifications for construction companies and having a higher-grade classification allows construction firms to work on a larger variety of construction projects. Government regulations for construction may be found in the *Regulation on Administration of Foreign-Invested Construction Enterprises.* Engineers and constructors use the International Federation of Consulting Engineer's contract forms that are issued by the Ministry of Construction and National Commerce and Administration Bureau. Quantities are measured using the Royal Institute of Chartered Surveyors measurement methods. Construction is subject to the *Regulation on Examining Occupational Safety and Health Management System* issued by the National Economic and Trade Commission. There are no unions.

Names on business cards should be in the local dialect and English. Rank and hierarchy are essential in business and lower-ranking managers do not do business with higher-ranking managers. Be thoroughly prepared for meetings-including bringing technical experts and translators, know specific details about proposals, be familiar with a company before trying to conduct business, and be familiar with the competition. Answers to questions always should be consistent no matter how a question is being asked by someone (questions may seem repetitive).

Business is not rushed because individuals and companies have to establish their credibility as they participate in business meetings before making business deals. Contracts indicate a formal agreement and documenting what happens during negotiations helps to ensure that agreements reflect what was discussed during contract negotiations.

If a foreigner wants to live in the People's Republic of China, residency permits are required along with visas. Medical tests are required to obtain a driver's license and to swim in public swimming pools. Within China, citizens must have a residency permit to live in a particular region, and if they do not, they are paid lower salaries. There are no independent unions, and workers are not allowed to organize. Access to internet Web sites and the media are regulated by the government (censorship).

Corporate cultures are casual and people may talk loudly because they are used to being in large groups. Ranks within organizations are clear and people at each level may be distinguished by the location of their desks, subtle differences in the way they dress, and their business titles. Social-emotional and negotiating relationships are used in business. Leaders are appointed by the government based on their experience and expertise or their political connections. There is great respect for people in authority, and it is difficult for subordinates to express their opinions or to disagree with their supervisors. Employees are sensitive to criticism from their supervisors, but they hide it for fear of losing their jobs.

Construction techniques vary throughout China. There are ultramodern high-rise structures, such as the ones that are in Shanghai, but in rural areas structures still may be made out of clay bricks. A rural Chinese village is shown in Figure 14.6.

Structures in major cities are constructed quickly, with little regard for building codes and safety. There are unique construction methods being used in China, such as drainage systems built on slopes that are a series of vertical Ys that are interconnected until they reach the bottom of the slope. During construction, entire structures are covered with bamboo scaffolding so that workers have access to the outside surface of every floor.

There is a severe shortage of skilled and unskilled labor in Taiwan, and foreigners may fill these positions. The Ministry of the Interior controls unions

Figure 14.6 Rural village, China.

and union organizing activities. The National Health Insurance program covers everyone that works in Taiwan, and each participant pays a monthly premium for medical treatment (socialized medicine).

Use of business agents and construction company classifications: Gifts could be interpreted as bribes, and tipping is discouraged in China. Someone with connections is called a *guanxi,* which means he or she has relationships of mutual obligation with others through the "back door."

Companies are required to work through a state agency to do business in the country and to have local partners. The required amount of local participation on projects depends on the government classification of a foreign company; there are classifications of special grade and grades I, II, and III. Grades of I, II, and III include primary contractors that are allowed to subcontract out some of the work and specialty contractors. Labor contractors are classified as grade I or II. Classifications are based on management expertise, number of engineers, the amount of fixed assets and circulating capital, mechanical equipment, and total output per year. There are approximately 100,000 registered construction companies, and fewer than 3000 are classified as grade I, which allows them to engage in any type of construction.

There are over 1000 joint-venture construction companies in China. Foreign companies may enter the construction market through:

• Joint ventures or cooperative engineering and construction companies
• Real estate development companies
• Doing construction supervision only

- A construction consulting firm
- Bidding on projects as a joint general contractor
- Supervising projects funded by foreign companies or overseas loans
- Setting up representative offices in the country

The Foreign Ministry helps members of foreign firms conduct business in China. There is a Ministry of Construction in China, which is a department of the State Council (there is a construction committee or construction department in every provence), and the Information Department provides official information about the country. There are four levels of government that develop regulations—the republic, provinces, regions, and local agencies.

The People's Republic of China joined the World Trade Organization (WTO) in 2001, it is a member of the Association of South East Asian Countries (ASEAN), and it participates in the Asian Free Trade Agreement (AFTA).

Industries and materials available: The People's Republic of China is the second-largest consumer of oil (the United States is first), and it is the largest consumer of copper, tin, zinc, platinum, steel, and iron ore (*Newsweek,* May 31, 2004). Confirmed reserves in China include oil, natural gas, oil shale, coal, iron, copper, aluminum, molybdenum, tin, zinc, mercury, uranium, graphite, phosphorus, and sulfur. Steel and cement may be obtained locally. Highways and railways are available throughout the country for transporting construction materials, but delays are encountered due to a deteriorating infrastructure and traffic congestion.

Social standards: The country is called the *People's Republic of China,* not *China,* and Taiwan is called the *Republic of China.* If the two names of the countries are reversed it may result in disastrous consequences. Conversations with foreigners may start with some type of a compliment that is replied to with a self-deprecating remark. Anyone who is not Chinese is considered to be a *nonperson* and they are not afforded the same respect as Chinese show other Chinese. China is a homogeneous society, with less than 10 percent of the population from minorities that are concentrated in specific regions of the country.

Chinese citizens are encouraged to have only one child if they live in a city and two if they live in a rural collective. If people have more than the government prescribed limit of children, their children may not receive medical benefits or education.

In some major cities, drivers of automobiles are only allowed to use their parking lights at night so that cyclists will not be blinded by automobile headlights, because there are millions of cyclists. The cities are crowded and vehicles are not always allowed to use the streets. Figure 14.7 shows a typical street scene in one of the smaller cities in the People's Republic of China.

14.3.3 South Korea

Languages spoken: Korean is the official language and English is taught in high school.

Figure 14.7 City street scene, China.

Religions: Forty-four percent of citizens have no religious affiliation, 26 percent are Buddhist, 1 percent are Confucian, and 1 percent other.

Climate: The climate is temperate with heavy rains in the summer and occasional typhoons (high winds and flooding).

Culture and customs: Confucianism dominated the country until recent years, and it is more a philosophy rather than a religion. Confucianism emphasizes the importance of humanity. Elders are respected, and people in power and their citizens have faith in each other.

Third-party introductions are preferred; therefore, it is appropriate to wait to be introduced to other people at gatherings. A junior person will bow first, but senior people will be the first ones to offer their hand for a handshake. Weak handshakes, and a slight nod of the head, are an acceptable greeting. To greet each other, South Koreans might use a slight bow and a handshake. When shaking hands with an elder, the right forearm is held by the left hand. People also bow at the ends of meetings. Western women have to initiate handshakes with Korean men, but if a man is of a higher status than the woman, he should initiate handshakes. Always greet elderly people first, because they are highly respected in the Korean culture. When anything is given to elderly people it should be extended with both hands and a slight bow.

Eye contact indicates sincerity and that you are being attentive to the speaker. If someone is an authority figure, people may not be able to maintain eye contact with him or her. Some men try to avoid eye contact with women. Do not touch people on the back or arm unless good friends. Physical contact is limited to good friends and family. People should never be criticized in public, because it will cause them to lose face (embarrassment). Smiles may be used to mask embarrassment. To beckon someone, the hand is extended with the palm facing down and the fingers are moved in and out because people do not beckon someone with an index finger.

Business practice standards: North Korea is closed to foreign investments. Foreign businesses were allowed into South Korea in the November Declaration of the Fourth World Trade Organization Ministerial Conference in 1994. Foreign firms that obtained construction licenses in the late 1990s returned their licenses to the government and started focusing exclusively on construction management rather than construction because they could not secure enough work.

Prior to the economic crisis, people expected lifetime employment in South Korea, and this may still be the case in North Korea because it is a Communist country. Korean firms have a short-term orientation, they use collective decision making, authority is concentrated at the top of firms, and firms are operated on the principles of humanism. Promotions are based on seniority.

Use of business agents: Connections are the most important way to secure business, and a few large companies control each industry. Nepotism (the hiring of family members) is practiced in Korea.

Construction materials available in the country: Steel and electronic components.

Social standards: Major industries are electronics, telecommunications, automobile production, chemicals, shipbuilding, and steel production. There is mandatory military service for 2 years or more at age 18. South Korean construction firms use military personnel to build projects, which makes them more competitive than firms from other countries in the global construction market.

The Chosun dynasty controlled Korea for hundreds of years before it became a republic. There are only a few surnames in Korea, and people with each surname are direct descendants of families that existed during the dynasty.

Korea has one of the highest per capita rates of alcohol consumption in the world. Scotch, beer, and *Soju* (a local drink that is 23 percent alcohol) are the drinks of choice. If people are urging others to drink, they will continue drinking so that they do not lose face.

14.4 Eastern Europe

The official name for the 10 nations of Eastern Europe that were formerly part of the Soviet Union is the Commonwealth of Independent States (CIS). Information written about Eastern European countries may be inaccurate or intentionally deceptive in nature, because government officials in these regions try to disguise economic problems from the outside world and from their own citizens.

Massive construction projects were undertaken by the former Soviet Union during the early part of the twentieth century that employed millions of forced laborers (imprisoned dissidents) and military personnel, and these projects pushed the limits of technology. During the 1930s, Russia promoted its projects all over the world as examples of Russian superiority. Many of these projects failed shortly after they were constructed, or they were never actually completed by the government. Examples include the Dnieper dam (Dveprastro), a steel city in Magnitogorsk, a coal city in Kuznetsk, and the White Sea Canal. In the 1950s, 59 percent of government officials in Russia were engineers and by the 1980s, the government was run by 89 percent engineers (a technocratic government).

In the 1980s, the Soviet Union was dismantled and regions of the country were divided into 10 separate nations that formed new governments. Prior to the 1980s, the Communist government controlled the economy; therefore, after the 1980s the governments of the newly formed countries had to quickly establish private industries. The "mafia" (secret criminal organizations) stepped in to fill the void left by the Communist government, and in most Eastern European countries construction is controlled by organizations with connections to the Russian mafia. Obtaining construction materials requires working through the mafia and paying bribes. Since government funds are limited many Eastern European countries are still relying on foreign aid to pay for construction projects.

Over 30 percent of college students in Russia major in engineering, but engineering degrees are highly specialized compared with engineering programs in other parts of the world. Instead of being awarded a degree in mechanical engineering, a Russian engineer could be a ball-bearings engineer for paper mills. There are hundreds of engineering specialties, including agricultural machinery, machine tools, casting equipment, automobiles, tractors, aircraft, a specialization in copper and alloys, drilling wells for oil, bridge design, large-scale buildings, hydraulic structures, and erecting industrial buildings.

To obtain a degree in civil, electrical, chemical, or mechanical engineering, students have to attend one more year of school beyond the engineering specialty degree program. Technical degrees are 3-year programs, and bachelor's degrees require 5 years of study. Eighty-eight percent of engineering degrees are awarded through technical institutes rather than through universities such as the University of Leningrad or the University of Moscow. Engineering programs stress science, materials, and descriptive technology.

The government provides planning and standardization details on "how to prepare sites, what bulldozers to use, what earthmovers and trucks to use, what cranes to employ, etc." (Graham, 1993, p. 69). Sometimes specifications ignore site characteristics, groundwater levels, population densities, types of soil or rock for foundations, and seismic considerations.

14.4.1 Bosnia and Herzegovina

Languages spoken: Bosnian, Serbian, and Croation. The languages are written in either Cyrillic or Latin alphabets. Jewish folk songs are sung in Ladino, a fifteenth-century Spanish dialect.

Religions: Islam, Catholicism, Orthodox Christian, and Jewish.

Holidays: There are combinations of religious, secular, and family holidays, including the Orthodox holiday *Slavia* (family saint's name day). Specific holidays include January 1—Nova Godina (New Years Day); January 9—Dan Republike (Republic Day); March 1—Dan Nezavisnosti (Independence Day); May 1—Me Unarodni Praznik Rada (May Day); November 25—Dan dr avnosti (National Day); Muslimanska Nova Godina (New Year)—first day of Muharram, Mevlud (Prophet's Anniversary); twelfth day of Rabee'ul Awwal; Kurban Bajram (Had Ibajram, Big Feast); tenth day of Hajj, Ramazanski Bajram (Little Feast); and the first day of Shawwal. There are 16 different Orthodox and Roman Catholic holidays that are celebrated on different dates, including Easter, Christmas, All Soul's Day, and named saints' days (Christmas and Easter are celebrated on different days by Catholics and Orthodox Christians), and there are 17 Jewish holidays.

Climate: Hot summers and cold winters (except the coastline where there are mild rainy winters).

Culture and customs: In business, handshakes are used for greetings and when someone leaves a meeting. In social relationships people touch their friends' arms or backs and kiss each other during greetings. Croats kiss each other twice, and Serbs kiss each other three times. People are used to a strong government, and they are still adjusting to free-market enterprise.

Business practice standards: Engineering designs should include a seismic analysis because of the earthquake activity in this location. Environmental issues are another concern because of the pollution created by industrial plants. Sites for the disposal of urban wastes are limited and the infrastructure was severely damaged during the civil war in the 1990s. Productivity is 80 percent less than it was before the civil war. A black market economy is flourishing, along with money-laundering activities and political corruption, due to weak law enforcement, that is not reflected in national statistics. Funds for construction and basic services come from reconstruction assistance and humanitarian aid from other countries. Some of the money designated for infrastructure repair is being siphoned off for other purposes.

The construction industry employs only 2.5 percent of the population, and it has a reputation for producing high-quality products that was established when native companies were building projects in Germany, Iraq, Kenya, Kuwait, Libya, Pakistan, and Russia before the dismantling of the Soviet Union. Engineering firms are either partially privatized or fully privatized, and there are firms that specialize in bridges, roads, and tunnel construction. Some firms provide both design and supervisory services to clients. The housing market is growing because people are now allowed to own land and homes instead of being assigned to government housing, as was done during the Communist regime.

The four main universities that award engineering and construction degrees are the University of Sarajevo, the University of Mostar, the Univeristy of Banja Luka, and the University of Tuzla. There are 3-year degree programs (associate degrees), 5-year programs (engineering degrees), master's degree programs (2 years), and PhD programs (4 to 5 years). Construction workers attend specialized secondary schools. Engineers and construction workers expect to be paid more by foreign firms than what is paid locally.

Construction companies are capable of constructing large industrial projects, but 50 percent of their equipment was lost during the civil war and the skilled workforce migrated to safer countries. There are only 24 insurance companies that operate in the country, and they are Austrian or Austrian-joint ventures. Therefore, it is difficult to obtain insurance for large construction projects. It is a civil-law country, and there are both state laws and entity laws. Since the legal system is not based on precedence law, contracts describe everything in great detail. Conflicts are solved by arbitration or mutual agreement. A major problem for construction firms is unidentified minefields, as the country has the second highest number of minefields in the world.

Use of business agents: To secure work on private projects, personal contacts are needed, but government projects are bid competitively. There are no government alliance requirements.

Construction materials available in the country: Steel is manufactured, but timber is hard to locate because of the deforestation that took place during the civil war. Stone, cement products, and prestressed reinforced concrete elements are available locally.

Social standards: Official names are the Federation of Bosnia and Herzegovina and the Republic of Srpska. The country gained sovereignty in 1991 when the Bosnians (48 percent) and Croats (14 percent) agreed to create a joint Bosniak-Croat Federation to end the fighting between ethnic groups and to stop the ethnic cleansing of Muslims (Bosniaks) by Serbs (37 percent), which live dominately in the eastern part of the country.

A joint multiethnic and federal democratic republic was created in 1995, along with the Office of High Representative (OHR), which oversees civilian aspects of the 1995 agreement. The country is unusual in that it has two first-order administrative divisions and one internationally supervised district (the Republic of Srpska). As of 2006, the country is still occupied by European Union peacekeeping forces that maintain peace and stability. Before the civil war, it was a centrally planned economy, and this is still evident throughout the country because companies are still overstaffed. Bosnia is the center of the former Yugoslavia, and it is ranked as one of the poorest republics in the former Yugoslavian states, with 44 percent unemployment. Most of its industries were destroyed in the civil war in the 1990s.

There are water shortages; therefore, water is rationed at certain times of the day. The primary social unit is the family. It is a patriarchal society, and mental illnesses are never discussed outside of a family. Architecture reflects both the Ottoman and Austro-Hungarian empires; and Islamic art, including calligraphy, fine metalworking, and carved wooden gates and screens, adorn Bosnian homes and public buildings. When a couple gets married, the wedding couple's initials are woven into carpets and these carpets are highly valued in Bosnia.

14.4.2 Romania

Languages spoken: The official language is Romanian and some people also speak Hungarian and German.

Religions: Predominately Eastern Orthodox.

Climate: Temperate continental with four seasons. Cold, cloudy winters with snow and fog and warm, sunny summers. There are extreme temperatures in the Baragan region ($-36°$F to $110°$F) and over 55 inches of rain in the Carpathian Mountains. A *cravat* is a cold northeasterly wind that blows in from the Eastern European plain. Earthquakes are a problem in the south and southwest regions.

Culture and customs: Romanians are used to deprivation. Obtaining products is still difficult in Romania because not many products are manufactured locally. As a result of living under Communist rule, Romanians are frugal and they invest their money rather than spending it for instant gratification. Privatization has increased home ownership and housing costs are increasing rapidly.

Romanians may be reserved in their interactions with others and self-deprecating when it comes to their abilities. They may speak softly as a sign of respect to others.

Business practice standards: Romania operates under a bicameral parliament with a president, a senate, and a chamber of deputies.

Engineering and metalworking are 26 percent of all industrial production. Construction firms are transitioning from state owned to privately owned and as a consequence, the industrial sector has been ignored in the early part of the 2000s. If a company is publicly owned, decisions are made by "office borders" who are part of the government.

There are five main universities—Bucharest, Cluj, Brasov, Timisoara, and Isasi—and they offer degrees in civil engineering, building science, railways, roads and bridges, hydromechanics, construction equipment, topography, and engineering economics. There are 3-year degrees, but it takes 5 years to obtain a full engineering degree. At construction job sites, engineers are either called *structural engineers* or *field engineers.*

Employees are formal with their superiors (they use the polite form of the pronoun *you* for them) but informal with each other. Firms are hierarchical, which makes it easy to distinguish who is in charge. The primary causes of construction delays are weather, scheduling conflicts, poor-quality work, inefficient equipment, dealing with the bureaucracy, and smoking breaks. Bonuses that are offered to workers are products rather than cash, such as cars, cell phones, or paying an employee's rent.

The Romanian Builders and Contractors Association (ARACO) has 1000 members, and members perform 80 percent of the construction work in Romania. Foreign contractors that operate in Romania are from France, Italy, Israel, Austria, and the United States.

Use of business agents: Romania has not had alliance requirements since 1998, but business agents help to expedite transactions. Foreign firms have to register with the government to operate in Romania.

Construction materials available in the country: Lumber, aluminum, cement (cement plants are owned by the French company LaFrance and the German company Holderbank Cement, and these two companies own 70 percent of the cement market), reinforcing steel, river aggregate, and petroleum products are produced in Romania. Romania is the largest oil producer in Central and Eastern Europe, and it has the fourteenth largest refining capacity in the world. Most structures are built out of reinforced concrete, which does not perform well in seismic zones such as Romania. Construction equipment is produced in Romania, but most construction firms cannot afford to buy it. Other construction materials are supplied by Germany or Italy.

14.4.3 Russian Federation (Russia)

Languages spoken: Russian and about 130 other languages. Not many people in Russia know how to speak English.

Religions: Orthodox Christian. A law was passed in 1997 that limits religious freedom and that made it more difficult for foreign missionaries to obtain visas.

Climate: The climate varies throughout the country with temperatures dropping below $-50°F$ in Siberia.

Culture and customs: Citizens were not allowed to demonstrate individual initiative while the Communist Party was in power, but individual initiative survived through criminal activities. There was a hidden, shadow economy during Communism, and the enterprises that thrived in the shadow economy were able to transition to free market enterprises.

Business practice standards: Computers may be brought into the country, but computers and software are searched when someone is leaving the country and encryption software or Global Positioning Satellite (GPS) software might be confiscated when someone goes through customs as they exit Russia.

Bartering and countertrade are used to pay for services such as electrical energy, natural gas, oil, and raw materials. Organized crime rings are powerful, and they have seized property from citizens through intimidation. Even state and private firms collaborate with organized crime organizations to produce raw materials and parts, to sell their products for a higher price than they would get through legal methods, or to avoid government taxes.

The government does not allow large firms to declare bankruptcy. Before operating in the country, it is essential to understand the legal and tax system, as well as technical standards and regulations. Organizations have some elements of self-government, but they still belong to state authorities.

A few firms produce over 95 percent of the construction equipment, and there is a shortage of heavy equipment on construction projects because of its high cost. The construction industry used to be 100 percent mechanized, but now manual labor is used when construction equipment is not available. Construction equipment may have only two or three attachments (e.g., dippers, hydraulic hammers, and rippers) that have to be used for all applications.

Excavation and site preparation are accomplished using *stampers* and *bush-saws* (scarifiers attached to trucks or row-crop tractors). Bulldozers and scrapers are only used on large industrial projects and irrigation projects. Compaction is done using impact compactors called *lift hitches* or crawler-mounted excavators, and complete compaction is seldom achieved on projects. Soil nailing is not used in construction, because it requires equipment that has to be imported from Italy.

Construction workers complete 2 years of technical training and then serve 2 to 3 years in the armed forces before joining the civilian workforce. Engineers study for 5 years and technicians study for 4 years. There are doctorate of science degrees and candidate of science degrees that are equivalent to doctorate of philosophy degrees in other countries.

Modularization is a common construction method, and many housing complexes contain apartments that are either model A or model B, because there are no other choices. Turkish, Finnish, Italian, and German contractors have more success in collaborating with firms in Russia than do U.S. firms. Local contractors are experienced in building large-scale industrial projects and power plants. Air-conditioning is not used, and radiators are the main source of heat.

Russian multilingual, multinational, and multicultural issues are barriers to effective project management. Some project managers set up "ethnic competitions" on projects to increase productivity. Businesses are autocratic, and managers may not know about laws protecting intellectual property, patents, cadastral surveying (legal rights associated with land), import and export regulations, or about organizational structures other than autocratic ones.

Use of business agents: Business agents are required by the Ministry of Foreign Trade, and this agency directs members of foreign firms to the appropriate foreign trade organization (FTO). Embezzlement is a problem in the construction industry due to low wages.

Construction materials available in the country: Reinforced concrete is the primary construction material, and it is finished by hand, owing to a lack of power tools, and women perform finishing tasks as well as plastering, painting, and installing floor finishes. The construction industry is comprised of 50 percent women.

Social standards: Money does not motivate citizens of Russia, because there is nothing to buy in many regions of the country. There is a high level of alcoholism and divorces are easy to obtain in Russia. The literacy rate is high, and many citizens have college degrees.

Glasnost (openness) and *perestroika* (restructuring) were introduced in the 1980s, along with more open relations with other countries, but an attempted coup d'etat in 1991 transferred power in all but name, and former Communist Party members were in power again. In the 2000s, state control is resurfacing and some of the signs of increasing state control include the suspension of elections for governors and the dismissal of the prime minister and the cabinet in 2004.

There are three major political and economic regions in Russia: (1) The European states are allied with the West, (2) Middle Asian countries and Azerbaijan are oriented to the Islamic world, and (3) the Far Eastern industrialized districts rely on Japan, South Korea, and other Far Eastern industrialized countries. Communist ideas of common equality are still being promoted in Russia. In the 2000s, there are high deficits, high inflation, a weak unconvertible currency, and political instability. In 2005, Russia signed a military treaty with China to balance the power being wielded by the United States and to try and establish political hegemony.

14.5 The Near and Middle East

The will of Allah influences life in Islamic countries, and some people may not feel that they have self-determination unless they are able to leave their native country. Countries in the Near and Middle East are either ruled by kings (Saudi Arabia, Qatar, Oman, Jordan, Kuwait, and Jordan), by *shii'a* clerics (Iran), by emirs (Djibouti), by sultans (Bahrain), or by dictators (formerly Iraq and Libya). Many Middle Eastern countries are dependent on oil or foreign aid to keep their economies stable or thriving, because there are few other industries (oil is 80 percent of exports in Iran). The funds received from the oil industry do not always trickle down to the masses through infrastructure projects because it is concentrated in the hands of the monarchy or other rulers.

There has also been a "brain drain" in Middle Eastern countries, because the educated classes are leaving the Middle East and settling in foreign nations (200,000 people per year leave Iran). Citizens have tried to challenge their governments, but government officials quickly suppress uprisings or the uprisings dissolve due to a lack of organization. Unemployment rates are high in the more populous countries and drug addiction is becoming a social problem, because there is easy access to drugs from Afghanistan, a major supplier of opium.

Islam is the official religion in most of the countries in the region, but there are also Christians and Jews. Mosques, and in some countries vehicles, have loud speakers that call the faithful to prayer five times a day, starting at first light.

14.5.1 Iran

Languages spoken: Persian, Persian dialects, and Shushtari (1750 years ago captured engineers were taken to Shustar to build dams, water mills, artificial canals, and irrigation systems), which is an older version of the Persian language.

Religions: The official religion is Shiism, a sect of Islam. There is a theocratic regime in power, which means that the government is run by clerics. There are also Christians and Jews.

Holidays: The most important holiday is *Nowurz*—New Year's Day—and it is on the first day of the equinox in spring. This holiday begins precisely at the equinox, so each year it starts at a different time, and it is a secular holiday. Other holidays include *Charshanbeh-soori* (Festive Wednesday) to bid farewell to the old year and *Sizdah-bedar*, the thirteenth day of the new year, which is a day to banish bad fortune by throwing sprouted greens into flowing water. Islamic holidays are observed (see Yemen).

Climate: Varies throughout the country.

Culture and customs: The New Year's celebration takes weeks of preparation, because everyone cleans out their house, buys or makes new clothes, and bakes traditional pastries and sweets, including sugar-coated almonds, baklava, almond cookies, and rice cookies. The ceremonial setting is called a *haftseen,* and *haft* means "seven." Seven symbolic items that represent sin and that begin with the "s" sound are displayed along with other meaningful objects such as a mirror, colored eggs, and goldfish in a bowl. The objects represent "health, renewal, prosperity, fertility, and the usual universal hopes shared by all people at any New Year's celebration" (DuMas, 2003, p. 106). Other items that are used are "olives for abundance, gold coins for prosperity, hyacinth for beauty, and vinegar and garlic for life's bitter moments" (Asayesh, 1999, p. 204). It is customary to slaughter a lamb on festive occasions and roast it whole. During Muharram, households have *rowzehs,* which are a "gathering where the Koran and stories of martyrdom are chanted by mullahs in turbans and flowing brown robes" (Aseyesh, 1999, p. 74). If someone uses the expression *gharb-zadeh,* it means that someone is enamored with Western culture.

Business practice standards: Eighty percent of exports are oil-related products. There is wide disparity between the rich and the poor, because oil funds are used for *cosmetic* change projects rather than urban and rural infrastructure. Cities suffer from overurbanization, as a result of 64 percent of the population migrating to cities in the past 45 years. Construction projects are needed to support the infrastructure. The economy is controlled by the state and there is corruption.

Unemployment was over 25 percent in 2005, which means that 16 million people earn a living for 63 million people. A large segment of the population is in their twenties because there was a "baby boom" in the 1980s. One fourth of the college educated citizens of Iran move to other countries. Foreign companies may rely on having well-educated workers, because jobs that require an education are scarce.

The government provides financial incentives to attract foreign businesses, but there is taxation discrimination. The inflation rate is over 16 percent, and there are continuous currency devaluations even though the government tries to stabilize its currency.

Iran requires 200 billion rials (UD$28 million) per year to build enough houses to meet the needs of its housing sector, and if that money is not available many people build their own homes. Land distribution policies are dictated by the Ministry of Housing and Urban Development. Commercial, educational, and industrial buildings make up only 12 percent of the total number of construction sites, but they are valued at 32 percent of total construction investment. Construction technology has not changed in Iran in 50 years due to a lack of construction materials and construction-management techniques.

Use of business agents: Business agents are required in Iran.

Construction materials available in the country: Residential structures are built using clay bricks that are made locally, because they are the least expensive construction material. Brick structures collapse during earthquakes. The capital of Iran, Tehran, is built on three earthquake faults and many of the buildings in Tehran collapsed during earthquakes in the latter part of the twentieth century. Steel beams for commercial structures have to be imported from Asian countries, and up to 20 percent of the beams may be defective. If other construction materials were available locally, energy savings would be over 50 percent.

Social standards: The government of Iran considers its subjects to be infants based on a concept known as *Valayat-e Faqih,* which translates to "guardianship of the jurist." The practice of *Valayat-e Faqih* was designed originally to protect orphans, the insane, and abandoned property. During the rein of Ayatollah Khomeini, who was called *Faqih,* which means the "supreme ruler," the *Valayat* was extended to include everyone in the country, and the reining leader is considered to be the proxy father of everyone.

Iran was a monarchy until the shah was deposed in 1979, during the Iranian revolution. Prior to the revolution, the only women who wore the *chador,* which translates into "veil" or "tent," were older women and villagers, but now signs warn against *Badhejabi.* The signs state that women should be adequately covered if they are over 9 years of age.

Iranians should not be called *I-raynians,* because the proper pronunciation is "E-rahn-ians," and the country name is pronounced "Erahn."

In locations where there are no sanitation facilities, *aftabehs* are used (steeping stones are placed on both sides of a pit, and a metal pail of water is placed next to it). Oil is delivered to homes by the *nafti* (the "oil man"), and it is usually stored in private yards called *hayat khalvat.* People are jailed if they are members of the People's Mujahedeen (an Islamo-Marxist group), and foreigners need to be careful of guilt by association. If *Marg bar Amrika* is being chanted, people from the United States should leave the area immediately, because it means death to Americans.

14.5.2 Israel

Languages spoken: Hebrew and Arabic. English is taught in middle and high school. Other languages spoken are Turkish and Asian languages.

Religions: Eighty percent Jewish. Islam and Christianity are also practiced in Israel. The Sabbath is from sundown on Friday to sundown on Saturday. Public transportation stops during the Sabbath and stores are closed.

Holidays: The following holidays are observed in Israel, but the dates change yearly based on the lunar-solar calendar: Rosh Hashanah (New Year), Yom Kippur (Day of Atonement), Sukkoth (Festival of Tabernacles), Simchat Torah (Rejoicing of the Law), Hanukkah (Festival of Lights), Fast of Tenth of Tevet, Tu B'Shevat (New Year of Trees), Fast of Ester, Purim, Sushan Purim, Passover, Yon Hashoah (Holocaust Day of Remembrance), Yon Hazikaron (Memorial Day), Yom Haatzma'ut (Independence Day), Tom Yerushalayim (Jerusalem Day), Shavuot, and the Feast of Tisha B'Av.

Climate: Hot, dry summers and cool, mild winters, but areas close to the Dead Sea are humid.

Culture and customs: No matter what religion someone follows, they do not have to work on Jewish holidays. Jewish people worship in synagogues rather than churches, and men wear a small hat on the back of their head called to a yarmulke when they are in a synagogue or if they are Orthodox Jews. Jews are either kosher or nonkosher (see social standards). Orthodox Jews, called Hasidic Jews, do not operate machinery or wear leather products on the Sabbath and they may wear hair locks that are long curls on each side of the head. They wear Eastern European clothing in the fashion of the nineteenth century and a white apron under their jackets. To celebrate the start of the Sabbath on Friday night, families light candles. The Torah, which is in the form of a scroll when it is used in Temples, contains passages from the Old Testament of the Bible.

Business practice standards: There are labor laws and building codes, but they are not rigidly enforced by the government. Only half of the businesses in Israel are privately owned, because the government owns a quarter of them and the other quarter are owned by *Histagrut,* which is the General Federation of Labor. The state had the exclusive right to generate power until 2006, but starting in 2006 20 percent of the power is generated by private industry. The state-owned Bezeq Israel Telecom provides the communication infrastructure.

There are only a few large construction companies in Israel, because most firms have less than 100 employees. Manuel labor is performed by 43 percent foreigners, 37 percent Israelis, and 20 percent Palestinians, and foreigners are paid less than Israelis. Construction supervisors are Israeli, Turkish, Arabic, Jewish, or Asian. Middle- or upper-class Jewish families own most of the construction firms. Workers are Ethiopian, Arabic, Jewish, and Turkish. Palestinian workers are not allowed to live in Israel; therefore, they have to cross the border into and out of Israel every day to work.

Laborers are paid by the hour, and they do not receive any benefits. Hourly wages are tied to output instead of the number of hours worked, with bonuses given for exceeding productivity targets. If workers are employed by large construction firms, they receive higher wages, a cost-of-living allowance, and severance pay if they are laid off. In the public sector there are unions, but unions do not exist in the private sector.

Use of business agents: Connections.

Construction materials available in the country: Chemicals, electronic equipment, fertilizer, paper, plastics, and textiles are available in Israel. Iron and steel are imported from other countries. Israel has no hydroelectric power or coal deposits, so it is dependent on foreign oil. Concrete and masonry are the most common building materials. Stone is used in facades and for indoor flooring because it is available locally.

Social standards: If someone is *kosher,* it means that they do eat meat and poultry if they are prepared in the same kitchen as dairy products. Kosher restaurants do not serve dairy and meat products that are prepared in the same kitchens. Other foods that are not eaten include pork, shellfish, and seafood other than fish. There also may be separate seating areas in restaurants for people eating dairy and nondairy foods.

Most of the disputes between foreign nations and Israel are over land that has access to water, because Israel has limited access to fresh water. The border of Israel has changed each time there has been a war between Israel and other nations.

14.5.3 Jordan

Languages spoken: Arabic.

Religions: Sunni (Islam) 92 percent, Christianity 6 percent, and some Greek Orthodox, Roman Catholicism, Syrian Orthodox, Coptix Orthodox, Armenian Orthodox, a small population of Shi'a (Islam), and Druze (which is a secret religion from Lebanon).

Holidays: See the Islamic holidays in Yemen.

Climate: Arid desert with a rainy season in the west that lasts from November to April. Jordan experiences droughts and earthquakes.

Culture and customs: A standard greeting is *alaykum,* which means "peace be upon you," and the response is *alaykum issallam,* which means "upon you, peace also." If someone is not Muslim, they will say *marhaba* and respond *marhabtayn,* which mean "welcome" and "welcome to you." Muslims do not eat food that is prepared by non-Muslims.

Business practice standards: Unemployment was over 16 percent in 2005. The economy is dependent on foreign assistance and remittances from Jordanians who work abroad.

Construction projects are concentrated in the mining sector, telecommunications, power generation, and water transportation systems. The government of Jordan is encouraging oil and gas exploration.

The capital requirement for investing in Jordan of 100,000 Jordanian dinars was removed in 1997, and the Encouragement of Foreign Investment Regulation of 1995 removed the requirements for foreign firms to obtain permits through licensed brokers and approval of the cabinet to operate in Jordan. This regulation also provides tax exemptions for investing in rural regions of Jordan. For additional information consult the website www.inforprod.co/il/country/jord2f.htm. Foreign firms are no longer restricted to only 50 percent ownership in Jordanian companies in the financial market, the transportation sector, insurance companies, banking, telecommunications, and the agricultural sector. There is a 50 percent foreign ownership restriction in the construction industry for trading companies and for the mining sector.

All fixed assets used in construction projects are exempt from custom's duties and taxes, including equipment, machinery, tools, and spare parts if they do not exceed 15 percent of the value of the fixed asset they will be used to repair. Net profits from projects are exempt from taxes for up to 10 years from the start of commercial production.

There are free-trade zones that are supervised by an autonomous agency that was established by the Free Trade Zone Corporation Law. To qualify to operate in a free-trade zone, firms have to be applying new technology, using local raw materials, improving local labor skills, and reducing imports. Firms that receive free-trade-zone status are exempt from taxes for 12 years, they may repatriate their profits, the items that they import are not taxed, employees do not pay income or social service taxes, and structures are exempt from licensing fees and taxes.

Use of business agents: Business agents are required in Jordan.

Construction materials available in the country: Cement and glass are available, but machinery and crude oil are imported from other countries. Jordan mainly conducts business with firms from Saudi Arabia, China, Kuwait, Germany, Iraq, and the United States.

Social standards: Jordanians are demonstrative, loud voices are an indication of emotion, and talking quietly is a sign of respect and maturity. People stand close to others while they are speaking to them. Men hold hands with other men, and there is no sexual connotation attached to it. Public displays of affection between the sexes are not acceptable in Jordan.

14.5.4 Turkey (The Republic of Turkey)

Languages spoken: The official language is Turkish. Other languages spoken in Turkey include Kurdish, Arabic, Armenian, and Greek. The Turkish language is written using the Latin alphabet, not the Arabic alphabet.

Religion: Islam.

Holidays: National Sovereignty and Children's Day—April 23; Ataturk Commemoration and Youth and Sports Day—May 19; Victory Day—August 30; Republic Day—October 29; Seker Bayrami (Sugar Feast Day)—a 3-day festival at the end of Ramadan; and Kurban Bayrami (Slaughter Feast)—a 4-day festival. Additional Islamic holidays are observed, and they are listed under Yemen.

Climate: There are three climate zones: (1) Mediterranean, with hot and dry summers and mild and wet winters, (2) Continental, with hot and dry summers and cold and harsh winters, and (3) the Black Sea area, which is temperate and rainy all year round. Turkey is located in an active earthquake zone.

Culture and customs: Turkey is a male-dominated society, and people do not understand what the terms *politically correct* are intended to convey in society. It is an informal country where business is discussed at work and during meals. Turks are proud of their country and their heritage. A large percentage of the population is under the age of 20, which increases the economic burden of those over 20. Traffic congestion influences the daily life of Turks, and it causes delays on construction projects when materials are being delivered.

Turkey does not officially recognize the Kurdish state in the southeast region that has a population of 20 million people; therefore, citizens of this region are not subject to the same legal rights as Turkish citizens.

Business practice standards: Turkey will be joining the European Union in 2007, and to accommodate the requirements of being a European Union member, it is upgrading its laws and regulations, many of which will affect the construction industry. Alliance requirements conform to the European Union Unity of Customs agreement of 1994. The inflation rate in Turkey has been approximately 15 percent per year, and the Turkish economy was unstable during the early 2000s. Turkish construction firms perform work in 55 other nations. Some businesses in foreign nations stopped accepting letters of credit from Turkish firms, and this is limiting their ability to secure work in other nations. Construction of a 1000-mile oil pipeline from Baku, Azerbaijan, to the Mediterranean port city of Ceyhan was started in 2002.

Many of the construction firms in Turkey have been family owned for three generations. People in Turkey may speak two or three languages. There are 29 universities that offer civil engineering programs in Turkey, and there are many government officials in Turkey who have earned engineering degrees.

The hierarchy in construction organizations is flat, and there are no employees between engineers and workers; therefore, engineers perform both engineering and management functions. Productivity rates depend on individual workers and how motivated they are to perform their tasks.

International standards are used for engineering and construction projects. There are limited safety requirements, firms pay minimal compensation to injured workers, and safety is left up to individuals.

Use of business agents: Business agents are required in Turkey.

Construction materials available in the country: Turkey produces iron ore, billet, sheet iron and steel, petroleum products, coal, bricks, thermal insulation, glass, aluminum, marble, cement, and lumber.

Social standards: The eastern region of Turkey is known as *Anatolia,* or *Asia Minor,* and the western region is called *Thrace.* Turkey is strategically located at the center of Eastern European and Middle Eastern commerce, and it has access to four seas: (1) the Mediterranean in the south, (2) the Aegean in the west, (3) the Sea of Marmara, between the European and Asian land masses, and (4) the Black Sea in the north.

14.6 North America

14.6.1 Canada

Languages spoken: English and French. The official language in Quebec is French, and the rest of the country speaks English.

Religions: Roman Catholic in French Canada; United Church in western Canada; and Catholics, Baptists, and Anglicans in eastern Canada.

Holidays: There are different holidays in each province of Canada. Quebec has Roman Catholic holidays, as well as secular holidays.

Climate: There are weather extremes, and the coasts have milder weather than the interior part of the country.

Culture and customs: When speaking to French Canadians use the polite *vous* for pronouns instead of *tu*, unless they are a family member or a close friend.

Business practice standards: There has been both a British and a U.S. influence on business practices in Canada, but most businesses operate more like European businesses. In business, be direct and provide data to clarify what is stated in verbal discussions.

Use of business agents: Business agents are not required in Canada.

Construction materials available in the country: Lumber, cement, steel, stone, tile, and aggregate.

Social standards: There are nationalist feelings of people of French descent versus people of English descent. Canadians follow basic rules of etiquette, and they may be conservative and reserved in their interactions with foreigners. They are direct in their speech, what they say is what they mean, and there are no hidden meanings.

14.6.2 United States of America

Languages spoken: English.

Religions: Protestantism, Catholicism, and various other religions (see Appendix C).

Holidays: Secular holidays: New Years' Day—January 1; Presidents Day (in honor of former presidents Lincoln and Washington), which may be a week-long holiday in some states that takes place in February; Easter in March or April; St. Patrick's Day—March 17; Memorial Day—the last Monday in May;

Fourth of July (Independence Day); Labor Day—the first Monday in September; Columbus Day—October; Halloween—October 31; Veterans Day—November; Thanksgiving—November; and Christmas—December 25. Martin Lurther King Day (who was a civil rights activist who was assassinated in 1968) is observed on January 16. In addition, there are regional holidays, such as San Jacinto Day in Texas (the day Texas defeated Mexican troops to regain Texas); Mardi Gras in Louisiana, celebrated in February.

Climate: The climate varies throughout the country from (1) hot and humid in the south (90°F to 100°F) with mild winters, (2) Mediterranean climate in California with mild winters, (3) hot and dry in the Southwest, (4) moderate temperatures and wet in the winter in the Northwest, (5) snow and cold in the winter (below 50°F) and hot summers in the Midwest, (6) moderate temperatures and snow in the Rockies, and (7) hot and humid summers and cold and snowy winters in the Northeast region.

Culture and customs: Customs and culture vary throughout the country, but there are common stereotypes including (1) people from the Northeast appear to be businesslike and competitive (in-your-face) until someone gets to know them (in New York City people move at a frenzied pace); (2) in the South people practice *southern hospitality,* which means "being polite and gracious"; (3) in the Midwest people are known for their integrity, their ethics, and being straightforward; (4) in the Southwest people have a more leisurely pace of life; (5) in the Northwest people are more concerned about their environment (and preserving it); and (6) in California people are under stress from the high cost of living and traffic congestion that causes long commutes to and from work.

People only shake hands at work, and greetings are usually just a *Hello* and *Goodbye* (a sign of the abrupt business culture) or *Hello, how are you?,* which is usually answered with *Fine, thank you.*

Business practice standards: In the United States, businesses and business decisions are driven by tax laws, which allow businesses to deduct their expenses from their taxable income and depreciate capital expenditures over several years and deduct them from their taxes. The official business-tax rate is over 30 percent, but most businesses pay less than 10 percent in taxes after writing off expenses and depreciation. Large companies incorporate in other countries to avoid paying federal taxes in the United States.

In business, principles are stressed, along with hard measures. Objective facts and scientific processes are used to analyze data, and scientists are notorious for measuring results and developing methods to quantify accomplishments. Experiments and specific information lead to generalizations about large groups of people. Little attention is paid to philosophy, which is common in other cultures. Companies are goal-oriented, and people are defined by their work.

People accept honest and constructive criticism, and it does not cause them shame or harm their pride; therefore, trial and error are acceptable in the U.S. culture. People may disagree openly with each other and their superiors in meetings. Promotions are based on achievements (or politics), and individuals are rewarded not groups. People form cabals (small groups) for self-promotion or to discredit others within organizations. Deals are made quickly with little socializing, and business is conducted even during meals. Work hours are 9 A.M. to 5 P.M. on the East Coast and 8 A.M. to 5 P.M. on the West Coast.

Contractors need to have a contractor's license in some states, which requires an 8-hour examination, but some states are lax about enforcing licenses. Engineering professional licenses are issued by individual states, not the federal government, and they require graduation from an Accreditation Board for Engineering and Technology (ABET)–accredited university, an 8-hour fundamentals of engineering examination (sometimes referred to as the engineering in training, or EIT exam), a 2- or 4-year internship in a company where there are professional engineers (the title used for *licensed engineers*), and a professional engineering–license examination that is 8 hours long. To obtain a structural engineering or geotechnical engineering license requires two additional 8-hour examinations. A professional license is required only if someone is designing institutional buildings such as schools, hospitals, prisons, and court houses. Civil engineers are required to consider public safety above other considerations by their professional society, as stated in their code of ethics.

Federal and state governments regulate safety through the Occupational Safety and Health Administration (OSHA), which inspects job sites for safety violations and issues fines for violations. Permits are required for construction, and they are issued through municipal governments or counties, and bribing government officials is illegal. Unions are strong on the East Coast; they are nonexistent due to open-shop (nonunion) laws in the South (the 12 states that seceded from the Union during the Civil War in the 1860s), and in the middle of the United States and in the Western states there are both union and nonunion workers, with the demarcation line running through the middle of Iowa. Unions are national organizations that negotiate labor contracts on behalf of workers, ensure that job conditions are safe for workers, and help to settle jurisdictional disputes (when more than one union declares that a specific type of work is performed by their craft). The National Labor Relations Board (NLRB) settles disagreements between labor and management.

Construction is managed by either construction managers (sometimes called *job superintendents*), project managers, or project-management teams. Universities offer degrees in civil engineering (ABET-accredited engineering programs), construction engineering and management (ABET-accredited engineering program), construction management (ABET-accredited technology programs), project management (nonaccredited programs), and mechanical, electrical, chemical, aerospace (ABET-accredited programs), and architectural engineering and architecture. Undergraduate degrees require 4 years of study, master's degrees require 5 years, and PhDs require 2 to 10 years beyond a master's degree. Workers may become construction managers by working their way up through the construction trades.

Construction sites are highly regulated by the federal and state governments. There are laws on (1) the allowable amounts of pollution and the generation of hazardous wastes that were enacted to protect the environment; (2) safety regulations such as having workers wear harnesses attached to a structure when they are working above ground level and having a construction site that is safe for workers; (3) antidiscrimination laws that prohibit discrimination in employment based on race, nationality, sex, or sexual orientation; (4) wage laws that set the minimum wage and the maximum work hours, along with overtime rates (time and a half or double time); (5) laws on disability accommodations, such as having wheel chair ramps and signs written in Braille (Braille is a written language where the letters and numbers are represented by small bumps); (6) negligence laws called *tort laws* that allow people to sue other people for negligence; and (7) laws governing contracts and the execution of contracts.

Contracts are detailed, and blank standard-form contracts for engineering and construction services may be purchased from the American Institute of Architects, the Associated General Contractors, and other organizations. A high proportion of jobs have claims for changes in the scope of work, and claims are settled through negotiation, mediation (conciliation), arbitration, dispute-review boards, or litigation (the legal system), and Chapter 7 discusses these dispute-resolution methods. Lawsuits are prevalent in the United States, and liability issues (being sued for negligence, which is willfully or accidentally allowing someone to be injured or killed on a job site) influences engineering designs and construction.

Projects are managed by either project managers or construction managers that may be called job-site superintendents and if a firm is large enough there will be project-management teams that include a project manager, an assistant project manager, a project engineer, a project-controls supervisor, schedulers, cost-control engineers, estimators, field engineers, a quality-control manager, and a safety manager.

In the United States, there are only a few major multinational E&C firms that perform overseas work, and some of them are the same firms that built landmark projects during the 1930s through the Work Projects Administration and the Civilian Conservation Corps, such as the Hoover Dam, which is shown in Figure 14.8, and the Grand Coulee Dam. Most of the construction projects in the United States are built by regional or local contractors. Construction projects in the U.S. are 30–35 percent residential construction; 40–50 percent nonresidential (institutional, educational, light industrial, commercial, religious, and recreational); 20–25 percent engineered construction using heavy materials such as concrete, steel, and earth; and 5–10 percent industrial construction (power plants, refineries, etc.).

Use of business agents: Business agents are not required in the United States.

Construction materials available in the country: Cement, lumber (large wood elements may be made out of several layers of thin wood fused together with epoxy to form beams called *glue lams*), plywood and plystrand (wood chips compressed between a thin veneer of plywood), bricks, masonry blocks, asphalt, aggregate, sheetrock, and nonstructural steel (steel produced using recycled cars). Most large structural steel elements are imported from foreign countries.

Social standards: Many people are defined by what they do, not by their heritage or religion. People from the United States may come across to foreigners as too

Figure 14.8 Hoover Dam, Nevada-Arizona border, United States.

strong, too fast, intimidating, phony, and insincere due to cultural differences. They also provide too much information too quickly compared with people from other cultures.

People are concentrated in the cities (there are 27 cities with populations of over 1 million people in each city), with vast amounts of open space between cities. People from the United States are used to having a lot of choices in their products, their housing, their education, their entertainment, and their recreation. There is a high literacy rate in the United States. Over 90 percent of the population attends kindergarten (a grade before the first grade) through the twelfth grade, because education is free until the twelfth grade. Community colleges and state universities charge lower tuition than private universities, which allows more people to attend college. The college graduation rate was over 35 percent in the 1980s, but it has declined to below 25 percent in the 2000s.

Electronic devices that are popular include large-screen televisions (called *big-screen TVs*), HDTV (high-definition televisions), cell phones, hard-line telephones, computers, and digital cameras. Most families own at least one car. There are only subway mass transit systems in a few cities, including Boston, Chicago, and New York. The United States is an open society where people are free to practice any religion, but that does not mean that prejudice does not exist. Many cities and schools are affectively segregated by nationality and race. Discrimination is more covert than overt, because the government legally de-segregated the country and passed nondiscrimination laws in the 1960s.

State laws vary throughout the country, and there are thousands of laws still on the books that could be enforced if the legal establishment decided to enforce them. The statistic for the high divorce rate (50 percent) is deceptive, because some people marry and divorce several times, which skews the percentage. Some states have "no-fault" divorces, whereas others require one spouse to "show cause" in order for a divorce to be granted. Liquor laws are set by states, and some parts of states may be "dry" (no liquor is sold), whereas others are "wet" (liquor is sold). Some states require liquor to be sold in clubs, only where food is sold, or not before noon on Sunday (Texas). Some southern states do not allow contracts to be formed on Sundays due to religious beliefs.

14.7 Oceana

14.7.1 Australia

Languages spoken: English, called *strine,* is a mixture of British English, Aborigine, and invented words.

Religions: Christianity (Anglican, Roman Catholic, and Protestant).

Holidays: Christmas—December 25; New Year's Day—January 1; Easter—March or April (the Great Exodus takes place at Easter); Australia Day—January 26.

Climate: Reverse seasons from the northern hemisphere, with the summer in November through February and the winter in June through August. The largest region of Australia is semiarid, and 40 percent of the land mass is covered with sand dunes. The north and east regions are tropical rain forests, and there is snow in the Alps.

Culture and customs: Australia is called the *Land Down Under, Ozzie Land,* or *Oz,* and Australians are called *diggers.* Beer is called *neck oil,* and *mate* is a male-to-male greeting. If someone is performing well at work they are called a *bottler,* the person in charge is called a *bloke,* and someone who is not performing well is called a *bludger.* If someone is from Perth, they are called *Westalians.* The interior region is called the *Outback.* Football (soccer) is referred to as *footie.* *PAME* is the term for "person of mother England." If someone is referred to as a *bludger,* it means that they appear to be working, but they are actually doing very little work, and this is a trait that is admired in Australia.

Business practice standards: The pace of work is slow. Workers are meticulous and specialists in their chosen field. The client-contractor relationship is informal, and people are self-deprecating. People work hard, but they hide it. Shorts may be worn to work. Being a manual laborer is prestigious in Australia. Australia is a patriarchal society; but if a woman demonstrates expertise in her area, workers will respect her.

Punctuality is expected in relation to meetings and work activities. Handshakes should be firm. People get to the point quickly in meetings, and in business transactions people are direct but informal. Showing confidence is a way to gain respect. Bargaining is not a standard business practice, and contracts are enforced as written.

Educational programs for engineering students require 4 years for an undergraduate degree, and university courses are structured so that students may attend college part-time. Foreign engineering degrees may not be valid in Australia.

Use of business agents: Business agents are not required, but having connections is helpful.

Industries and materials available: The main industries are wool; mining coal and metal ore; steel; wheat, barley, and sugar exportation; tourism; and technology. Steel is produced locally, and the steel industry competes with Japan and South Korea

Social standards: Australia is the least densely populated country in the world. It contains six states and several territories. It is a mixture of capitalism and socialism. The victory sign should not be used (index and third finger upright and spread out), because it is an obscene gesture unless it is done with the palm away from the person doing it. Thumbs up is another obscene gesture. People do not wink at each other, especially not at women. When males ride in taxies they sit in the front seat. It is a classless society and it is not proper to boast. Relationships are informal and casual. If Australians are praised they think they are being set up for something bad.

Do not discuss business at social functions. People do not like to have their time wasted; therefore, meetings and meals start on time. Condescending people are not well tolerated. Australians are pragmatic and do not like to be compared to people in other countries (especially the United States or Great Britain).

Local animals include kangaroos, koalas, Tasmanian devils, bandicoots, wallabies, and wombats.

14.8 South America

14.8.1 Brazil

Languages spoken: Portuguese.

Religions: Mainly Roman Catholicism.

Holidays: Catholic holidays including Christmas, Easter, and Good Friday.

Climate: The climate varies throughout the country because it is a large country.

Culture and customs: Inquire as to whether someone minds having his or her first named used in conversations before using it. Last names are a combination of both parent's names. Titles are used such as *Senhor* for a man or *Senhora* for a woman. Spanish is not spoken, so an interpreter should be used who speaks Portuguese.

Business practice standards: There is a paternalistic business climate, where family ties are strong and a person's first loyalty is to his or her family. Businesses are controlled or partially controlled by the government. Working relationships are not merely for one project or one transaction, they are continuous relationships. Business cards are given to business associates. There is a social aspect to business before starting transactions. Expect to bargain in business.

The construction industry is sophisticated as a result of major industrial, multinational projects, such as power plants and oil refineries, having been built in the country. Brazil is a cosmopolitan country with a mixture of nationalities including native Brazilians, Spanish, Europeans, and other nationalities.

Use of business agents: Despachantes should be hired to assist in business transactions.

Social standards: Brazilians do not like to be called Latin Americans, because they are independent from the rest of South America. It is an informal culture, but people like to be greeted individually instead of as a group. It is an affront to their dignity if they are not greeted individually or not said goodbye to when someone leaves. Being expressive when talking is normal. If "American time" or "airport time" is being used, this means to arrive on time, but for other situations being late is acceptable (and expected if it is a dinner at someone's home).

Construction materials available in the country: Most construction materials may be obtained in Brazil.

14.8.2 Peru

Languages spoken: Spanish.

Religions: Roman Catholicism. In the Andes, Roman Catholicism mixed with an old Inca religion.

Holidays: New Year's Day—January 1; Three Kings Day—January 6; Maundy Thursday and Good Friday; Easter week—March or April; Labor Day—May 1; National Days—July 28 and 29; Saint Rose Day—August 30; Battle of Angamos—October 7; All Saint's Day—November 1; Feast of the Immaculate Conception Day—December 8; and Christmas—December 25.

Climate: Peru has 28 of the 32 different types of climates, including hot deserts, dry forests, humid savannas, rain forests, cold plateaus, cool steppes, and icy mountains.

Culture and customs: Peruvians are predominately *Mestizos,* which means a mix of Spanish and Incan, but there are also Germans, Japanese, and Chinese. There are two types of culture: (1) Latin in the coastal regions and (2) the mystical and spiritual Andean culture.

Business practice standards: Major construction projects are in the areas of mining, hydroelectric power, water systems, and gas production. Twenty major roadways will be built that will connect Peru with Brazil, Chile, Colombia, and Bolivia in the 2000s. Sixty percent of construction workers have high school diplomas, and 40 percent have technical degrees from colleges. Sixty percent of universities are public, civil engineering is the most common engineering major, and engineering programs require 5 years. The most prestigious universities are the Universidad Nacional de Ingeniaria (public) and the Universidad Catolica (private). Only people who score in the top 20 percent on general examinations attend college.

Since workers are highly productive, labor costs are only 30 percent of the total costs on construction projects. During the summer months of December, January, and February, lunch breaks at construction job sites last 2 to 3 hours so that workers are not working in the hot sun during the hottest part of the day. Heavy construction equipment is not used as frequently in Peru as it is in other developed countries.

Peru belongs to the Andean Countries Union, which is a free-market community, as well as the South American Community of Nations (CSN). There is always the threat of new political movements taking over the government in Peru. There is widespread corruption, there are political parties that oppose current governments, and it takes a long time to pass new laws. And justice is slow. Peru is one of the fastest growing economies in South America, but Brazil, Argentina, and Chile have larger economies.

The *National Construction Regulations* are a compendium of safety laws, and they are enforced at construction job sites. Bureaucratic processes delay most activities in Peru.

Seventy percent of workers depend on public transportation, or they prefer to walk to work to avoid congested roadways. Travel in the Andes is treacherous when there are high winds.

Use of business agents: Business agents are not required, but family connections are important in Peru.

Construction materials available in the country: Cement, aggregate, and steel. Cement is produced by four major companies: (1) Andino, (2) Cementos Pacasmayo, (3) Cemento Atlas, and (4) Cemento Sol. Aggregate is available close to construction sites, because there are many rivers. Hierro Peru and Aceros Arequipa produce A36 carbon steel, A572 high-strength steel, and A588 corrosion-resistant high-strength steel. Construction equipment has to be imported from other countries.

Social standards: Major exports are oil, natural gas, coffee, minerals, and fish. People like to debate politics, talk about soccer, and discuss the government, but they do not talk about their personal lives at work.

14.9 Southeast Asia

Southeast Asia is a study in contrasts, because there are crowded major cities, such as the ones in India, the Philippines, and Indonesia, as well as remote islands and small villages throughout the region. Figure 14.9 is a photo of one of the Fiji Islands, which is one of the more remote locations in Southeast Asia. Cities in India have populations that exceed the populations of any other city on earth. This section describes several of the countries located in Southeast Asia.

14.9.1 The Republic of India

Languages spoken: Hindi is the national language, although English is the language of the Union and is used for business. There are 14 other languages recognized by the Indian constitution: (1) Bengali, (2) Telugu, (3) Marathi, (4) Tamil, (5) Urdu, (6) Gujarati, (7) Malyalam, (8) Kannada, (9) Oriya, (10) Punjabi, (11) Assamese, (12) Kashmiri, (13) Sindhi, and (14) Sanskrit.

Religions: Hinduism, Islam (34 percent), Christianity (3 percent), Sikhism (2 percent), Buddhism (1 percent), and Jainism (0.5 percent). People who practice the Hindu faith do not eat meat, eggs, or fish (animals are sacred); people who practice Islam do not eat pork, ham, or bacon, nor do they drink alcohol. Alcohol is illegal in some parts of the country or a special license is required that may be obtained from travel agents or at Indian airports.

Holidays: Independence Day—August 15; Republic Day—January 26; Mahatma Gandhi Jayanthi—October 2; Amgedkar Jayanthi—April 15; and other festivals such as Deepavali, Binayak Chaturthi, Baisakhi, Dassera, Kite Flying, Ramzan,

Figure 14.9 Remote island, Fiji Islands.

Bakrid, Ugadi, and Sankranthi (pongal) are celebrated on different days each year because they are dependent on the position of the moon. The Hindu festival of Dussera takes place in October, Diwali is 20 days later, and Holi is in March.

Climate: Eight different climate zones, all of which have monsoon rains and tropical weather. The highest temperatures in some regions reach 45°C.

Culture and customs: Religion influences both personal and business activities. There is no self-determination because everything is in the hands of God or Allah. It is considered irreligious to plan for the future. Social harmony is important. There is no nonverbal cue for saying no in India. Shaking the head from right to left merely means that someone understands what a speaker is saying to them.

People are addressed as *Srimathi* and *Sri* (Mrs. and Mr.), and if someone has a title, it is used before their name. A suffix of *ji* after the last name can be added as a form of respect. Always greet the eldest person in a group first.

Shoes are removed when entering someone's home. One form of greeting is a *namaste,* which requires that both hands be placed together with the fingers pointing toward the sky, and a slight bow of the head. Saris are still worn by some women (brightly colored long pants with a midcalf-length dress over the pants with a matching scarf worn around the neck).

Unusual Indian expressions include *eve teasing,* which means the "harassment of women"; *Himalayan blunder* is a serious mistake; *hotel* is a generic term for a place to eat; *tops* are earrings; *rubber* is an eraser; a *bearer* is a waiter; *good name* is used to inquire as to someone's name; and *cent percent* means 100 percent.

Business practice standards: The national government has no power over the state governments in India in the matters of foreign investment. There are no bankruptcy laws in India. Firms need government permission to close down. Foreign investors are not allowed to invest in housing or real estate except for company property used to conduct business. Titles to land are unclear because there is no reliable system for recording interest in property. Land cannot be used as collateral for loans. The oil industry controls many of the decisions made in the government, and the Indian "mafia" (secret criminal organizations) controls some industries.

At work, the boss gives orders, and the orders are not questioned by subordinates because the bosses are the ones who are paid to make decisions. In discussions, only senior members speak, but this does not mean that everyone present agrees with them.

The Payment Wage Act of 1936 and the Minimum Wage Act of 1948 are still in effect, and the Industrial Disputes Act requires permission to lay off workers or close a business if there are more than 100 employees; therefore, it is easier for firms to downsize using voluntary retirements.

Two-dimensional AutoCAD and STADD-3 are used in engineering firms, along with manual drafting methods and physical models. Project scheduling is done on spreadsheets rather than commercial scheduling programs.

Structural engineers are required to have professional registration (called *licensure*) through the Institution of Engineers, but licenses are not required for other areas of civil engineering.

Architects and engineers design projects, and they also manage construction. When applying for jobs or during job interviews, potential job candidates may supply school records and copies of diplomas and awards.

Construction workers are recruited through intermediaries (labor agents) on a daily basis. Construction sites are described as war zones because of the high number of accidents and deaths per year (5000 deaths per year). Workers do not receive paid holidays, sick leave, health care, pensions, or any other benefits. There are no standards for construction materials.

Standard work hours are 9 A.M. to 5 P.M. five days a week for private companies and 10 A.M. to 1 P.M. six days a week for government offices. Government workers are guaranteed lifetime employment, so there is little incentive to work efficiently or quickly.

Use of business agents: Use local business agents. Connections with powerful families are helpful. Permits are required for most business transactions, and if someone tries to operate without the proper permits, he or she will be fined by the government. Corruption is a problem, and bribery speeds up the government procurement system for licenses, permits, and completion certificates.

The government regulates the price of grain, sugar, edible oils, medicines, energy, fertilizer, and water. There is a dual pricing system for food—there is a fixed price at government distribution outlets while market prices control the open market.

Construction materials are available in the country: Steel, cement, masonry, brick, concrete, and asphalt.

Social standards: The Indus Valley civilization is one of the oldest in the world (5000 years old). It was invaded by Aryans in 1500 B.C., and there were Arab incursions in the fifteenth century. Britain assumed control of the country in the nineteenth century, and India became a free nation in 1947. Pakistan split off from India in 1947, and Bangladesh was formed in 1971 in the eastern part of Pakistan. There are 27 states and 7 Union territories in India. India and Pakistan have fought three wars over the Kashmir, and citizens of both sides are taught that they won the wars. Terrorist acts are a daily occurence in India due to disputes over land that have continued for the last 50 years. India does not acknowledge the fact that Pakistan gave parts of the country to China in 1965 and disputes continue with Pakistan over the Indus River.

Standardized tests are administered to children to determine who will move forward in the educational system, so there is limited self-determination on the choice of careers. Large companies provide housing, medical facilities, and schooling for their employees and their families.

14.9.2 Indonesia

Languages spoken: Bahasa Indonesia (or Indonesian) is the official government language. Chinese, Dutch, and English are spoken by some people, and there are many local languages. In Bahasa Indonesian there is no conjugation of verbs. To indicate past tense the word *sudah* is inserted anywhere in a sentence. To indicate that a word is plural it is said twice. An example is the word for "houses," which in Bahasa Indonesian would be *ruma ruma*.

Religions: Islam, Hinduism, Confucianism, Buddhistism, Christianity, and Wayang. Wayang is a mystical veneration of ancestors that has 150 different deities. Wayang includes the practice of "shadow puppets" that are either carved from wood, or made of leather, and used behind a sheet with a light shining on them so that the audience sees only the shadows of the puppets. The puppet shows are used to teach people about the religion and the deities.

Holidays: Islamic holidays (see Yemen).

Climate: Hot and humid with monsoons.

Culture and customs: Indonesia is a diverse country because there are so many different islands (over 13,000 in the archipelago). Men are called *bapak* and women are called *ibu* when they have reached a certain status or age. People of the same sex hold hands, link their small fingers together, kiss, and hug each other. Prostitution is controlled by the government, and government officials test prostitutes every month for sexually transmitted diseases.

Only spoons and forks are used to eat. Spoons are held in the left hand upside down and forks are held in the right hand. Spoons are used to cut food as there are no knives and most food comes in small pieces.

There are no public restrooms in the country, and people relieve themselves in public in discrete places or in local rivers, and the smell permeates the air. It is polite to turn away or look the other way while someone is relieving himself or herself. If homes have toilets they will be Eastern-style toilets. People shower in public shower areas that may conceal the lower half of their bodies. In some areas of Indonesia people only wear a sarong around the lower half of their body.

Families may have 15 to 20 children, and Muslim men may have up to four wives. Many people live with their families in small, one-room houses or in public places. All male relatives may be called *brother* or *uncle,* and all female relatives may be called *sister* or *auntie.* Fingers or feet should never be pointing at anyone, but pointing with a thumb is acceptable.

Business practice standards: Arguing or shouting at someone at work or during negotiations will cause him or her to lose face. Things are accomplished by establishing relationships with other people, by gaining influence, and by doing favors for other people. In business relations, people are concerned about the feelings of others, and matters are dealt with indirectly to preserve working relationships. Appendix B contains a case study that includes information or business practices in Indonesia.

Indonesia is one of the few remaining countries where there is a major craft industry and manual metalworking. Figure 14.10 shows a metal shop that is the last place in the world where handmade gongs are forged (gongs are used in the music of Indonesia). There are special laws that govern the exporting of

Figure 14.10 Metalworking shop, Indonesia.

handicrafts, and some foreign countries, such as the United States, have agreements not to tax handicrafts when they are imported into their countries.

Use of business agents: Business standards vary by regions because there are different cultures throughout Indonesia. Business agents are used to expedite business transactions. Many people operate in the *nonmonetary* economy where barter is used to secure food and services.

Construction materials available in the country: Teak wood, lumber, and natural gas; everything else has to be imported into the country.

Social standards: In 1965, there was a coup, and millions of people were killed throughout the country. Another coup occurred in the early 2000s. There is no self-determination in Indonesia, life is left up to the will of Allah or God, and people do not plan for the future because of their beliefs. Citizens have to obtain transmigration permits from the government to move anywhere.

Shoes are never worn in a mosque or other sacred locations. Never embarrass someone in public (losing face).

On some islands, foreigners are unfamiliar, especially if it is an outer island, and they are welcomed with no reservations. Foreigners may be called Mister even if they are women. Blond or red hair and blue or green eyes are unusual, and people may attempt to pet or touch unusually colored hair to see if it is real. Outside large cities, there may not be any electricity, running water, or sanitation facilities. Even without running water, people maintain good hygiene.

Water is purchased in 5-gallon gas containers. If possible, cities that house thousands of people are built over the water on wooden docks so that the human wastes and garbage are washed out to sea when the tide moves out.

A native instrument, called an *anklong,* is carved out of bamboo, and each instrument plays a different note in the scale. *Anklongs* may be played while they are resting on a stand by one person, or each person holds one instrument to produce a symphony.

Even though there are over 13,000 islands in Indonesia, the majority of people do not know how to swim.

Head hunting was outlawed in 1965, but on remote islands it is still practiced because the taking of a head is a rite of passage for males. In the cities, most foreigners live in compounds and hire guards to protect their homes. In rural areas, the military may provide protection for job sites.

14.9.3 The Philippines

Languages spoken: Filipino, English, and eight other major dialects—(1) Tagalog, (2) Cebuano, (3) Ilocano, (4) Hiligaynon or Lionggo, (5) Bical, (6) Waray, (7) Pampango, and (8) Pangasinense—and 111 other dialects. The Philippines is the third-largest English-speaking country in the world.

Religions: Predominately Roman Catholicism, but there are other religions including Islam, Protestantism, and Buddhistism.

Holidays: New Year's Day—January 1; EDSA Revolution Day (People's Power Day)—February 22; Heroes' Day (Day of Valor), Baatan Day (Araw ng Kagitingan)—April 9 (this day commemorates the death march that took place in Baatan during World War II, where prisoners of war were forced to march to Bataan without food or water and many of them died along the way); Labor Day—May 1; Independence Day—June 12; Manila Day—June 24; National Heroes' Day—August 31; Barangay Day—September 11; National Thanksgiving Day—September 21; All Saints' Day—November 1; All Souls' Day—November 2; Bonifacio Day—November 30; Christmas—December 24 and 25; and Rizal Day—December 30. Other holidays that are on different days each year are Holy Thursday, Good Friday, and Easter.

Climate: Tropical—hot and dry from March to May, wet and rainy from June to October, and a cool, dry season from November to February. The country has earthquakes, typhoons, volcanic eruptions, floods, and landslides.

Culture and customs: Filipinos are famous for *bayanihan,* which is a spirit of kinship and camaraderie. There are different cultures throughout the country, including the Ilocanos in the north, the Tagalogs in the central plains, the Visayans in the central islands, and the Muslims in Mindanao.

Business practice standards: The Philippines is a member of the Association of South East Asian Nations (ASEAN), and it participates in the Asian Free Trade Agreement (AFTA). The legal system is a combination of Anglo-American, Roman, and Spanish laws. Sources of law are the Civil Code, the Penal Code, the National Internal Revenue Code, the Labor Code, and the Code of Commerce. The Philippines is politically unstable, and military actions against insurgents are a concern to businesses.

Bamboo scaffolding is used at construction sites, which requires additional time to set up and dismantle. Construction equipment is not used at construction projects, because manual labor is still inexpensive. Road conditions are deteriorating, and at times they are inaccessible to large construction equipment. In the cities, it is difficult to locate space for laying down materials.

The government does not interfere with the construction industry, and there are currently no alliance requirements for foreign firms. Private clients finance construction projects, and they prefer build-own-transfer projects.

Religion affects all aspects of life in the Philippines, including work and construction, which is affected by the numerous religious and secular holidays observed in the country.

Use of business agents: Business agents are required in some areas of the country.

Construction materials available in the country: Most construction materials have to be imported from other countries.

Social standards: People from the Philippines are called *Filipinos,* and they are descendants of Malaysians and Indonesians.

The Philippines is an archipelago of over 7000 islands. The four main island groupings are the Luzon, the Visayan, the Mindanao, and the Sulu. Transportation is inexpensive because there are buses and trains in the cities and on the outer islands there are jeepneys, multicabs, and tricycles (a motorcycle with an attached passenger cabin with a third wheel).

14.9.4 Thailand

Languages spoken: Thai (Siamese), the elite speak English, and there are local dialects throughout the country. The alphabet has 44 letters, and it is based on Sanskrit and Pali (it was developed in 1283 by King Ramkamheng). The language uses monosyllabic words, and each has a complete meaning. It is a tonal language with five different tones—low tone, high tone, falling tone, rising tone, and middle tone. An example is the word *mai,* which means "new" if pronounced with a low tone, "wood" if a high tone, "not" with a falling tone, and "silk" with a rising tone. There is no conjugation of verbs and present, past, and future tenses are expressed by adding one word anywhere in a sentence.

Religions: Buddhism with some Islam, Hinduism, and Christianity.

Holidays: New Year's Day—January 1; Makhabucha Day—in February or March; Chakri Day—April 6; Songkhran Day—May 1; Visakhabucha Day—in May or June; Asalahucha Day—in July or August; Queen's birthday—August

12; Chaulalongkorn Day—October 23; King's Birthday—December 5; Constitution Day—December 10; and New Year's Eve—December 31.

Climate: Rainy tropical weather in the summer and dry and cool in the remaining months of the year.

Culture and customs: A *wai* is a standard greeting, and it is done with the hands together with the fingers pointing up at chest level while the head is bent down. A *wai* is used to show respect, humbleness, and friendliness and as a way of saying thank you. *Sawasdee* means "good morning" or "good evening." Social standing is an important element of society, and it is based on connections, lineal heritage, income, and education.

Business practice standards: The Alien Occupation Law requires that all foreigners secure a work permit from the government before they perform work in Thailand. There are exemptions for military personnel, consular personnel, officials of the United Nations, personal servants who work for the three previously listed types of people, persons who work under an agreement from the king, and people who are benefiting the local educational system, the arts, culture, or sports. In order to obtain a permit the following documents are required: a valid passport, a nonimmigrant visa, evidence of educational qualifications, medical certificates, and a job description from employers. It is illegal for people to perform work that is not listed on their work permit. Terminations in employment, transfers, or changes in employment have to be reported to the government within 15 days.

The Occupation Safety and Health (OSH) agency is the government agency responsible for safety at construction job sites, and it is supported by the International Safety Organization. Safety is the responsibility of owners and contractors. There are regulations on excavation, gas cutting and welding, electric arc welding, scaffolding and ladders, lifting supervision, cranes, tools and equipment, safety training, pile driving, and construction environment.

Outside of major cities the architecture in Thailand is different than what is found in other countries, and it is modeled after the royal palace in Bangkok. Figure 14.11 shows a picture of one of the buildings located on the grounds of the royal palace.

Use of business agents: Connections are used to secure business along with bribery.

Construction materials available in the country: Tungsten and tin.

Social standards: There are elements of both Hindu and Chinese culture in the Thai culture. Thailand was called Siam until 1939. The government is a constitutional monarchy, with a prime minister who is selected by a majority coalition of members of the House of Representatives, who are elected officials. It is a bicameral parliament with a National Assembly (*Rathasapha or Wuthisapha*) and a House of Representatives (*Sapha Phuthaen Ratsadon*).

Figure 14.11 Thai architecture, Bangkok, Thailand.

14.10 Western Europe

14.10.1 France

Languages spoken: French and some people also speak English.

Religions: Catholicism and Islam.

Holidays: There are 6 weeks of paid vacation in most companies, and most people go on vacation in August and return by mid-September. In addition, there are many holidays throughout the year.

Climate: Four seasons with mild summer and winter temperatures, unless there is a heat wave.

Culture and customs: People are referred to as *Monsieur, Madam,* or *Mademoiselle* with or without last names.

Business practice standards: The bureaucratic culture of France makes it difficult to quickly accomplish tasks because there are complicated procedures and paperwork. The government is involved in private-sector businesses, but businesses are not state-owned or state-controlled. Wages are taxed at rates between 40 and 50 percent, and social services are supplied with tax revenue. Unemployment rates are high—21 percent in 2005 (36 percent for men under age 25)—in areas called *sensitive urban zones,* which are areas where ethnic minorities reside in housing projects (12 million people are counted in these zones).

If someone does not speak French, knowing something about French culture or history is helpful in establishing business relationships. Decisions are not made quickly, so patience is required to work in France or with French people.

Use of business agents: Use a local agent or a local representative of a firm when dealing with French firms. It is useful to hire someone who is called a *defogger,* because their job is to reduce barriers for companies.

14.10.2 The Federal Republic of Germany

Languages spoken: German and some people speak English.

Religions: Protestantism and Catholicism.

Climate: Four seasons with mild summer and winter temperatures, unless there is a heat wave.

Culture and customs: The country is called the Federal Republic of Germany rather than referred to as Germany. People are called *Herr, Frau* (a married woman), or *Fraulein* (an unmarried woman), along with their last name. If someone has a doctorate degree in law or medicine, they are called *Herr Doctor* with their last name, and professors are called *Herr Professor* with the last name. People do not use first names unless they are asked to use them. Always find out a person's proper title.

Business practice standards: Germany is a paternalistic society. Germans are adept at negotiations because they do not give much away while they are negotiating, and this may slow down negotiations for contracts. There is a strong sense of loyalty to employers, and many people work for the same company most of their adult lives. Companies are required by the government to pay 2 years' worth of salary to any employee who is laid off from work.

The work atmosphere in the country is ordered, and work is planned in advance, including meetings. Casual dress is not acceptable in the work environment, and business associates are restrained in their interactions. Always be well prepared for meetings, and provide factual technical data that can be backed up with statistics if necessary. Presentations should be well-organized and authoritative to demonstrate confidence.

Status is important, and it is referred to as *wichtig tun* ("acting important"). Maintaining a detached attitude achieves more than being pushy. If product comparisons are used to sell a product or service, a firm could be taken to court to prove implied or stated superiority (Copeland and Griggs, 2001).

Contracts contain detailed information, but the German Civil Code may take precedence over contract clauses, so it helps to be familiar with the German legal system and the German Civil Code. There are two types of authority related to signing contracts: (1) p.p. or p.p.a., which means *per procura,* and it indicates that someone has restricted authority when signing contracts; and (2) i.V, which means *in Vertretung,* which indicates that someone is an executive who has full authority to sign all legal documents. Workers cannot be laid off unless they are paid 2 years salary (a government regulation), so temporary workers may be used on construction projects or other projects that are of a short duration. There is a high unemployment rate, the workforce is older, and it is mostly younger people who are unemployed. Unions are strong and strikes are common.

Germans are known for their machine work (milling) and precision in their work, and they are one of the top exporters in the world. Situations are thoroughly analyzed before decisions are made, and decisions are made by top executives, not delegated to lower-level managers.

East Germany is still transitioning from a centrally planned state-owned economy to a private enterprise economy, so there is still a great deal of government influence in industries.

Use of business agents: Not required but useful in former East Germany.

Social standards: Germans are reserved with strangers and formal in their relationships, which could be misinterpreted as being unfriendly. It may take a great deal of time to get to know some Germans, and foreigners may not be able to get to know them at all on a personal level. If Germans are bilingual they may be able to switch quickly between English and German. If foreigners speak German they will be welcomed in the country, but if they only speak English, their reception may be cool.

14.10.3 Italy

Languages spoken: Italian is the official language.

Religions: Roman Catholicism.

Holidays: Catholic and national holidays.

Climate: Hot and humid in the summer and mild winters in the southern region and mild summers and colder winters in the northern region.

Culture and customs: Italians are expressive when they speak and in their gestures. Talking loudly and arguing are part of the culture. Families are important, and extended families might live in the same house or in the same neighborhood. Italians should be referred to by the name of their city rather than as Italians, such as Florentines, Romans, and Sicilians. Italian is not pronounced "Eye-talian"; the proper pronunciation is "Eee-talian."

Vatican City (Holy See) is an individual state even though it resides in Italy in the city of Rome. The pope has full legal, executive, and judicial powers in Vatican City.

Business practice standards: Italy is a republic, and its major trading partners are Germany, France, the United States, the United Kingdom, Spain, the Netherlands, and Belgium. Italian contractors have been working throughout the world since World War II. Italian craftsman are known for producing high-quality products. Italian marble is used throughout the world for building facades, columns, and flooring.

Strikes (when people refrain from working) are a frequent occurrence in Italy, and several unions may be on strike at the same time. Strikes may last several days or for weeks or months.

Many Italian construction firms are family owned and operated, and nepotism (hiring family members) is prevalent within companies. Certain families control the industry, and their influence extends to other countries, including the United States.

Use of business agents: Family connections are required to do business in the construction industry.

Construction materials available in the country: The country exports construction equipment, engineering services, and marble. Italy imports energy products such as petroleum.

Social standards: Italians are warm, informal, and welcoming to foreigners. Family members do not move far when they marry, and the residents of towns all could be related in some manner. Italians are proud of their individual heritage, such as being Sicilian, Roman, or Florentine.

14.10.4 Spain

Languages spoken: Castilian Spanish.

Religions: Roman Catholicism.

Holidays: Catholic and national holidays.

Climate: Summers are hot in the interior and moderate and cloudy on the coasts. Winters are cloudy and cold in the interior and partly cloudy and cool along the coast.

Culture and customs: A persons last name will include the last name of their mother and their father. Men are called *Señor,* and women are *Señorita* with their first names, and married women retain their maiden names when they marry. Supervisors are not addressed by their first name. People dress up for business and for social gatherings.

Business practice standards: The economy is a mixed capitalist and socialist. There is high unemployment (11 percent in 2005). The government is in the process of privatizing many industries. Spain is part of the European Union. One in four people in Spain work in the construction industry. People prefer to invest in construction rather than to pay taxes to the Hacienda (the government tax agency). Interest rates for homes are artificially low due to government intervention. The Spanish master plan calls for the refurbishment of the airport system, seaports, and rail system by 2010, along with the building of a high-speed rail system.

A few major companies dominate the construction industry. Most of them are part of SEOPAN, which is a group of construction companies that control about 22 percent of the construction market, and they specialize in government development and environmental projects. Most construction projects are performed by subcontractors and supervised by contractors.

Labor costs are higher in Spain than they are in other countries because of the social welfare system. Government regulations for hiring and firing workers are stringent.

Use of business agents: Not required, but connections and family relationships are important.

Construction materials available in the country: Most construction projects use concrete and steel because steel and cement are produced in Spain.

Social standards: Social welfare is stressed in Spain. Afternoon siestas are from 1:30 to 4:30 P.M., during the heat of the day and people return to work after the siesta period. Everyone is treated with respect, and pride and honor are still important in Spain.

14.11 Summary

This chapter presented specific information on countries throughout the world that helps to demonstrate the issues that E&C personnel may encounter when they are working in a foreign country. If engineers and constructors have a basic understanding of why people behave differently in foreign cultures, it helps them work more effectively in foreign environments.

REFERENCES

Engineers and constructors from countries throughout the world provided material that was used in this chapter, and they are listed in the Acknowledgment section.

Asayesh, G. 1999. *Saffron Sky.* Boston, Mass.: Beacon Press.

Bosrock, M. 2005. Put Your Best Foot Forward. *International Education Systems.* St. Paul, Minnesota.

Bruner, B. 2005. *TIME Almanac and Information Please.* 2005. Needham, Mass.: Pearson Education Company.

Copeland, L., and L. Griggs. 2001. *Going International: How to Make Friends and Deal Effectively in the Global Marketplace.* New York: Random House.

Central Intelligence Agency (CIA). 2005. *Central Intelligence Agency Fact Book.* Washington, D.C.: CIA.

Dumas, F. 2003. *Funny in Farsi.* New York: Random House Trade Publications.

Graham, L. 1993. The Ghost of the Executed Engineer: Technology and the Fall of the Soviet Union. Boston, Massachusetts: Harvard University Press.

Kirkbride, P., Tang, S., and Westwood, R. 1993. The Chinese Concept of Face. *American Anthropologist,* vol. 46, no. 1, pp. 45–64.

Clinton, Bill. Newsweek. 2006. A Global Menace, *Newsweek,* New York, May 15.

Regulation on Administration of Foreign Invested Construction Enterprises. 2006. *Chinese Government Publication.* Ministry of construction, joint ordinance of ministry of construction and ministry of foreign trade and economic cooperation. Beijing, China.

Regulation on Examining Occupational Safety and Health Management System. 2005. *Chinese Government Publication.* National economic and trade commission. Beijing, China.

Appendix A

Glossary

This appendix provides definitions for terms that are used in the book that may be unfamiliar to engineering and construction (E&C) professionals. Terms are defined using common definitions that reflect their usage in the global E&C marketplace.

Alliances   Two or more companies join together to perform an objective such as building a construction project.

Arbitration   A panel of experts (usually three) that review information from a construction project and that provide a decision on a claim for a change-order payment.

Build-own-transfer (BOT)   One firm is responsible for all phases of a construction project, including the feasibility study, financing, land acquisition, design, engineering, construction, and operation.

Capital intensive   Requiring the purchase of extensive amounts of assets or equipment.

Cofinancing   When two parties are providing different parts of the financing for one project.

Commonwealth   A nation or state that has self-government.

Competitiveness   The act of trying to secure work when there are other companies that also are trying to secure the same work.

Completion guarantees   A completion date is guaranteed or the parent company will cover the cost of project overruns with equity investments.

Construction engineering   A branch of civil engineering that emphasizes the construction aspects of civil engineering.

Construction failures   During the construction of a project, when an element in the structure does not perform its function, and the structure collapses.

Construction management   Managing the construction phase of a project, including monitoring and controlling construction, costs, the schedule, safety quality, labor, resources, and other aspects of construction.

Contract   When two or more parties agree to perform work for compensation, the contract is the legally binding written document that explains the responsibilities of all parties, the scope of work, and the amount of compensation to be paid for the work.

Conciliation   When someone is persuaded to accept the opinion or facts presented by someone else.

Communist   State-planned and state-controlled society, where the state owns and operates all property and business.

Cooperative strategies   The methods for working together that detail how work will be performed and by whom.

Constitution   A system of laws and methods for organizing a group or a nation.

Countertrade   If a company or country cannot pay for a project with hard currency, it may trade goods for the project.

Cross-cultural   Between two or more cultures.

Cultural   Pertaining to the customs of a society.

Delays   When a task or some unit of work is not completed by the time it was scheduled to be completed.

Democracy   Government by people or through elected officials.

Developed nations or countries   Countries that rely on manufacturing or service as their primary industry and as the base of their economies.

Developing nations or countries   Countries that rely on agriculture or other non-industrial enterprises as the base of their economies.

Dictatorship   A country where one ruler has absolute power and authority over citizens.

Domestic   Operating within a native country.

Eastern culture   Intellectual and artistic tastes of citizens of Asia, East Asia, Southeast Asia, and the Middle East.

Economic   The monetary policies of an organization, a region, or a country.

Engineering and construction (E&C)   The process of designing and building structures.

Environmental   Pertaining to the world surrounding someone, including air, water, land, and the atmosphere.

Expatriate   Someone working in a country that is not his or her native country.

Expropriation   When a government takes private land for public use.

Feasibility studies   Research and economic calculations performed to estimate whether a project that is being considered will have a positive rate of return.

Federation   When states or provinces unite and are controlled by a central authority.

Financing   Providing funds to pay for a project that will be repaid with interest or other products.

Foreign   Countries or people from nations.

Foreign national   Someone who is working outside his or her native country in another country.

Global   Pertaining to the entire world.

Inspections   Reviewing work that has been performed by others for conformance to standards.

International   Between two or more nations.

International Standards Organization (ISO) The body of individuals that develops standards for use throughout the world, located in Geneva, Switzerland.

Islamic Practicing Islam (or the Muslim) religion, of which Mohammed was the prophet (the law of Islamic nations is called *sharia*).

Joint ventures When two or more companies come together to perform a project as one entity.

Judicial Pertaining to the legal process that involves decisions by judges.

Jurisdiction Legal span of control of a particular court.

Kyoto Protocol Environmental treaty for limiting the creation of toxic greenhouse gas emissions, developed by the United Nations and ratified in 2005 by countries throughout the world.

Labor When human beings perform work.

Legal jurisdiction A particular place where certain laws are followed, and the laws of the jurisdiction are used to settle legal cases (same as legal system).

Labor intensive When a project uses human beings to perform work instead of machines.

Legal systems Systems set up by governments to enforce laws and regulations.

Litigation When two or more parties want a legal definition of their rights in a dispute, in a situation where a contract has been breached, or when someone is harmed by another.

Management The process of achieving objectives through others.

Management-information systems (MIS) Computerized systems used for tracking documents, costs, schedules, materials, or personnel.

Mediation When a neutral third party assists in relaying information to two or more parties involved in a dispute.

Mitigation strategies Techniques and processes for solving a problem or reducing delays.

Modalities When a financial institution provides a loan and also participates in another loan by providing expertise or information.

Monarchy Ruled by a king or a queen.

Negotiation When two or more parties bring in another neutral party to help solve a dispute.

Newly emerging nations (NECs) Nations that are in the process of moving toward industrialization to decrease reliance on agriculture as the economic base of their economies.

Newly industrialized nations (NICs) Nations that have reached a level of manufacturing and production where industry substantially contributes to the economic base of their economies.

Nonrecourse financing A customer is willing to make advance payments on output of a product that will be delivered in the future; there is a guaranteed sale of the product; or the project is using known technology that reduces the risks involved with the project.

Operations level   The level where work is performed, a project is constructed, or a plant is being operated.

Partnerships   When two or more entities come together to form an organization, and all parties are responsible for the organization and achieving its objectives.

Plans   The drawings created by engineers for structures that are used to build structures.

Political   Pertaining to the politics of a country.

Privatization   When a project that is normally designed and built by a government is built by a private organization.

Productivity   The amount of time it takes to perform a task.

Program management   In charge of managing a program from inception to completion.

Project   A series of sequential, nonrepetitive tasks that when completed achieve a stated objective.

Project financing   When a project is considered for financing on its own merits not the merits of a company.

Project management   The managing of a project from inception to completion.

Republic   Power is in the hands of those who vote, and it is exercised through their representatives.

Social   Pertaining to the society of a country.

Socialist   When the citizens of a country collectively own and operate enterprises rather than having businesses owned by private citizens.

Specifications   The technical documents that describe how a project will be built, the materials to be used for the project, the methods that will be used to build the project, and other detailed project requirements.

Standards   An ideal level of performance or quality that other items are measured against for comparison.

Strategic level   The management level within an organization where executives make long-term decisions and perform planning.

Technical level   The level at which short-term decisions and plans are made within an organization.

Technologic   Relating to technology.

Technology transfer   When members of one organization teach members of another organization how to perform technical tasks, how to design a structure, or how to construct a structure.

Terrorism   Acts of violence against others for political or economic gain.

Unicameral parliment   Having only one legislative body.

Western culture   Intellectual or artistic tastes of citizens of North America, South America, the Mediterranean, and Europe.

Appendix B

Case Study: Managing Projects in Rural Developing Countries

A.1 Introduction

The construction techniques used in rural areas of Indonesia do not reflect the advancements that have taken place in industrialized nations. Even though many developing countries have imported modern construction technology for larger building or industrial projects, there is a marked absence of this technology on smaller projects.

This appendix discusses the types of difficulties that project-management personnel might encounter on construction projects in rural areas. Information for the case study was collected during a rural canal-construction project in Indonesia. Although some of the difficulties cited in this appendix are unique to Indonesia, many of them also apply to construction projects in other developing countries. The intent of this case study is to provide insight into the types of challenges that expatriate engineering and construction (E&C) personnel face in developing countries.

This case study is divided into several areas reflecting the different divisions within project management. The first section provides an introduction to the project that was used for this case study. The second section presents information on project-management issues, including labor practices, planning and scheduling, cost control, safety, materials management, and technical processes. The third section provides a discussion on the types of management issues that need to be addressed on rural projects in developing countries. The name and location of the project are not mentioned to respect the confidentiality of the firm responsible for building the project.

A.2 Project Introduction

Adjacent to the canal project used in this case study, there is an ultramodern, liquefied natural gas (LNG) plant. The LNG plant was built using industrialized technology and techniques combined with local Indonesian labor. Surrounding the liquefied natural gas plant is an area that was originally jungle terrain that was transformed into a thriving community by a variety of local construction projects. Figure A.1 is an aerial view of the LNG plant.

One example of the types of rural construction projects built in this community was a *canal* project that was built to alleviate flooding in the community adjacent to the LNG plant. Unfortunately, the surrounding community still flooded even after the drainage canal was built.

The objective of the canal project was to produce a storm-drainage canal that would carry rainwater out of a residential community. Storm water did flow through the drainage canal, but the outlets for the canal were not large enough; therefore, the adjacent community still flooded during wet monsoons. The canal was 8 feet wide and 8 feet deep, but the outlets that carried water under the roadway and out of the community were only 18 inches wide (Figure A.2), which caused water to back up behind the outlets as shown in Figure A.3.

This case study was developed by conducting research, field-data collection, and photo documentation.

A.3 Historical Background

Indonesia has only recently joined the ranks of what are considered to be *developing nations*. After World War II, Indonesia was still an underdeveloped nation com-

Figure A.1 Aerial view of a liquefied natural gas plant.

Figure A.2 Water canal–drainage system outlet under roadway.

Figure A.3 Community flooding after the construction of the water canal.

prised of over 13,000 islands that was struggling to overcome the image of being a *third-world country* (May, 1981). Indonesia did not gain independence from Holland until 1945. During the 1970s, natural gas was discovered in Indonesia, but the country lacked the expertise necessary to extract natural resources (Shaplon, 1970). During the 1970s, the Indonesian government did not have the funds required to train people in technical fields, as the country was still struggling to develop a viable form of government after a coup in 1965. For many years, a significant part of the population experienced starvation or near starvation. Once the Indonesian government discovered that Indonesia possessed vast reserves of oil, gas, and mineral resources, the country moved into a period where industries were developed to help the government achieve self-sufficiency and feed its citizens.

A.4 Project Management Issues

A.4.1 Labor

Indonesia has a completely different demographic structure than developed countries, because 90 percent of its population of over 238 million people are under the age of fifteen (Bruner, 2005). Having so many young people in a country creates a labor pool of predominately unskilled workers, which results in a situation where millions of people do not have jobs, and yet firms cannot find skilled workers.

The system used in Indonesia to recruit labor requires the following. For large projects (projects larger than US$1 million in labor costs), advertisements are placed in major national newspapers. Prospective employees submit job applications to the head or regional office of the firm building the project, which is usually located in Jakarta, on the island Java. After employees are selected, transmigration permits have to be issued by the government before employers may transport workers to job sites. The policy of requiring transmigration permits forces people to have employment before they can relocate or discourages them from seeking employment outside of the area where they are currently residing. Projects may be delayed while the government is processing permits for workers, and companies could be faced with a shortage of laborers in a country that is flush with unemployed workers. For small projects no advertisement other than word of mouth is required to locate employees, because workers go to job sites seeking employment opportunities.

In developing countries, the construction industry is labor intensive because there will be subtle or overt pressure by the government to use a large labor force, and labor rates are low when compared with industrialized nations. The state of the economy in Indonesia has contributed to low labor rates, because the amount of money required for survival is much less than what is required in developed countries. A large portion of the population is still in a nonmonetary economy, where barter and trading are used to secure the necessities of life. Most of the wealth in Indonesia is concentrated in the upper echelons of society, which reduces the actual per capita income of ordinary citizens. Heavy construction equipment is rarely used in Indonesia, because it is expensive to transport overseas. Most of the islands in Indonesia do not have docking facilities for container ships. There is a lack of trained mechanics, and a shortage of spare parts to repair equipment.

A.4.2 Training Programs

Large-scale construction projects (projects costing over US$100 million). For large projects, such as LNG plants, where the labor force is well defined long before the project starts, contractors set up their own training programs to train workers, because most workers cannot afford to pay for technical training programs.

Training programs start 2 to 6 months before actual craftwork commences, and both expatriate and national teachers teach the training sessions. Training programs are financed by clients, and while they are being trained, workers are paid at a salary that is less than the normal craft rate paid during construction. Training courses last from 6 to 26 weeks, depending on the craft and the level of talent among the students. At the end of training session, workers receive a certificate of completion.

For nonmanual positions (i.e., field office staff such as engineers, accountants, etc.), companies use two general guidelines for hiring: (1) people with no experience but who have a college degree or (2) people who do not have a college degree but have good work experience. Training is not provided for nonmanual workers before they are sent to field job sites, but after 3 months, they are eligible to be nominated to return to the home office for a 3- to 12-month period in which they participate in *technology transfer*–training programs. If a job site is large enough, technology-transfer programs are offered at job sites, along with courses in the language of the contractor (it is interesting to note that foreign nationals are not being trained to speak the local language; rather, local people are being taught the language of expatriate contractors).

At the end of most projects, manual workers are laid off, but the contractor tries to absorb as many nonmanual workers as possible into the national home office. If an employer is able to retain some of the professional, technical, or administrative staff after a project is completed, it makes it easier for the employer to develop a more accurate schedule and bid estimates on future projects.

A.4.3 Planning and Scheduling

Trying to set up and maintain schedules for construction projects is challenging in industrialized countries and almost impossible in developing countries. Two items that have adverse effects on schedules in developing countries are low wages and labor-intensive industries. Another area that contributes to scheduling difficulties is that workers are paid in accordance with the number of days they work and hired on a day-to-day basis, so it is advantageous to workers to have a project continue for as long as possible.

Schedules may also be delayed if workers are operating in industrialized nations, also under "rubber time" rather than by real time. Rubber time indicates that flexibility is tolerated in situations where people from industrialized nations would demand punctuality. People may show up hours late for work or appointments. The apparent lack of concern over time originates out of a *spiritual orientation* rather than being rationally or materially oriented, as is found in other societies (May, 1981). This spiritual orientation also creates another situation that hampers efforts to schedule or update projects, it is known as *asal bapak senang*, which loosely translates to "as long as the boss is happy." This means that if a subordinate thinks that his or her supervisor wants to hear that a task is being accomplished the supervisor will be told that it is being accomplished even if it is not being accomplished (Fodor's, 1990). The only way to elicit accurate responses is to re-phrase yes or no questions into a format that requires another type of answer, such as an explanation or a date. If yes or no questions are asked, workers will answer yes because they want to please their supervisor. This situation fosters false and inaccurate information that is incorporated into construction schedules.

Since most rural projects do not use critical-path method (CPM) schedules or bar charts the workers are faced with a lack of short-term goals, or milestones, that they can use to judge the progress of their work. As a consequence, there are no incentives for workers to complete tasks quickly. With no schedules to adhere to, situations arise whereby workers must stop their work and wait for other work-ers to deliver materials and equipment or to finish their tasks, which results in lengthy delays.

A.4.4 Cost Control

The cost aspects of rural Indonesian projects have also been adversely influenced by inaccurate or nonexistent scheduling techniques. If projects are contracted as lump-sum contracts, owners are financially protected from inadequate cost-control procedures since contractors absorb cost overruns. By using lump-sum contracts, the Indonesian government is trying to force contractors into situations where they must become cost conscious in order to stay in business. Indonesian contractors are beginning to realize that they will have limited profit margins if projects go beyond a normal duration for that particular type of project. However, in a country that has high unemployment, high levels of underemployment, and no labor un-ions to regulate wages, contractors still may make large profits on lump-sum jobs merely by paying workers lower salaries if the project continues beyond its sched-uled completion date.

Indonesia does not allow unions, nor are there any set minimum wages or craft scales. There is a Department of Manpower and Transmigration that issues gov-ernment guidelines that are followed by government agencies and major firms. Contractors are able to avoid following government guidelines by using subcon-tractors, because subcontractors do not have to follow government regulations.

A.4.5 Safety

In rural locations, safety issues on construction projects may be ignored because there are always new workers to replace workers who are injured or who die on projects. Employers do not have to provide any type of compensation to workers who are injured on the job or to the families of workers who are killed on the job. Workers' compensation insurance is available only on projects that are sponsored directly by the Indonesian government. Subcontractors are not required to provide workers' compensation benefits.

On nongovernment or multinational projects, workers rarely use hard hats, safety glasses, steel-toed shoes, gloves, or safety harnesses. The use of bamboo poles for bracing and scaffolding is still common in Indonesia. The government does not regulate safety on construction projects.

All grievances related to labor conditions are handled by the government Directorate of Labor Relations, which has representatives only in densely populated areas of the country. Without unions or other labor organizations, Indonesian workers find that there is little they can do to alter their present working conditions. Workers feel that it is a privilege merely to have a job, and they do not worry about the conditions under which they must work. The spiritual orientation of the Indonesians teaches them to appreciate what they have rather than always aspiring to possess more.

A.4.6 Materials Management and Methods

Materials management was not used during the case study project. Materials were acquired locally or were "borrowed" (taken without permission) from other construction sites. No scheme exists for small contractors to import materials and most contractors do not have enough capital to stockpile materials for several jobs.

Contractors use whatever materials are available in their local region, be it bamboo, straw, mud, wood, rattan, cardboard, etc. Materials and small hand tools are not kept at job sites. Construction materials and hand tools are scarce commodities and difficult to secure; therefore, materials and tools are brought to the job site on a daily basis, which reduces productivity because workers are idle why they are waiting for materials and tools.

If a batch plant is available, concrete will be brought to job sites in concrete trucks, but the concrete chutes will be constructed out of wood scraps, with boulders as supports, as is shown in Figure A.4. To spread concrete while it is being placed, workers stand barefoot in the freshly poured concrete and spread it with their hands or makeshift tools, such as pieces of scrap lumber.

A.4.7 Labor Practices

Construction projects in Indonesia are hampered by direct and indirect actions that result from the presence of a uniquely structured *labor class system*. If someone has thumbnails 1 to 3 inches long, this indicates that the person uses intelligence rather than strength to earn a living, and it advertises the fact that the person is not a common laborer. Class distinctions affect productivity, because workers will not perform tasks that jeopardize their status, and additional workers must be hired to perform jobs these workers consider demeaning. One practice is for each

Figure A.4 Concrete pumping operation (chute supported by rocks and lumber).

engineer and the other professional staff to hire *tea boys,* whose only job is to make coffee or tea and to keep the office clean. When the Dutch occupied Indonesia, they created the titles *tea boys* or *yard boys,* and these titles are still used for grown men, which is insulting to Indonesians.

Because of the importance of the class system, there are large separations in the work force on rural projects. It is not a common practice for an engineer to work as a direct supervisor of field laborers, because project foreman perform this function, whether they are qualified or not, and field engineers only make brief daily visits to job sites. This puts the burden of making field decisions onto the project foreman (*leadman*). Leadmen usually do not have formal education or training to prepare them to make engineering judgments or decisions, but because of the importance of status, the leadman will make engineering decisions without seeking the advice of an engineer because they do not want to "lose face" (be embarrassed) by showing that they do not know what to do. The concept of losing face prevents supervisors from directly criticizing their subordinates. A reprimand for an incorrect action should always be done in private, or workers will not return to work.

Another area that influences performance on a construction project is that "people are acutely aware of their proper place and do not aspire to change. To try to change one's position or status would be considered forward, if not presumptive or tempting fate. Such behavior would be discouraged by others" (May, 1981, p. 352). Foreigners misinterpret this behavior as a lack ambition on the part of Indonesians, but Indonesians are taught to appreciate the fact that they have a job.

There may be racial tensions at job sites between workers from different native islands and between foreigners and native workers. The first loyalty of workers is to other *island nationals,* which are workers from the same island. Some workers will work with people from other islands, but they will only socialize with people from their native region. Even though the national language is *Bahasa Indonesia,* many different languages are spoken throughout the country. It is possible to have a worker from an island such as Boogie who only speaks *Booganise,* who is not able to communicate with anyone else on the project.

A.5 Management Issues

Capital intensive projects in developing countries are undertaken only if it can be determined that they will generate profits. But sometimes decisions to build construction projects could be based solely on the atmosphere prevailing in the current government. Projects of social importance may be overlooked in favor of *reckless industrial adventures* or those that provide *aesthetic improvements* (May, 1981).

Centralized government agencies control which projects will be built, and this puts local governments in a position of not having any influence on the types of projects that will be built in their region. Decisions to build projects are made at the upper echelons of government, with no input from local citizens. This system has contributed to a lack of essential services in rural areas, such as running water, electricity, paved roads, and sewer and sanitation facilities. Indonesian villages and most of the areas in large cities would be considered *slums* by industrialized standards because of the lack of public works projects.

Workers who live in rural areas constitute the main body of unskilled laborers, and when they see the inequity of their surroundings as compared with project environments, resentments result that lead to workers taking materials from projects and using them to improve their home environments. Perceived inequalities might cause workers to slow down production to make their jobs easier to perform.

Another difficulty that hampers construction progress is the widespread use of *graft* (bribery) and political favors. Construction companies may have to pay *fees* to a variety of different government agencies to procure licenses, equipment, materials, or services. In some cases, because of these fees, the cost of projects might exceed original estimates by as much as 100 percent or more. "Often bribery is regulated by tradition, and an official is considered to transgress only if he oversteps accepted limits under what is known in one area as the TST, *tahu, sama, tabu,* 'lets keep it to ourselves'" (May, 1981). This could create problems for members of foreign firms, if their native governments prohibit the bribing of foreign government officials (bribing foreign government officials is illegal in the United States because of the Foreign Anticorruption Practices Act and the United Nations Convention Against Corruption). One method used to circumvent anticorruption laws is the use of *business agents* who take care of TST for foreign firms.

A.6 Indonesian Educational Programs

In Indonesia, only members of the upper class have opportunities for advanced education. The government realizes that it needs to provide additional education to meet the growing demands of development, but it is promoting vocational schools as opposed to university programs. Less than 1 percent of the population of Indonesia graduates from college.

Even though a great deal of money was spent during the 1960s to build colleges, by 1965 there were only 45 state universities, 85 academies at the higher education level, and 200 private institutions of higher learning. There has been only a small increase in the number of colleges since the 1960s (*World Almanac*, 2005). If laborers do not receive training prior to starting a job, they merely follow the example of their fellow workers. This results in a situation where no one on a project has ever been taught to execute the work properly.

A.7 Case Study: Conclusions

Citizens of developing nations desire to see the designation of a *developing country* removed from their country and replaced with *emergent nation* or newly industrialized nation, but they still lack the technical expertise to achieve this goal. Middle and lower-level managers and supervisors in developing countries try to implement Western technology and techniques on rural construction projects, but it is difficult to implement technology unless it is *appropriate technology* that fits the needs of rural environments. If appropriate technology is not used on construction projects, the groundwork to support proposed redevelopment does not exist.

The shortcomings of the construction techniques used in the rural areas of developing countries could be alleviated by the introduction of the concepts inherent in project management. The areas that need to be stressed include the following: estimating and cost control, scheduling, field supervision, material management, safety, quality control, and site organization. When these types of project-management concepts are introduced to developing countries, managers have to be careful not to apply industrialized formulas to countries that are devoid of industrialized motives. Managers need to adapt basic project-management concepts to fit the particular needs of a country and stress methods for planning, organizing, and controlling projects. Even the introduction of rudimentary drawings, estimates, bar charts, and site-organization techniques would prevent cost overruns and low-quality products. Productivity rates would increase if less time were spent waiting for new materials or new job assignments, and efficient site organization would directly affect the safety aspects of projects. Accidents could be prevented if workers were aware of their proper job assignments or when they are not in their proper work environment.

One important step for members of developing countries is to teach these concepts to small contractors. Large contractors provide their own training programs, which only reach a small proportion of the population. The majority of public works projects in developing countries are being built by small contractors or subcontractors; therefore, the necessity of educating this segment is essential.

Anyone who is sent to work in rural areas of a developing country should take the time to study the country, its culture, its traditions, its demographics, and its political and social systems in order to be able to accurately plan and monitor the project. If industrialized standards are used to plan and control projects in developing countries, there may be severe delays and cost overruns during construction due to cultural differences or political interference. Being aware of these types of issues during the planning phases of projects helps to ensure that projects will be built more efficiently, in less time, and for a lower cost.

REFERENCES

Bruner, B. 2005. *TIME Almanac and Information Please.* Needham Mass.: Pearson Education Company.

Copeland, L., and L. Griggs. 2001. *Going International.* New York: Random House.

Fodor's Modern Guides, 1990. *Fodor's Southeast Asia,* Distributed by David McKay Company, New York.

May, Brian. 1981. *The Indonesian Tragedy.* Singapore: Graham Brash Ltd.

Shaplon, Robert. 1969. *Time Out of Hand.* New York: Harper and Row Publishers.

Webster's New Unabridged Dictionary. 2005. New York: Simon and Schuster.

World Almanac, 2005, World Almanac Education Group, Inc., New York, New York.

Appendix C

Definitions of Religions

Some of the following definitions are from *Newsweek,* August 29–September 5, 2005; *U.S News and World Report,* January 31–February 7, 2005; and *Webster's Unabridged Dictionary,* 2005:

- *Anglican*—The Church of England (Protestant Episcopalian).
- *Baptist*—A member of a Protestant denomination that believes that baptism should be given only to adult believers and by immersion in water rather than only being sprinkled with water.
- *Buddhism*—Religions and philosophy founded in India in the 6th or 5th century B.C. by Siddhartha Gautama, call the Buddha. An Asian religion that teaches the practice of meditation and the observance of moral precepts (The Concise Columbia Encyclopedia, 1994, p. 122).
- *Charismatic*—Neo-Pentecostal Christians whose worship emphasizes the gifts of the spirit (charismata) listed by St. Paul in Corinthians, especially healing and speaking in tongues (The Concise Columbia Encyclopedia, 1994 p. 160).
- *Episcopalian*—Part of the Angelican Communion in the United States.
- *Evangelical United Brethren Church*—Protestant denomination created in 1946 by the union of the Evangelical Church and the United Brethren in Christ. The church has an Episcopal government and stresses prayer, devotion to Jesus, and individual responsibility. In 1967 it became part of the United Methodist church (The Concise Columbia Encyclopedia, 1994, p. 287).
- *Fundamentalism*—In Protestantism, conservative religious movement that arose among members of various denominations early in the 20th century. Its aim is to maintain traditional interpretations of the Bible and what its adherents believe to be the fundamental doctrines of the Christian faith (The Concise Columbia Encyclopedia, 1994, p. 322).
- *Hasidism*—A Jewish movement founded in Poland in the 18th century by Baal-Shem-tov. It encourages joyous religious expression through music and dance and teaches that purity of heart is more pleasing to God than learning from the Torah (The Concise Columbia Encyclopedia, 1994, p. 377).
- *Hinduism*—Western term for the religious beliefs and practices of innumerable sects to which the vast majority of the people of India belong. Hinduism was originally a synthesis of indigenous religions and the religion brought to India by the Aryans in 1500 B.C. (The Concise Columbia Encyclopedia, 1994, p. 391).
- *Judaism*—The religious beliefs and practices and the way of life of the Jews. Central to these is the notion of monotheism, adopted by the biblical Hebrews

(The Concise Columbia Encyclopedia, 1994, p. 453). Followers of the Jewish religion study the Torah.

- *Kabbalah*—A form of Jewish mysticism focused on uncovering the hidden meaning of the Torah through meditation and study rooted in the Zohar, a thirteenth-century Aramaic text. Traditional practitioners do not usually wear the red string bracelet favored by new devotees.
- *Lutheran*—Protestant denomination founded by Martin Luther.
- *Islam*—Submission to, or having peace with, God. The religion of which Mohammad was the prophet. Its salient feature is its devotion to the Koran, or Qu'ran, a book believed to be the revelation of God to Mohammad. (The Concise Columbia Encyclopedia, 1994, p. 322). Within Islam there are Sunnis and Shiites.
- *Orthodox Eastern Church*—Community of Christian churches, independent but mutually recognized, originating in Eastern Europe and Southwest Asia through a split with the Western church (The Concise Columbia Encyclopedia, 1994, p. 650).
- *Pentecostalism*—A form of Protestantism in the United States that emphasizes baptism in the Holy Spirit speaking in tongues and faith healing.
- *Protestant*—Christian not of a Catholic or Orthodox Eastern church faith includes Lutherans and Anglicans and other denominations.
- *Roman Catholic*—Christian church headed by the pope, the bishop of Rome. All members of the church accept the gospel of Jesus as handed down by the church, the teachings of the Bible, and the church's interpretations of those teachings (The Concise Columbia Encyclopedia, 1994, p. 750).
- *Sufism*—Muslim philosophical and literary movement. It emerged among the Shiites in the late 10th and early 11th century, borrowing ideas from Neo-Platonism, Buddhism, and Christianity (The Concise Columbia Encyclopedia, 1994, p. 844).
- *Wayang*—Mystical religion that includes 150 different deities that is practiced in Indonesia.
- *Zen Buddhism*—Buddhist sect of Japan and China, based on the practice of meditation rather than on adherence to a particular scriptural doctrine (The Concise Columbia Encyclopedia, 1994, p. 969).

REFERENCES

The Concise Columbia Encyclopedia, 1994. Columbia University Press, New York, New York.
Newsweek, August 29–September 5, 2005.
U.S. News and World Report, January 31–February 7, 2005.
Webster's New Unabridged Dictionary, 2005. Simon and Schuster, New York, New York.

Index